Argument-Driven Inquiry in Third-Grade Science

Three-Dimensional Investigations

Argument-Driven Inquiry in Third-Grade Science

Three-Dimensional Investigations

Victor Sampson and Ashley Murphy

NSTA press
National Science Teachers Association
Arlington, Virginia

Claire Reinburg, Director
Rachel Ledbetter, Managing Editor
Andrea Silen, Associate Editor
Jennifer Thompson, Associate Editor
Donna Yudkin, Book Acquisitions Manager

ART AND DESIGN
Will Thomas Jr., Director

PRINTING AND PRODUCTION
Catherine Lorrain, Director

NATIONAL SCIENCE TEACHERS ASSOCIATION
David L. Evans, Executive Director

1840 Wilson Blvd., Arlington, VA 22201
www.nsta.org/store
For customer service inquiries, please call 800-277-5300.

22 21 20 19 4 3 2 1

Library of Congress Cataloging-in-Publication Data

Names: Sampson, Victor, 1974- author. | Murphy, Ashley, 1988- author.
Title: Argument-driven inquiry in third-grade science : three-dimensional investigations / by Victor Sampson and Ashley Murphy.
Description: Arlington, VA : National Science Teachers Association, [2019] | Includes bibliographical references and index.
Identifiers: LCCN 2018041212 (print) | LCCN 2018049891 (ebook) | ISBN 9781681405186 (e-book) | ISBN 9781681405179 | ISBN 9781681405179¬q(print)
Subjects: LCSH: Science--Methodology--Study and teaching (Primary)--Handbooks, manuals, etc. | Science--Experiments. | Inquiry-based learning.
Classification: LCC Q182.3 (ebook) | LCC Q182.3 .S24 2019 (print) | DDC 372.35/044--dc23
LC record available at *https://lccn.loc.gov/2018041212*

Contents

Contents

Contents

Preface

There are a number of potential reasons for teaching children about science in elementary school. Some people, for example, think it is important to focus on science in the early grades to get students interested in science early so that more people will choose to go into a science or science-related career. Some people think that science is important to teach in the early grades because children ask so many questions about how the world works and the information included as part of the science curriculum is a great way to answer many of their questions. Others think it is important to focus on science in elementary school because children need a strong foundation in the basics so they will be prepared for what they will be expected to know or do in middle or high school. Few people, however, emphasize the importance of teaching science because it is useful for everyday life (Bybee and Pruitt 2017).

Science is useful because it, along with engineering, mathematics, and the technologies that are made possible by these three fields, affects almost every aspect of modern life in one way or another. For example, people need to understand science to be able to think meaningfully about policy issues that affect their communities or to make informed decisions about what food to eat, what medicine to take, or what products to use. People can use their understanding of science to help evaluate the acceptability of different ideas or to convince others about the best course of action to take when faced with a wide range of options. In addition, understanding how science works and all the new scientific findings that are reported each year in the media can be interesting, relevant, and meaningful on a personal level and can open doors to exciting new professional opportunities. The more a person understands science, which includes the theories, models, and laws that scientists have developed over time to explain how and why things happen and how these ideas are developed and refined based on evidence, the easier it is for that person to have a productive and fulfilling life in our technology-based and information-rich society. Science is therefore useful to everyone, not just future scientists.

A Framework for K–12 Science Education (NRC 2012; henceforth referred to as the *Framework*) is based on the idea that all citizens should be able to use scientific ideas to inform both individual choices and collective choices as members of a modern democratic society. It also acknowledges the fact that professional growth and economic opportunity are increasingly tied to the ability to use scientific ideas, processes, and ways of thinking. From the perspective of the *Framework*, it is important for children to learn science because it can help them figure things out or solve problems. It is not enough to remember some facts and terms; people need to be able to use what they have learned while in school. We think that this goal for science education not only is important but also represents a major shift in what should be valued inside the classroom.

Preface

The *Framework* asks all of us, as teachers, to reconsider what we teach in grades K–5 and how we teach it, given this goal for science education. It calls for all students, over multiple years of school, to learn how to use disciplinary core ideas (DCIs), crosscutting concepts (CCs), and scientific and engineering practices (SEPs) to figure things out or solve problems. The DCIs are key organizing principles that have broad explanatory power within a discipline. Scientists use these ideas to explain the natural world. The CCs are ideas that are used across disciplines. These concepts provide a framework or a lens that people can use to explore natural phenomena; thus, these concepts often influence what people focus on or pay attention to when they attempt to understand how something works or why something happens. The SEPs are the different activities that scientists engage in as they attempt to generate new concepts, models, theories, or laws that are both valid and reliable. All three of these dimensions of science are important. Students not only need to know about the DCIs, CCs, and SEPs but also must be able to use all three dimensions at the same time to figure things out or to solve problems. These important DCIs, CCs, and SEPs are summarized in Table P-1.

When we give students an opportunity to learn how to use DCIs, CCs, and SEPs to make sense of the world around them, we also provide an authentic context for students to develop fundamental literacy and mathematics skills. Students are able to develop literacy and mathematics skills in this type of context because doing science requires people to obtain, evaluate, and communicate information. Students, for example, must read and talk to others to learn what others have done and what they are thinking. Students must write and speak to share their ideas about what they have learned or what they still need to learn. Students can use mathematics to measure and to discover trends, patterns, or relationships in their observations. They can also use mathematics to make predictions about what will happen in the future. When we give students opportunities to do science, we give students a reason to read, write, speak, and listen. We also create a need for them to use mathematics.

To help students learn how to use DCIs, CCs, and SEPs to figure things out or solve problems while providing them a context to develop fundamental literacy and mathematics skills, elementary teachers will need to use new instructional approaches. These instructional approaches must give students an opportunity to actually do science. To help teachers who teach elementary school make this instructional shift, we have developed a tool called argument-driven inquiry (ADI). ADI is an innovative approach to instruction that gives students an opportunity to use DCIs, CCs, and SEPs to construct and critique claims about how things work or why things happen. As part of this process, students must talk, listen, read, and write to obtain, evaluate, and communicate information. ADI, as a result, creates a rich learning environment for children that enables them to learn science, language, and mathematics at the same time.

TABLE P-1

The three dimensions of *A Framework for K–12 Science Education*

<table>
<tr>
<td colspan="3">Science and engineering practices (SEPs)
• SEP 1: Asking Questions and Defining Problems
• SEP 2: Developing and Using Models
• SEP 3: Planning and Carrying Out Investigations
• SEP 4: Analyzing and Interpreting Data
• SEP 5: Using Mathematics and Computational Thinking
• SEP 6: Constructing Explanations and Designing Solutions
• SEP 7: Engaging in Argument From Evidence
• SEP 8: Obtaining, Evaluating, and Communicating Information</td>
<td colspan="3">Crosscutting concepts (CCs)
• CC 1: Patterns
• CC 2: Cause and Effect: Mechanism and Explanation
• CC 3: Scale, Proportion, and Quantity
• CC 4: Systems and System Models
• CC 5: Energy and Matter: Flows, Cycles, and Conservation
• CC 6: Structure and Function
• CC 7: Stability and Change</td>
</tr>
<tr>
<td colspan="6">Disciplinary core ideas</td>
</tr>
<tr>
<td colspan="2">Earth and Space Sciences (ESS)
• ESS1: Earth's Place in the Universe
• ESS2: Earth's Systems
• ESS3: Earth and Human Activity</td>
<td colspan="2">Life Sciences (LS)
• LS1: From Molecules to Organisms: Structures and Processes
• LS2: Ecosystems: Interactions, Energy, and Dynamics
• LS3: Heredity: Inheritance and Variation of Traits
• LS4: Biological Evolution: Unity and Diversity</td>
<td colspan="2">Physical Sciences (PS)
• PS1: Matter and Its Interactions
• PS2: Motion and Stability: Forces and Interactions
• PS3: Energy
• PS4: Waves and Their Applications in Technologies for Information Transfer</td>
</tr>
</table>

Source: Adapted from NRC 2012.

This book describes how ADI works and why it is important, and it provides 14 investigations that can be used in the classroom to help students reach the performance expectations found in the *Next Generation Science Standards* (NGSS Lead States 2013) for third grade.[1] The 14 investigations described in this book will also enable students to develop the disciplinary-based literacy skills outlined in the *Common Core State Standards* for English language arts (NGAC and CCSSO 2010) because ADI

1 See *Argument-Driven Inquiry in Fourth-Grade Science* (Sampson and Murphy, forthcoming) and *Argument-Driven Inquiry in Fifth-Grade Science* (Sampson and Murphy, forthcoming) for additional investigations for students in elementary school.

gives students an opportunity to give presentations to their peers, respond to audience questions and critiques, and then write, evaluate, and revise reports as part of each investigation. In addition, these investigations will help students learn many of the mathematical ideas and practices outlined in the *Common Core State Standards* for mathematics (NGAC and CCSSO 2010) because ADI gives students an opportunity to use mathematics to collect, analyze, and interpret data. Finally, and perhaps most important, ADI can help emerging bilingual students meet the *English Language Proficiency (ELP) Standards* (CCSSO 2014) because it provides a language-rich context where children can use receptive and productive language to communicate and to negotiate meaning with others. Teachers can therefore use these investigations to align how and what they teach with current recommendations for improving science education.

References

Bybee, R. W., and S. Pruitt. 2017. *Perspectives on science education: A leadership seminar.* Arlington, VA: NSTA Press.

Council of Chief State School Officers (CCSSO). 2014. *English language proficiency (ELP) standards.* Washington, DC: NGAC and CCSSO. *www.ccsso.org/resource-library/english-language-proficiency-elp-standards.*

National Governors Association Center for Best Practices and Council of Chief State School Officers (NGAC and CCSSO). 2010. *Common core state standards.* Washington, DC: NGAC and CCSSO.

National Research Council (NRC). 2012. *A framework for K–12 science education: Practices, crosscutting concepts, and core ideas.* Washington, DC: National Academies Press.

NGSS Lead States. 2013. *Next Generation Science Standards: For states, by states.* Washington, DC: National Academies Press. *www.nextgenscience.org/next-generation-science-standards.*

Sampson, V., and A. Murphy. Forthcoming. *Argument-driven inquiry in fifth-grade science: Three-dimensional investigations.* Arlington, VA: NSTA Press.

Sampson, V., and A. Murphy. Forthcoming. *Argument-driven inquiry in fourth-grade science: Three-dimensional investigations.* Arlington, VA: NSTA Press.

Acknowledgments

We would like to thank Dr. Linda Cook, director of science for Coppell Independent School District (ISD) in Texas, and the following individuals from Coppell ISD for piloting these lab activities and giving us feedback about ways to make them better.

Kristina Adrian
Teacher
Richard J. Lee Elementary

Jennifer Baldwin-Hays
K–5 Instructional Coach
Pinkerton Elementary

Julie Bowles
Teacher
Valley Ranch Elementary

Heidi Brown
Teacher
Wilson Elementary

Kelsea Burke
Teacher
Cottonwood Creek Elementary

Kasey Cross
Teacher
Wilson Elementary

Denise Danby
Teacher
Valley Ranch Elementary

Cindy Daniel
Teacher
Richard J. Lee Elementary

Andi Feille
K–5 Instructional Coach
Wilson Elementary

Ohlia Garza
Teacher
Denton Creek Elementary

D'Ann Green
K–5 Instructional Coach
Valley Ranch Elementary

Nathan Harvey
Teacher
Austin Elementary

Rachelle Hendricks
Teacher
Lakeside Elementary

Kristan Hruby
K–5 Instructional Coach
Town Center Elementary

Jacque Johnson
K–5 Instructional Coach
Richard J. Lee Elementary

Kasey Kemp
Teacher
Town Center Elementary

Cassandra Knight
Teacher
Richard J. Lee Elementary

Brittney Krommenhoek
Teacher
Austin Elementary

Sarah Leishman
K–5 Instructional Coach
Denton Creek Elementary

Rachel Lim
Teacher
Town Center Elementary

Kelly Matlock
Teacher
Wilson Elementary

Katie Nelson
Teacher
Mockingbird Elementary

Acknowledgments

April Owen
Teacher
Mockingbird Elementary

Dawn Rehling
Teacher
Lakeside Elementary

Karli Reichert
Teacher
Richard J. Lee Elementary

Jody Reynolds
K–5 Instructional Coach
Mockingbird Elementary

Frankie Robertson
Teacher
Lakeside Elementary

Renee Rohani
Teacher
Town Center Elementary

Liliana Rojas
Teacher—Dual Language Program
Wilson Elementary

Maureen Salmon
K–5 Instructional Coach
Lakeside Elementary

Meredith Schaaf
Teacher
Town Center Elementary

Priscilla Shaner
Teacher
Richard J. Lee Elementary

Jennifer Stepter
Teacher
Denton Creek Elementary

Liz Tanner
Teacher
Lakeside Elementary

Meghan Tidwell
Teacher
Cottonwood Creek Elementary

Casey Wagner
Teacher
Town Center Elementary

About the Authors

Victor Sampson is an associate professor of STEM (science, technology, engineering, and mathematics) education at The University of Texas at Austin (UT-Austin). He received a BA in zoology from the University of Washington, an MIT from Seattle University, and a PhD in curriculum and instruction with a specialization in science education from Arizona State University. Victor also taught high school biology and chemistry for nine years. He is an expert in argumentation and three-dimensional instruction in science education, teacher learning, and assessment. Victor is also an NSTA (National Science Teachers Association) Fellow.

Ashley Murphy attended Florida State University and earned a BS with dual majors in biology and secondary science education. Ashley taught biology and integrated science at the middle school level before earning a master's degree in STEM education from UT-Austin. While in graduate school at UT-Austin, she taught courses on project-based instruction and elementary science methods. She is an expert in argumentation and three-dimensional instruction in middle and elementary classrooms and science teacher education.

Introduction

A Vision for Science Education in Elementary School

The current aim of science education in the United States is for *all* students to become proficient in science by the time they finish high school. *Science proficiency,* as defined by Duschl, Schweingruber, and Shouse (2007), consists of four interrelated aspects. First, it requires an individual to know important scientific explanations about the natural world, to be able to use these explanations to solve problems, and to be able to understand new explanations when they are introduced to the individual. Second, it requires an individual to be able to generate and evaluate scientific explanations and scientific arguments. Third, it requires an individual to understand the nature of scientific knowledge and how scientific knowledge develops over time. Finally, and perhaps most important, an individual who is proficient in science should be able to participate in scientific practices (such as planning and carrying out investigations, analyzing and interpreting data, and arguing from evidence) and communicate in a manner that is consistent with the norms of the scientific community. These four aspects of science proficiency include the knowledge and skills that all people need to have to be able to pursue a degree in science, be prepared for a science-related career, and participate in a democracy as an informed citizen.

This view of science proficiency serves as the foundation for *A Framework for K–12 Science Education* (the *Framework;* NRC 2012). The *Framework* calls for all students to learn how to use disciplinary core ideas (DCIs), crosscutting concepts (CCs), and scientific and engineering practices (SEPs) to figure things out or solve problems as a way to help them develop the four aspects of science proficiency. The *Framework* was used to guide the development of the *Next Generation Science Standards* (*NGSS;* NGSS Lead States 2013). The goal of the *NGSS,* and other sets of academic standards that are based on the *Framework,* is to describe what *all* students should be able to do at each grade level or at the end of each course as they progress toward the ultimate goal of science proficiency.

The DCIs found in the *Framework* and the *NGSS* are scientific theories, laws, or principles that are central to understanding a variety of natural phenomena. An example of a DCI in Earth and Space Sciences is that the solar system consists of the Sun and a collection of objects that are held in orbit around the Sun by its gravitational pull on them. This DCI not only can be used to help explain the motion of planets around the Sun but also can be used to explain why we have tides on Earth, why the appearance of the Moon changes over time in a predictable pattern, and why we see eclipses of the Sun and the Moon.

The CCs are ideas that are important across the disciplines of science. The CCs help people think about what to focus on or pay attention to during an investigation. For example, one of the CCs from the *Framework* is Energy and Matter: Flows, Cycles, and Conservation. This CC is important in many different fields of study, including astronomy, geology, and meteorology. This CC is equally important in physics and biology. Physicists use this CC when they study how things move, why things change temperature, and the behavior of circuits, magnets, and generators. Biologists use this CC when they study how cells work, the growth and development of plants or animals, and the nature of ecosystems. It is important to highlight the centrality of this idea, and other CCs, for students as we teach the subject-specific DCIs.

The SEPs describe what scientists do as they attempt to make sense of the natural world. Students engage in practices to build, deepen, and apply their knowledge of DCIs and CCs. Some of the SEPs include familiar aspects of what we typically associate with "doing" science, such as Asking Questions and Defining Problems, Planning and Carrying Out Investigations, and Analyzing and Interpreting Data. More important, however, some of the SEPs focus on activities that are related to developing and sharing new ideas, solutions to problems, or answers to questions. These SEPs include Developing and Using Models; Constructing Explanations and Designing Solutions; Engaging in Argument From Evidence; and Obtaining, Evaluating, and Communicating Information. All of these SEPs are important to learn because scientists engage in different practices, at different times, and in different orders depending on what they are studying and what they are trying to accomplish at that point in time.

Few students in elementary school have an opportunity to learn how to use DCIs, CCs, and SEPs to figure things out or to solve problems. Instead, most students are introduced to facts, concepts, and vocabulary without a real reason to know or use them. This type of focus in elementary school does little to promote and support the development of science proficiency because it emphasizes "learning about" science rather than learning how to use science to "figure things out." This type of focus also reflects a view of teaching that defines *rigor* as covering more topics and *learning* as the simple acquisition of more information.

We must think about rigor in different ways before we can start teaching science in ways described by the *Framework*. Instead of using the number of different topics covered in a particular grade level as a way to measure rigor in our schools (e.g., "we made third grade more challenging by adding more topics for students to learn about"), we must start to measure rigor in terms of the number of opportunities students have to use DCIs, CCs, and SEPs to make sense of different phenomena (e.g., "we made third grade more challenging because students have to figure out how to predict the motion of ball once is dropped into a half-pipe"). A rigorous class, in other words, should be viewed as one where students are expected to do science, not just learn about science. From this perspective, our goal as teachers should be to help our students learn how to use DCIs and CCs as tools to plan and carry out investigations, construct and evaluate explanations, and question how we know what we know instead of just ensuring that we "cover" all the different DCIs and CCs that are included in the standards by the end of the school year.

We must also rethink what learning is and how it happens to better promote and support the development of science proficiency. Rather than viewing learning as an individual process where children accumulate more and more information over time, we need to view learning as a social and then an individual process that involves being exposed to new ideas and ways of doing things, trying out these new ideas and practices under the guidance of more experienced people, and then adopting the ideas and practices that are found to be useful for making sense of the world (NRC 1999, 2008, 2012). Learning, from this perspective, requires children to "do science" while in school not because it is fun or interesting (which is true for many) but because doing science gives children a reason to use the ideas and practices of science. When children are given repeated opportunities to use DCIs, CCs, and SEPs as a way to make sense of the world, they will begin to see why these ideas, concepts, and practices are valuable. Over time, children will then adopt these ideas and practices and start using them on their own. We therefore must give our students an opportunity to experience how scientists figure things out and share ideas so they can become "socialized to a greater or lesser extent into the practices of the scientific community with its particular purposes, ways of seeing, and ways of supporting its knowledge claims" (Driver et al. 1994, p. 8).

It is important to keep in mind that helping children learn how to use the ideas and practices of science to figure things out by giving them an opportunity to do science is not a "hands-off" approach to teaching. The process of learning to use the ideas and practices of science to figure things out requires constant input and guidance about "what counts" from teachers who are familiar with the goals of science, the norms of science, and the ways things are done in science. Thus, learning how to use DCIs, CCs, and SEPs to figure things out or to solve problems is dependent on supportive and informative interactions with teachers. This is important because students must have a supportive and educative learning environment to try out new ideas and practices, make mistakes, and refine what they know and how they do things before they are able to adopt the ideas and practices of science as their own.

The Need for New Ways of Teaching Science in Elementary School

Science in elementary school is often taught though a combination of direct instruction and hands-on activities. A typical lesson often begins with the teacher introducing students to a new concept and related terms through direct instruction. Next, the teacher will often show the students a demonstration or give them a hands-on activity to complete to illustrate the concept. The purpose of including a demonstration or a hands-on activity in the lesson is to provide the students with a memorable experience with the concept. If the memorable experience is a hands-on activity, the teacher will often provide his or her students with a step-by-step procedure to follow and a data table to fill out to help ensure that no one in the class "gets lost" or "does the wrong thing" and everyone "gets the right results." The teacher will usually assign a set of questions for the students to answer on their own or in groups after the demonstration or the hands-on activity to make sure that everyone in the class "reaches the right conclusion." The lesson usually ends with the teacher reviewing the concept and all related terms to make sure that everyone in the class learned what they were "supposed to have learned." The teacher often accomplishes this last step of the lesson by leading a whole-class discussion, by assigning a worksheet to complete, or by having the students play an educational game.

Classroom-based research, however, suggests that this type of lesson does little to help students learn key concepts (Duschl, Schweingruber, and Shouse 2007; NRC 2008, 2012). This finding is troubling because, as noted earlier, one of the main goals of this type of lesson is to help students understand an important concept by giving them a memorable experience with it. In addition, this type of lesson does little to help students learn how to plan and carry out investigations or analyze and interpret data because students have no voice or choice during the activity. Students are expected to simply follow a set of directions rather than having to think about what data they will collect, how they will collect it, and what they will need to do to analyze it once they have it. These types of activities can also lead to misunderstanding about the nature of scientific knowledge and how this knowledge is developed over time due to the emphasis on following procedure and getting the right results. These hand-on activities, as a result, do not reflect how science is done at all.

Over the last decade, many elementary school teachers have adopted more inquiry-based approaches to science teaching to address the many shortcomings of these typical science lessons. Inquiry-based lessons that are consistent with the definition of *inquiry* found in *Inquiry and the National Science Education Standards* (NRC 2000) share five key features:

1. Students need to answer a scientifically oriented question.
2. Students must collect data or use data collected by someone else.
3. Students formulate an answer to the question based on their analysis of the data.
4. Students connect their answer to some theory, model, or law.
5. Students communicate their answer to the question to someone else.

Introduction

Teachers often use inquiry-based lessons as a way to give students a firsthand experience with a concept before introducing terms and vocabulary (NRC 2012). Inquiry-based lessons, as a result, are often described as an "activity before content" approach to teaching science (Cavanagh 2007). The focus of these "activity before content" lessons, as the name implies, is to help students understand the core ideas of science. Inquiry-based lessons also give students more opportunities to learn how to plan and carry out investigations, analyze and interpret data, and develop explanations. These lessons also give students more voice and choice so they are more consistent with how science is done.

Although classroom-based research indicates that inquiry-based lessons are effective at helping students understand core ideas and give students more voice and choice than typical science lessons, they do not do as much as they could do to help students develop all four aspects of science proficiency (Duschl, Schweingruber, and Shouse 2007; NRC 2008, 2012). For example, inquiry-based lessons are usually not designed in a way that encourages students to learn how to use DCIs, CCs, and SEPs because they are often used to help student "learn about" important concepts or principles (NRC 2012). These lessons also do not give students an opportunity to participate in the full range of SEPs because these lessons tend to be designed in a way that gives students many opportunities to learn how to ask questions, plan and carry out investigations, and analyze and interpret data but few opportunities to learn how to participate in the practices that focus on how new ideas are developed, shared, refined, and eventually validated within the scientific community. These important sense-making practices include developing and using models; constructing explanations; arguing from evidence; and obtaining, evaluating, and communicating information (NRC 2012). Inquiry-based lessons that do not focus on sense-making also do not provide a context that creates a need for students to read, write, and speak, because these lessons tend to focus on introducing students to new ideas and how to design and carry out investigations instead of how to develop, share, critique, and revise ideas. These types of inquiry-based lessons, as a result, are often not used as a way to help students develop fundamental literacy skills. To help address this problem, teachers will need to start using instructional approaches that give students more opportunities to figure things out.

This emphasis on "figuring things out" instead of "learning about things" represents a big change in the way we have been teaching science in elementary school. To figure out how things work or why things happen in a way that is consistent with how science is actually done, students must have opportunities to use DCIs, CCs, and SEPs at the same time to make sense of the world around them (NRC 2012). This focus on students using DCIs, CCs, and SEPs at the same time during a lesson is called *three-dimensional instruction* because students have an opportunity to use all three dimensions of the *Framework* to understand how something works, to explain why something happens, or to develop a novel solution to a problem. When teachers use three-dimensional instruction inside their classrooms, they encourage students to develop or use conceptual models, develop explanations, share and critique ideas, and argue from evidence, all of which allow students to develop the knowledge and skills they need to be proficient in science (NRC 2012). A large body of research suggests that all students benefit from three-dimensional instruction because it gives all students more voice and choice during a lesson and it makes the learning process inside the classroom more active and inclusive (NRC 2012).

We think investigations that focus on making sense of how the world works are the perfect way to integrate three-dimensional science instruction into elementary classrooms. Well-designed investigations can provide opportunities for students to not only use one or more DCIs to understand how something works, to explain why something happens, or to develop a novel solution to a problem but also use several different CCs and SEPs during the same lesson. A teacher, for example, can give his or her students an opportunity to figure out a way to predict the motion of a ball over time when it dropped into a half-pipe. The teacher can then encourage them to use what they know about Forces and Motion (a DCI) and their understanding of Patterns (a CC) to plan and carry out an investigation to figure out where the ball will be in the half-pipe each time it changes direction. In addition to planning and carrying out an investigation, they must also ask questions; analyze and interpret data; use mathematics; construct an explanation; argue from evidence; and obtain, evaluate, and communicate information (seven different SEPs).

Using a DCI along with multiple CCs and SEPs at the same time is important because it creates a classroom experience that parallels how science is done. This, in turn, gives all students who participate in the investigation an opportunity to deepen their understanding of what it means to do science and to develop science-related identities (Carlone, Scott, and Lowder 2014; Tan and Barton 2008, 2010). In the following section, we will describe how to promote and support the development of science proficiency through three-dimensional instruction by using an innovative instructional model called argument-driven inquiry (ADI).

Argument-Driven Inquiry as a Way to Promote Three-Dimensional Instruction While Focusing on Literacy and Mathematics

The ADI instructional model (Sampson and Gleim 2009; Sampson, Grooms, and Walker 2009, 2011) was developed as a way to change how science is taught in our schools. Rather than simply encouraging students to learn about the facts, concepts, and terms of science, ADI gives students an opportunity to use DCIs, CCs, and SEPs to figure out how things work or why things happen. ADI also encourages children to think about "how we know" in addition to "what we have figured out." The ADI instructional model includes eight stages of classroom activity. These eight stages give children an opportunity to *investigate* a phenomenon; *make sense* of that phenomenon; and *evaluate and refine* ideas, explanations, or arguments. These three aspects of doing science help students learn how to figure something out and make it possible for them to develop and refine their understanding of the DCIs, CCs, and SEPs over time.

Students will use different SEPs depending on what they are trying to accomplish during an investigation, which changes as they move through the eight stages of ADI. For example, students must learn how to ask questions to design and carry out an investigation in order to investigate a phenomenon, which is the overall goal of the first two stages of ADI. Then during the next stage of ADI, the students must learn how to analyze and interpret data, use mathematics, develop models, and construct explanations to accomplish the goal of making sense of the phenomenon they are studying. Students then need to evaluate and refine ideas, explanations, or arguments during the last five stages of this instructional model. Students, as a result, must learn how to ask questions; obtain, evaluate, and communicate information; and argue from evidence. These three goals therefore provide coherence to a three-dimensional lesson, create a need for students to learn how to use each of the SEPs (along with DCIs and CCs), and will keep the focus of the lesson on figuring things out. We will discuss what students do during each stage of ADI in greater detail in Chapter 1.

ADI also provides an authentic context for students to develop fundamental literacy and mathematics skills. Students are able to develop these skills during an ADI investigation because the use of DCIs, CCs, and SEPs requires them to gather, analyze, interpret, and communicate information. Students, for example, must read and talk to others to learn to gather information and to find out how others are thinking. Students must also talk and write to share their ideas about what they are doing and what they have found out and to revise an explanation or model. Students, as a result, "are able to fine-tune their literacy skills when they engage in science investigations because so many of the sense-making tools of science are consistent with, if not identical to, those of literacy, thus allowing a setting for additional practice and refinement that can enhance future reading and writing efforts" (Pearson, Moje, and Greenleaf 2010, p. 460). Students must also use mathematics during an ADI investigation to measure what they are studying and to find patterns in their observations, uncover differences between groups, identify a trend over time, or confirm a relationship between two variables. They also use mathematics to make predictions. Teachers can therefore use ADI to help students develop important literacy and mathematics skills as they teach science. We will discuss how to promote and support the development of literacy and mathematics skills during the various stages of ADI with greater detail in Chapter 1. We will also describe ways to promote and support productive talk, reading, and writing during ADI in the Teacher Notes for each investigation.

ADI investigations also provide a rich language-learning environment for emerging bilingual students who are learning how to communicate in English. A rich language-learning environment is important because emerging bilingual students must (1) interact with people who know English well enough to provide both access to this language and help in learning it and (2) be in a social setting that will bring them in contact with these individuals so they have an opportunity to learn (Lee, Quinn, and Valdés 2013). Once these two conditions are met, people are able to learn a new language through meaningful use and interaction (Brown 2007; García 2005; García and Hamayan 2006; Kramsch 1998). ADI, and its focus on giving students opportunities to use DCIs, CCs, and SEPs to figure things out, also provides emerging bilingual students with opportunities to interact with English speakers and opportunities to do things with language inside the classroom (Lee, Quinn, and Valdés 2013). Emerging bilingual students therefore have an opportunity to use receptive and productive language to communicate and to negotiate meaning with others inside the science classroom. Teachers, as a result, can promote and support the acquisition of a new language by simply using ADI to give emerging bilingual students an opportunity to *investigate* a phenomenon; *make sense* of that phenomenon; and *evaluate and refine* ideas, explanations, or arguments with others and then provide support and guidance as they learn how to communicate in a new language. We will discuss how teachers can use ADI to promote language development with greater detail in Chapter 1. We will also provide advice and recommendations for supporting emerging bilingual students as they learn science and how to communicate in a new language at the same time in the Teacher Notes for each investigation.

Organization of This Book

This book is divided into seven sections. Section 1 includes two chapters. The first chapter describes the ADI instructional model. The second chapter provides an overview of the information that is associated with each investigation. Sections 2–6 contain the 14 investigations. Each investigation includes three components:

- Teacher Notes, which provides information about the purpose of the investigation and what teachers need to do to guide students through it.
- Investigation Handout, which can be photocopied and given to students at the beginning of the lesson. The handout provides the students with

a phenomenon to investigate, an overview of the DCI and the CC that students can use during the investigation, and a guiding question to answer.

- Checkout Questions, which can be photocopied and given to students at the conclusion of the investigation. The Checkout Questions consist of items that target students' understanding of the DCI and the CC addressed during the investigation.

Section 7 consists of five appendixes:

- Appendix 1 contains several standards alignment matrixes that can be used to assist with curriculum or lesson planning.
- Appendix 2 provides an overview of the CCs and nature of scientific knowledge (NOSK) and nature of scientific inquiry (NOSI) concepts that are a focus of the different investigations. This information about the CCs and the NOSK and NOSI concepts are included as a reference for teachers.
- Appendix 3 lists some frequently asked questions about ADI.
- Appendix 4 provides of a peer-review guide and teacher scoring rubric, which can also be photocopied and given to students.
- Appendix 5 provides a safety acknowledgment form, which can be photocopied and given to students.

References

Brown, D. H. 2007. Principles of language learning and teaching. 5th ed. White Plains, NY: Longman.

Carlone, H., C. Scott, and C. Lowder. 2014. Becoming (less) scientific: A longitudinal study of students' identity work from elementary to middle school science. *Journal of Research in Science Teaching* 51: 836–869.

Cavanagh, S. 2007. Science labs: Beyond isolationism. *Education Week* 26 (18): 24–26.

Driver, R., Asoko, H., Leach, J., Mortimer, E., and Scott, P. 1994. Constructing scientific knowledge in the classroom. *Educational Researcher* 23: 5–12.

Duschl, R. A., H. A. Schweingruber, and A. W. Shouse, eds. 2007. *Taking science to school: Learning and teaching science in grades K–8.* Washington, DC: National Academies Press.

García, E. E. 2005. *Teaching and learning in two languages: Bilingualism and schooling in the United States.* New York: Teachers College Press.

García, E. E., and E. Hamayan. 2006. What is the role of culture in language learning? In *English language learners at school: A guide for administrators,* eds. E. Hamayan and R. Freeman, 61–64. Philadelphia, PA: Caslon Publishing.

Lee, O., H. Quinn, and G. Valdés. 2013. Science and language for English language learners in relation to *Next Generation Science Standards* and with implications for *Common Core State Standards* for English language arts and mathematics. *Educational Researcher* 42 (4): 223–233. Available online at *http://journals.sagepub.com/doi/abs/10.3102/0013189X13480524.*

Kramsch, C. 1998. *Language and culture.* Oxford, UK: Oxford University Press.

National Research Council (NRC). 1999. *How people learn: Brain, mind, experience, and school.* Washington, DC: National Academies Press.

National Research Council (NRC). 2000. *Inquiry and the National Science Education Standards.* Washington, DC: National Academies Press.

National Research Council (NRC). 2008. *Ready, set, science: Putting research to work in K–8 science classrooms.* Washington, DC: National Academies Press.

National Research Council (NRC). 2012. *A framework for K–12 science education: Practices, crosscutting concepts, and core ideas.* Washington, DC: National Academies Press.

NGSS Lead States. 2013. *Next Generation Science Standards: For states, by states.* Washington, DC: National Academies Press. *www.nextgenscience.org/next-generation-science-standards.*

Pearson, P. D., E. B. Moje, and C. Greenleaf. 2010. Literacy and science: Each in the service of the other. *Science* 328 (5977): 459–463.

Sampson, V., and L. Gleim. 2009. Argument-driven inquiry to promote the understanding of important concepts and practices in biology. *American Biology Teacher* 71 (8): 471–477.

Sampson, V., J. Grooms, and J. Walker. 2009. Argument-driven inquiry: A way to promote learning during laboratory activities. *The Science Teacher* 76 (7): 42–47.

Sampson, V., J. Grooms, and J. Walker. 2011. Argument-driven inquiry as a way to help students learn how to participate in scientific argumentation and craft written arguments: An exploratory study. *Science Education* 95 (2): 217–257.

Tan, E., and A. Barton. 2008. Unpacking science for all through the lens of identities-in-practice: The stories of Amelia and Ginny. *Cultural Studies of Science Education* 3 (1): 43–71.

Tan, E., and A. Barton. 2010. Transforming science learning and student participation in sixth grade science: A case study of a low-income, urban, racial minority classroom. *Equity and Excellence in Education* 43 (1): 38–55.

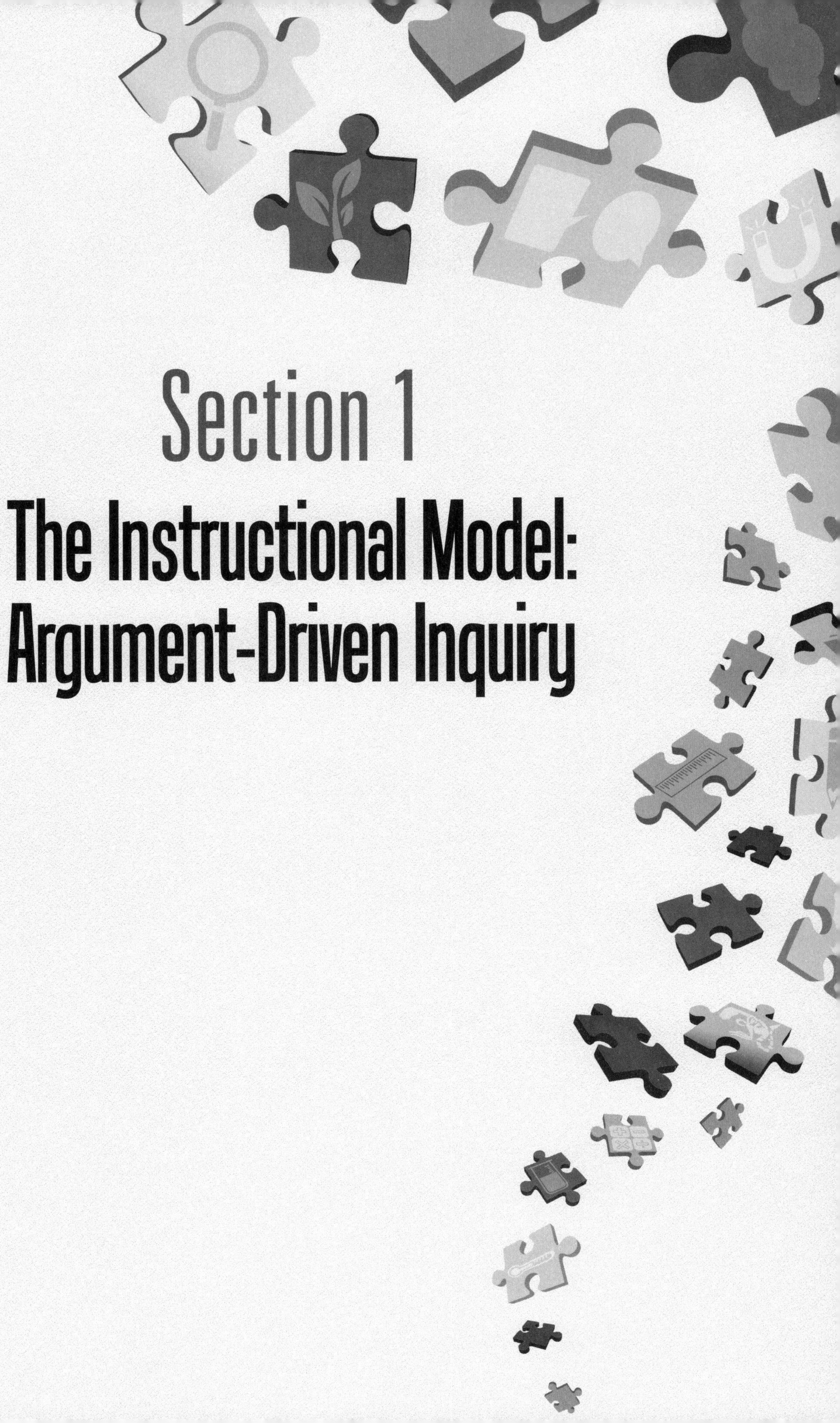

Section 1

The Instructional Model: Argument-Driven Inquiry

Chapter 1

An Overview of Argument-Driven Inquiry

The Argument-Driven Inquiry Instructional Model

The argument-driven inquiry (ADI) instructional model (Sampson and Gleim 2009; Sampson, Grooms, and Walker 2009, 2011) was created to help change the way science is taught in our schools. This instructional model includes eight stages of classroom activity. As children participate in each stage of ADI they have an opportunity to *investigate* a phenomenon; *make sense* of that phenomenon; and *evaluate and refine* ideas, explanations, or arguments. These eight stages of the instructional model provide a structure that supports children in third grade as they learn to plan and carry out an investigation to figure something out and enables them to develop and refine their understanding of the disciplinary core ideas (DCIs), crosscutting concepts (CCs), and scientific and engineering practices (SEPs) over time (NGSS Lead States 2013; NRC 2012). ADI also provides an authentic context for students to develop fundamental literacy skills and to learn or apply mathematical concepts and practices, and it provides a social setting that enables emerging bilingual students to acquire a new language as they learn science.

In this chapter, we will explain what happens during each of the eight stages of ADI. These eight stages are the same for every ADI investigation. Students, as a result, quickly learn what is expected of them during each stage and can focus their attention on learning how to use DCIs, CCs, and SEPs to figure out how something works or why something happens. Figure 1 provides an overview of the eight stages of the ADI instructional model.

FIGURE 1

The eight stages of the argument-driven inquiry instructional model

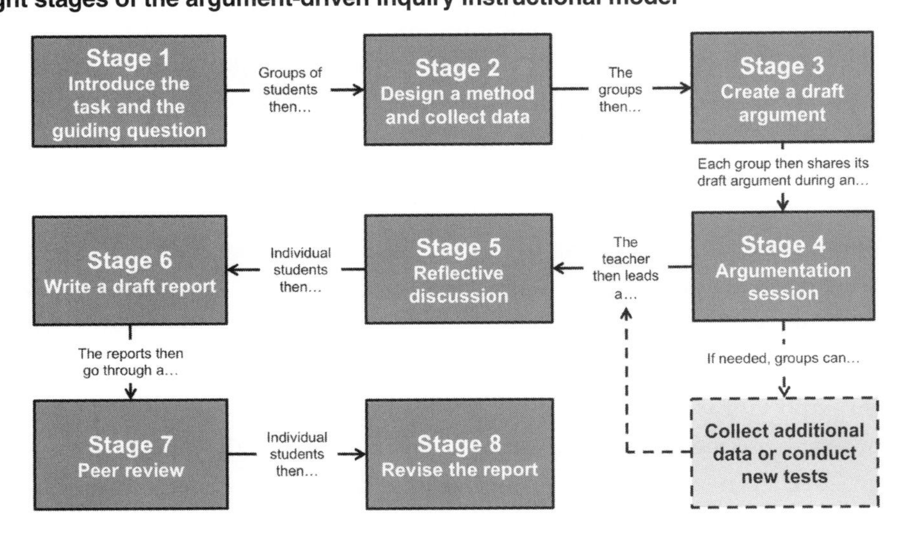

We will also provide hints (in boxes) for implementing each stage as part of our explanation of the ADI instructional model. To supplement the explanations and hints found in this chapter, Appendix 3 provides answers to frequently asked questions about ADI. These answers will help teachers encourage productive talk among students, support emerging bilingual students, improve students' reading comprehension, and improve the quality of feedback during the peer-review process; also included in these answers are techniques for helping students when they get stuck as they are developing their draft argument or writing their report.

Stage 1: Introduce the Task and the Guiding Question

An ADI activity begins with the teacher identifying a phenomenon to investigate and offering a guiding question for the students to answer. The goal of the teacher during this stage of the model is to create a need for students to use DCIs, CCs, and SEPs to figure something out. To accomplish this goal, teachers provide each student with a copy of an Investigation Handout. The teacher reads the first part of the handout "Introduction" out loud and then directs the students to explore a phenomenon for a few minutes. This exploration can be a firsthand or secondhand experience with a phenomenon. Examples of *firsthand experiences* include tinkering with several different magnets, dropping a marble into a bowl and watching how the motion of the marble changes over time, and using a rubber band to propel a marble into a cup. A *secondhand experience,* in contrast, involves watching a video of something that happens. An example of a video that provides a good secondhand experience with a phenomenon might be a pack of wolves hunting a musk ox or a beluga whale swimming under an ice sheet in the Arctic Ocean.

This brief exploration with a phenomenon is designed to encourage students to ask questions and create a need for them to figure something out. Students should be encouraged to record what they observe and any questions they might have during this brief exploration (see Figure 2). Emerging bilingual students should be allowed to use English, their native language, or some combination of the two (see Figure 2). The teacher should then give the students an opportunity to share their observations and questions with the rest of the class. At this point, students are interested and want to know more about the phenomenon. The teacher can then read the rest of the handout "Introduction" or have the students read it on their own. The handout also provides an overview of the DCI and the CC that the students will use during the investigation to figure things out, an overview of the task they need to complete, and a guiding question for the students to answer. This stage gives students an opportunity to learn how to ask questions (SEP 1) and obtain information (SEP 8) in the context of science.

It is also important for the teacher to hold a "tool talk" (Blanchard and Sampson 2018) during this stage, taking a few minutes to explain how to use the available materials and equipment. Teachers need to hold a tool talk because children in grade 3 are often unfamiliar with these materials and equipment. Even if the students are familiar with them, they may use them incorrectly or in an unsafe manner unless they are reminded about how they work and the proper way to use them. The teacher should therefore review specific safety protocols and precautions as part of the tool talk. Including a tool talk during this stage is useful because students often find it difficult to design aty method to collect the data needed to answer the guiding question (the task of stage 2) when they do not understand what they can and cannot do with the available materials and equipment.

Keep the following points in mind during stage 1 of ADI:

- The initial hands-on activity is important. It is designed to provide a phenomenon to explain, trigger students' curiosity, and "create a need to read." Do not skip it!
- There are many supports for helping students comprehend what they read (i.e., activating prior knowledge, providing a shared experience, making connections, synthesizing, and talking with peers) already embedded into this stage. You might not need to provide much extra support.
- Don't worry if students "don't get it" yet or struggle to comprehend what they are reading at this point; they will revisit the text later in the lesson.
- Students will likely use some or most of the information that they include in the "Things we KNOW from what we read…" box of the "know / need to figure out" chart to help justify their evidence in their arguments during stage 3 of the investigation.

Once all the students understand the goal of the investigation and how to use the available materials, the teacher should divide the students into small groups (we recommend three or four students per group) and move on to the second stage of the instructional model.

Stage 2: Design a Method and Collect Data

Small groups of students develop a method to gather the data they need to answer the guiding question and then carry out that method during

FIGURE 2

A student recording observations of a phenomenon in the Investigation Handout

this stage of ADI. The overall intent of this stage is to provide students with an opportunity to use a DCI and a CC to plan and carry out an investigation (SEP 3). It also gives children in third grade an opportunity to learn how to use appropriate data collection techniques, which include but are not limited to controlling variables, making multiple observations, and quantifying observations, and how to use different types of data collection tools, such as rulers to measure length, scales to measure mass, and graduated cylinders to measure volume. This stage of ADI also gives students a chance to see why some approaches to data collection and tools work better than others and how the method used during a scientific investigation is based on the nature of the question and the phenomenon under investigation. Students even begin to learn how to deal with the uncertainties that are associated with all empirical work as they discover the importance of attending to precision when they take measurements and attempt to eliminate factors that may change the results of their tests.

This stage begins with students discussing two questions that are designed to encourage the students to use a CC as a lens to determine what data they need to collect and how they should collect it (see the "Plan Your Investigation" section of the handout). For example, students might be asked to discuss questions such as

- What types of *patterns* might we look for to help answer the guiding question?
- What information do we need to find a relationship between a *cause* and an *effect*?
- What type of *cause-and-effect relationship* should we look for?
- How might the *structure* of what you are studying relate to its *function*?

FIGURE 3

A group of students working together to plan an investigation

Once students have discussed these questions in their small groups and shared their ideas with the rest of the class, each small group can begin to plan their investigation. To facilitate this process, the teacher should direct the students to fill out the graphic organizer in the "Plan Your Investigation" section of the handout. The graphic organizer helps guide students through the process of planning an investigation by encouraging them to think about what type of data they will need to collect, how to collect it, and how to analyze it. Figure 3 shows a group of students working together to plan an investigation using the graphic organizer.

If the students get stuck as they are planning their investigation, the teacher should not tell them what to do. Instead, the teacher should bring over some of the available materials and ask them probing questions, such as "I have all these materials, what data do I need to collect?" or "Now that we know what data we need, what should we do first?" Once the students create a plan for their investigation, the teacher should look it over and either approve it or offer suggestions about how to improve the investigation. If the teacher identifies a flaw in the investigation proposal, he or she should ask probing questions to help the students identify and correct the flaw rather than just telling them what to fix. For example, the teacher could ask the students questions such as "I'm not sure what you mean here, could you explain that another way?" or "Do you think you have all the information you need to answer the guiding question?"

Keep the following points in mind during stage 2 of ADI:

- Not all investigations in science are experiments. Scientists use different methods to answer different types of questions.
- The graphic organizer found in the "Plan Your Investigation" section of the Investigation Handout is designed to help students think about how to answer the guiding question of the investigation, what data they need to collect, and how they will need to analyze it; it is not a scientific method.
- The graphic organizer makes student thinking about investigation design visible so teachers can use it as an embedded formative assessment.
- Students can use some or most of the information that they include in the graphic organizer to help write their investigation report in stage 6.

It is important to remember that the student-designed investigations do not all need to be the same. The students will learn more about how to do science during the later stages of the instructional model when the groups use different methods to collect and analyze data. The groups should then carry out their plan and collect the data they need to answer the guiding question once their proposal is approved by the teacher (see Figure 4).

FIGURE 4

A group of students working together to collect data

Stage 3: Create a Draft Argument

The next stage of the instructional model calls for each group to create a draft argument. To accomplish this task, the students must first analyze the measurements (e.g., temperature and mass) and/or observations (e.g., appearance and location) they collected during stage 2 of ADI. Once the groups have analyzed and interpreted the results of their analysis (SEP 4), they will need to make sense of the phenomenon based on what they found out. They can then develop an answer to the guiding question based on what they figured out. Often, but not always, developing an adequate answer to the guiding question requires the students to construct an explanation (SEP 6). The students can then create a draft argument to share what they have learned with the other students in the class. The intent of having students craft arguments is to encourage them to focus not only on "what they figured out" but also on "how they know what they know." It also enables students to master the mathematical ideas and practices outlined in the *Common Core State Standards* in mathematics (*CCSS;* NGAC and CCSSO 2010).

The argument that the students create consists of a claim, the evidence they are using to support their claim, and a justification of their evidence. The *claim* is their answer to the guiding question and thus is often an explanation for how or why something happens. The *evidence* consists of an analysis of the data they collected and an interpretation of the analysis. The evidence often includes a graph that shows a difference between groups, a trend over time, or relationship between variables; it also includes a statement or two that explains what the analysis means (but not why it matters). The *justification of the evidence* is a statement that explains why the evidence matters. The justification of the evidence, in other words, is used to defend the choice of evidence by making the DCIs, CCs, and/or assumptions underlying the collection of the data, the analysis of the data, and interpretation of the analysis explicit so the other people can understand why the evidence is important and relevant. The components of a scientific argument are illustrated in Figure 5. Crafting an

FIGURE 5

The components of a scientific argument and some criteria for evaluating an argument

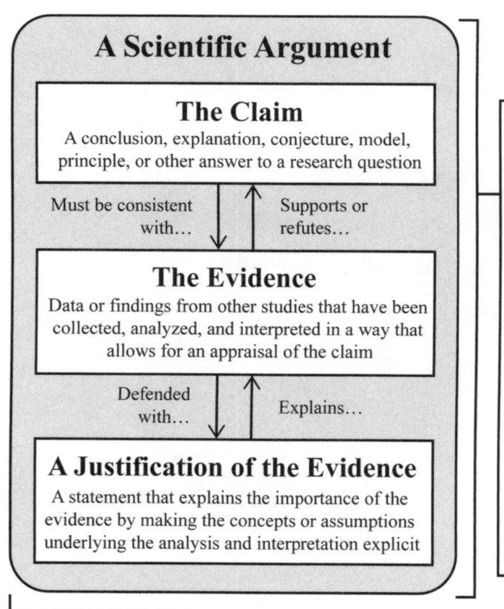

The quality of an argument is evaluated using...

Empirical Criteria

- The claim is consistent with the evidence.
- The amount of evidence is sufficient.
- The evidence is relevant.
- The method used to collect the data was appropriate and rigorous.
- The method used to analyze the data was appropriate and sound.

Theoretical Criteria

- The claim is sufficient (includes everything needed).
- The claim is useful (helps us understand the phenomenon we are studying).
- The claim is consistent with current theories, laws, or models.
- The interpretation of the data analysis is consistent with current theories, laws, or models.

The generation and evaluation of an argument are shaped by...

Discipline-Specific Norms and Expectations

- The theories and laws used by scientists within a discipline
- The methods of inquiry that are accepted by scientists within a discipline
- Standards of evidence shared by scientists within a discipline
- How scientists communicate with each other within a discipline

argument that consisting of these three components helps students learn how to argue from evidence (SEP 7).

It is not enough for students to be able to include all the components of an argument when they have an opportunity to argue from evidence. It is also important for students to understand that in science some arguments are better than others. Therefore, an important aspect of arguing from evidence in science involves the evaluation of the various components of the arguments put forward by others. The framework provided in Figure 5 highlights two types of criteria that students can and should be encouraged to use to evaluate an argument in science: empirical criteria and theoretical criteria. *Empirical criteria* include

- how well the claim fits with all available evidence,
- the sufficiency of the evidence,
- the relevance of the evidence,
- the appropriateness and rigor of the method used to collect the data, and
- the appropriateness and soundness of the method used to analyze the data.

Theoretical criteria refer to standards that allow us to judge how well the various components are aligned with the DCIs and CCs of science; examples of these criteria are

- the sufficiency of the claim (i.e., Does it include everything needed?);
- the usefulness of the claim (i.e., Does it help us understand the phenomenon?);
- how consistent the claim is with accepted theories, laws, or models (e.g., Does it fit with our current understanding of motion and forces?); and
- how consistent the interpretation of the results of the analysis is with accepted theories, laws, or models (e.g., Is the interpretation based on what we know about the relationship between structure and function?).

What counts as quality in terms of these different criteria varies from discipline to discipline (e.g., biology, geology, physics, chemistry) and within the specific fields of each discipline (e.g., the fields of ecology, botany, and zoology within the discipline of biology). This variation in what counts as quality is due to differences in the types of phenomena that the scientists within these disciplines or fields investigate (e.g., changes in populations over time, how a trait is inherited in a plant, an adaptation of an animal), the types of methods they use (e.g., descriptive studies, experimentation, computer modeling), and the different DCIs that they use to figure things out. It is therefore important to keep in mind that what counts as a quality argument in science is discipline and field dependent.

Each group of students should create their draft argument in a medium that can easily be viewed by the other groups. This is important because each group will share their draft argument with the other students in the class during the next stage of ADI (stage 4). We recommend that students use dry erase markers to create their draft argument on a

FIGURE 6

A group of students creating a draft argument on a whiteboard

2'× 3' whiteboard during this stage (see Figure 6). Students can also create their draft arguments using presentation software such as Microsoft's PowerPoint, Apple's Keynote, or Google Slides and devote one slide to each component of an argument. They can then share their arguments with others using a tablet or a laptop (see Figure 11 in the "Stage 4: Argumentation Session" section, later in this chapter). The choice of medium is not important as long as students are able to easily modify the content of their argument as they work and other students can easily see each component of their argument. Students should include the guiding question of the investigation and the three main components of an

FIGURE 7

The components of an argument that should be included on a whiteboard (outline)

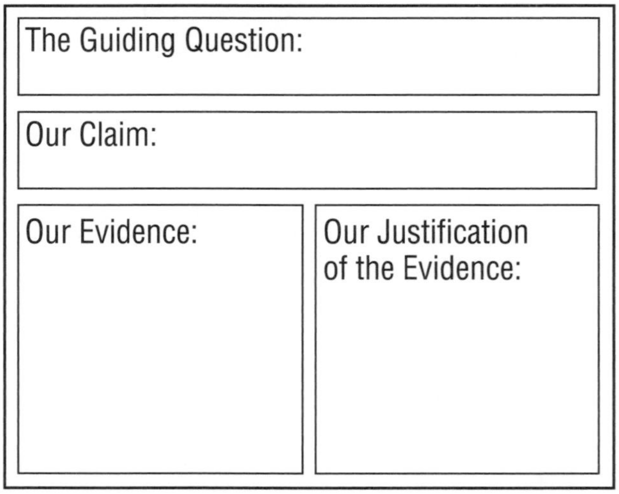

The Guiding Question:	
Our Claim:	
Our Evidence:	Our Justification of the Evidence:

Note: This outline is referred to as the "Argument Presentation on a Whiteboard" image in stage 3 of each investigation.

argument on the whiteboard. Figure 7 shows the general layout for a presentation of an argument.

Figures 8 and 9 provide examples of an argument crafted by students. Notice that the argument in Figure 8 is written in English and the argument in Figure 9 is written in Spanish. To help support emerging bilingual students in learning science and English at the same time, it is important to allow students to communicate their ideas in their first language, English, or a combination of the two. This is important because (a) it allows students to use their home language and culture as a resource for making sense of the phenomenon (Escamilla and Hopewell 2010; Goldenberg and Coleman 2010; González, Moll, and Amanti 2005), (b) students learn new languages through meaningful use and interaction (Brown 2007; García 2005; García and Hamayan 2006), and (c) students' academic language development in their native language facilitates their academic language development in English (Escamilla and Hopewell 2010; García and Kleifgen 2010; Gottlieb, Katz, and Ernst-Slavit 2009).

This stage of the model can be challenging for students in third grade because they are rarely asked to make sense of a phenomenon based on raw data, so it is important for teachers to actively work to support their sense-making. In this stage, teachers should circulate from group to group to act as a resource person for the students, asking questions that prompt them to think about what they are doing and why. To help students remember the goal of the activity, you can ask questions such as "What are you trying to figure out?" You can also ask

FIGURE 8

An example of a student-generated argument on a whiteboard

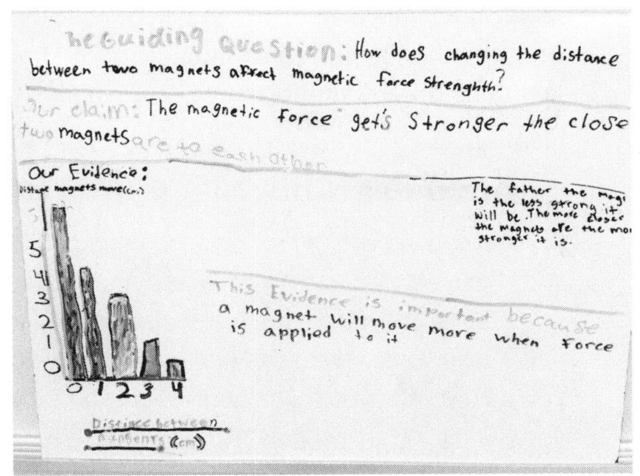

FIGURE 9

An example of a student-generated argument written in Spanish

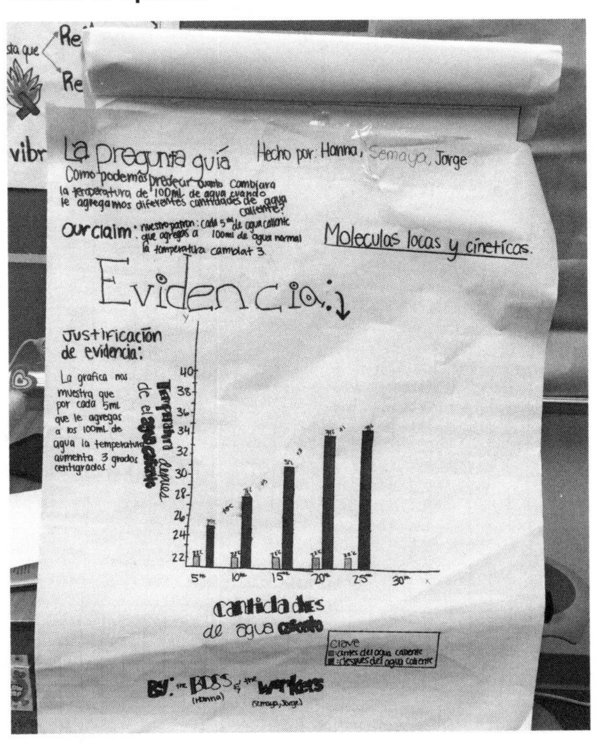

them questions such as "Why is that information important?" or "Why is that analysis useful?" to encourage them to think about whether or not the data they are analyzing are relevant or the analysis is informative. To help them remember to use rigorous criteria to determine if a claim is acceptable or not, you can ask, "Does that fit with all the data?" or (for an investigation concerning forces and motion) "Is that consistent with what we know about forces and motion?"

Keep the following points in mind during stage 3 of ADI:

- If you uncover a flaw or if something important is missing in an argument as you move from group to group, ask them probing questions such as "I see you put a table here, but you did not really explain what we should pay attention to. Is there a way to help your classmates understand what is really important in this table?" or "I'm not sure what you mean here, could you explain that another way?"
- All the arguments should not be the same (or perfect) at this point in the lesson. The argumentation session during stage 4 will be more interesting (and students will learn more about how to do science) if each group has a different claim, evidence, and justification of the evidence so there is something to discuss.
- The students will have an opportunity to revise their draft argument at the end of the argumentation session (stage 4 of the lesson).
- One of the best ways to ensure that important ideas spread through the class is to help one or two groups develop a very strong component of their argument. Other groups will see these examples and add a version of them to their own arguments during the argumentation session. For example, help one group make a perfect graph, another group write out a perfect interpretation of their analysis, and a third group include a core idea that the other groups are missing in their justification.

It is important to remember that at the beginning of the school year, students will struggle to develop arguments and will often rely on inappropriate criteria such as plausibility (e.g., "That sounds good to me") or fit with personal experience (e.g., "But that is what I saw on TV once") as they attempt to make sense of the phenomenon, construct explanations, and support their ideas with sufficient evidence and an adequate justification of the evidence. However, as students learn why it is useful to use evidence in an argument, what makes evidence valid or acceptable from a scientific perspective, and the importance of providing a justification for their evidence through repeated practice, *students will improve their ability to argue from evidence* (Grooms, Enderle, and Sampson 2015; Strimaitis et al. 2017). This is an important principle underlying the ADI instructional model.

Stage 4: Argumentation Session

The fourth stage of ADI is the argumentation session. In this stage, each group is given an opportunity to share, evaluate, and revise their draft arguments. The process of sharing their arguments with their classmates requires students to communicate their explanations for the phenomenon under investigation (SEP 8) and support their ideas with evidence (SEP 7). Other students in the class are expected to listen to presentations of these arguments, ask questions as needed (SEP 1), evaluate the arguments based on empirical and theoretical criteria (SEP 4 and SEP 7), and then offer critiques (SEP 8) along with suggestions for improvement. At the end of this stage the students are expected to revise their draft arguments based on what they learned from interacting with other students and seeing examples of other arguments.

This stage is included in the ADI instructional model because research indicates that students develop a better understanding of DCIs and CCs, learn how to argue from evidence, and acquire better critical-thinking skills when they are exposed to alternative ideas, respond to the questions and challenges of other students, and are encouraged to evaluate the merits of competing ideas (Duschl, Schweingruber, and Shouse 2007; NRC 2012). Research also suggests that students learn how to distinguish between alternative ideas using rigorous scientific criteria and are able to develop scientific habits of mind (such as treating ideas with initial skepticism, insisting that the reasoning and assumptions be made explicit, and insisting that claims be supported by valid evidence) when they have an opportunity to participate in these argumentation sessions (Sampson, Grooms, and Walker 2011). This stage provides the students with an opportunity to learn from and about the practice of arguing from evidence.

FIGURE 10

A group of students presenting their arguments to the other groups in the class during a whole-class presentation argumentation session

Teacher can use different types of formats during the argumentation session. One format is the *whole-class presentation format,* in which each group gives a presentation to the whole class (see Figure 10). This format is often useful when students are first learning how to propose, support, evaluate, challenge, and refine ideas in the context of science because the teacher can provide more support as students interact with each other. This format begins with the teacher asking a group to present their argument to the class. The teacher can then encourage students in the class to ask questions, offer critiques, and give the presenters suggestions about ways to improve. The teacher can also ask questions as needed. We also recommend that the presenters keep a record of the critiques made by their classmates and any suggestions for improvement. The students who listen to the presentation should also be encouraged to keep a record of good ideas or potential ways to improve their own arguments by recording them in the "Argumentation Session" section of the handout.

FIGURE 11

Students critiquing arguments and providing feedback during a gallery walk argumentation session. Students in this class created draft arguments using a tablet.

FIGURE 12

Some of the feedback that students gave to each other using sticky notes during a gallery walk argumentation session

Another format is called the *gallery walk.* In this format, each group sets up their argument so others can see it at their workstation. The entire group then moves to a different workstation. While at a different workstation, a group is expected to read the argument at that workstation (which was created by a different group) and offer critiques and suggestions for improvement (see Figure 11). Notice in Figure 11 that the students in this classroom created their draft argument on tablets and then used the tablets to share their argument with their classmates. The students should move to a different workstation after a few minutes so they can read and critique an argument written by a different group. Students should also be encouraged to keep a record of good ideas or potential ways to improve their own arguments as they move from workstation to workstation to offer critiques and feedback.

After repeating this process in the gallery walk three or four times, every group will have had an opportunity to read and critique three or four different arguments and receive feedback from three or four different groups. Students often use sticky notes to provide feedback to each other during a gallery walk argumentation session (see Figures 11 and 12). This type of format is useful because everyone in the classroom will be actively engaged during the argumentation session and will have a chance to see different arguments. This format, however, does not give students a chance to support, critique, and challenge the ideas, explanations, and arguments of other groups through talk, because two different groups are never at the same workstation at the same time.

A third format is called the *modified gallery walk.* We recommend that teachers use this format rather than a whole-class presentation format or gallery walk format whenever possible because it provides more opportunities for student-to-student talk and ensures that all ideas are heard and that all students are actively involved in the process. This is especially important for helping students develop speaking and listening skills. It also provides a context for emerging bilingual students to use both productive (speaking and writing) and receptive (listening and reading) language and learn science through meaningful interactions (Brown 2007; García and Hamayan 2006). In the modified gallery walk format (see Figure 13), one or two members of the group stay at their workstation to share their group's ideas (we call these students presenters) while the other group members go to different groups one at a time to listen to and critique the arguments developed by their classmates (we call them reviewers). We recommend that reviewers visit at least three different workstations during the argumentation session. We also recommend that the presenters keep a record of the critiques made by their classmates and any suggestions for improvement (see Figure 14). The reviewers should also be encouraged to keep a record of good ideas or potential ways to improve their own arguments as they travel from group to group. The presenters and the reviewers can record this information in the "Argumentation Session" section of the Investigation Handout.

FIGURE 13

An example of a modified gallery walk argumentation session. Notice that in this format there are multiple discussions going on at the same time.

FIGURE 14

An example of a modified gallery walk argumentation session. Notice that the presenter and a reviewer are recording good ideas that they can use to improve their argument.

It is important to note, however, that supporting and promoting productive interactions between students inside the classroom can be difficult because the practices of arguing from evidence (SEP 7) and obtaining, evaluating, and communicating information in science (SEP 8) are foreign to most students when they first begin participating in ADI. In a non-ADI classroom, students are not typically expected to think about or engage with the ideas, explanations, or argument of their classmates, so students are often reluctant to ask questions (SEP 1) in this type of context. To encourage more productive interactions between students, materials, and ideas, the ADI instructional model requires students to generate their arguments in a medium that can be seen by others. By looking at whiteboards, paper, or slides, students tend to focus their attention on evaluating evidence and how well the various components of the argument align with the DCI and the CC that students are using to figure things out rather than attacking the source of the ideas. This strategy often makes the discussion more productive and makes it easier for students to identify and weed out faulty ideas. It is also important for the students to view the argumentation session as an opportunity to learn. The teacher, therefore, should describe the argumentation session as an opportunity for students to collaborate with their peers and as a chance to give each other feedback so the quality of all the arguments in the classroom can be improved, rather than as an opportunity to determine who is right or wrong. This is why we ask students to keep track of what they see and hear so they can use the ideas of others to make their own argument better.

Just as is the case in earlier stages of ADI, it is important for the classroom teacher to be involved in (without leading) the discussions during the argumentation session. Once again, the teacher should move from group to group or from student to student to keep everyone involved and model what it means to argue from evidence. The teacher can ask the presenters questions such as "Why did you decide to analyze the available data like that?" or "Were there any data that did not fit with your claim?" to encourage students to use empirical criteria to evaluate the quality of the arguments. The teacher can also ask the presenters to explain how the claim they are presenting fits with a DCI or a CC or to explain why the evidence they used is important based on a DCI or a CC. In addition, the teacher can also ask the students who are listening to a presentation questions such as "Do you think their analysis is accurate?" or "Do you think their interpretation is sound?" or even (in the case of an investigation related to forces and motion) "Do you think their claim fits with what we know about forces and motion?" These questions can serve to remind students to use empirical and theoretical criteria to evaluate an argument during the discussions. Overall, it is the goal of the teacher at this stage of the lesson to encourage students to think about how they know what they know and why some claims are more valid or acceptable in science. This stage of the model, however, is not the time to tell the students that they are right or wrong.

Keep the following points in mind during stage 4 of ADI:

- Make sure to stress that the goal of the argumentation session is to share ideas and to help each other out. The goal of the argumentation session is not to see who has the best argument. Students need to see value in the ideas of others and view their classmates as a resource. This does not happen when students are trying to show others that they have the best argument.
- Be sure to remind students before the argumentation session starts that they will have an opportunity to revise their arguments based on what they learn from their peers at the end of the argumentation session. This is important because students need to have a reason to engage with the ideas of others.
- Encourage students to talk to each other during this stage. It is really important to give students voice and choice during the argumentation session.
- Use a note card to keep track of interesting ideas you hear or see, good examples, or important contributions that were made by different students when the students are sharing and critiquing the arguments. You can bring these ideas, examples, or contributions up at the end of this stage (before students begin revising their arguments) or during stage 5 (the reflective discussion).

At the end of the argumentation session, it is important to give students time to meet with their original group so they can discuss what they learned by interacting with individuals from the other groups and they can revise their draft arguments (see Figure 15). This process can begin with the presenters sharing the critiques and the suggestions for improvement that they heard during the argumentation session. The students who visited the other groups during the argumentation can then share their ideas for making the arguments better, based on what they observed and discussed at other stations. Students often realize that the way they collected or analyzed data was flawed in some way at this point in the process. The

FIGURE 15

Students discussing what they learned from other groups as they revise their draft arguments at the end of the argumentation session

teacher should therefore encourage students to collect new data or reanalyze the data they collected as needed. Teachers can also give students time to conduct additional tests of ideas or claims. At the end of this stage, each group should have a final argument that is much better than their draft one.

Stage 5: Reflective Discussion

The teacher should lead a whole-class reflective discussion during stage 5 of ADI. The intent of this discussion is to give students an opportunity to think about and share what they know and how they know it. This stage enables the teacher to ensure that all students understand the DCI and the CC they used during the investigation. It also encourages students to think about ways to improve their participation in scientific practices such as planning and carrying out investigations, analyzing and interpreting data, and arguing from evidence. At this point in the instructional sequence, the teacher should also encourage students to think about a nature of scientific knowledge (NOSK) or nature of scientific inquiry (NOSI) concept (Abd-El-Khalick and Lederman 2000; Lederman and Lederman 2004; Schwartz, Lederman, and Crawford 2004) (see Appendix 2).

It is important to emphasize that the reflective discussion is not a lecture; it is an opportunity for students to think about important ideas and practices and to share what they know or do not understand. The more students talk during this stage, the more meaningful the experience will be for them and the more a teacher can learn about student thinking.

The teacher should begin the discussion by asking students to share what they know about the DCI and the CC they used to figure things out during the investigation. The teacher can give a demonstration and/or show images as prompts and then ask questions to encourage students to think about how the DCI and the CC helped them explain the phenomenon under investigation and how they used the DCI and the CC to provide a justification of the evidence in their arguments. The teacher should not tell the students what results they should have obtained or what information should be included in each argument. Instead, the teacher should focus on the students' thoughts about the DCI and the CC by providing a context for students to share their views and explain their thinking. Remember, this stage of ADI is a *discussion,* not a presentation about what the students "should have seen" or "should have learned." We provide recommendations about what teachers can do and the types of questions that teachers can ask to facilitate a productive discussion about the DCI and the CC during this stage as part of the Teacher Notes for each investigation.

Next, the teacher should encourage the students to think about what they learned about the practices of science and how to design better investigations in the future. This is important because students are expected to design their own investigations, decide how to analyze and interpret data, and support their claims with evidence in every ADI investigation. These practices are complex, and students cannot be expected to master them without being given opportunities to try, fail, and then learn from their mistakes. To encourage students to learn from their mistakes during an investigation, students must have an opportunity to reflect on what went well and what went wrong. The teacher should therefore encourage the students to think about what they did during their investigation, how they chose to analyze and interpret data, how they decided to argue from evidence, and what they could have done better. The teacher can then use the students' ideas to highlight what does and does not count as quality or rigor in science and to offer advice about ways to improve in the future. Over time, students will gradually improve their abilities to participate in the practices of science as they learn what works and what does not. To help facilitate this process, we provide questions that teachers can ask students to help elicit their ideas about the practices of science and set goals for future investigations in the Teacher Notes for each investigation.

Keep the following points in mind during stage 5 of ADI:

- Keep this stage short—no more than 15 minutes.
- Make sure that students are doing most of the talking during this stage.
- Your goal during the reflective discussion is to figure out what students are thinking and build on their ideas as needed.
- Make sure that every student has an opportunity to contribute to the discussion and that they feel like their ideas and contributions are valued.

The teacher should end this stage with an explicit discussion of a NOSK or NOSI concept, using what the students did during the investigation to help illustrate one of these important concepts (NGSS Lead States 2013). This stage provides a golden opportunity for explicit instruction about NOSK and how this knowledge develops over time in a context that is meaningful to the students. For example, teachers can use the investigation as a way to illustrate the differences between observations and inferences, data and evidence, or theories and laws. Teachers can also use the investigation as a way to illustrate NOSI. For example, teachers might discuss the following concepts:

- How scientists investigate questions about the natural world
- The wide range of methods that scientists can use to collect data
- How science is a way of knowing
- The assumptions that scientists make about order and consistency in nature

Research in science education suggests that students only develop an appropriate understanding of NOSK and NOSI concepts when teachers *explicitly* discuss them as part of a lesson (Abd-El-Khalick and Lederman 2000; Lederman and Lederman 2004; Schwartz, Lederman, and Crawford 2004). In addition, by embedding a discussion of a NOSK concept or a NOSI concept into each investigation, teachers can highlight these important concepts over and over again throughout the school year rather than just focusing on them during a single unit. This type of approach makes it easier for students to learn these abstract and sometimes counterintuitive concepts. As part of the Teacher Notes for each investigation, we provide recommendations about which concepts to focus on and examples of questions that teachers can ask to facilitate a productive discussion about these concepts during this stage of the instructional sequence.

Stage 6: Write a Draft Report

Stage 6 is included in the ADI model because writing is an important part of doing science. Scientists must be able to read and understand the writing of others as well as evaluate its worth. They also must be able to share the results of their own research through writing. In addition, writing helps students learn how to articulate their thinking in a clear and concise manner, encourages metacognition, and improves student understanding of DCIs and CCs (Wallace, Hand, and Prain 2004). Finally, and perhaps most important, writing makes each student's thinking visible to the teacher (which facilitates assessment) and enables the teacher to provide students with the educative feedback they need to improve.

In stage 6, each student is required to write an investigation report using his or her Investigation Handout and his or her group's argument as a starting point. The report should address three fundamental questions:

1. What were you trying to figure out and why?
2. What did you do to answer your question and why?
3. What is your argument?

The format of the report is designed to emphasize the persuasive nature of science writing and to help students learn how to communicate in multiple modes (words, figures, tables, and equations). The three-question format is well aligned with the components of research articles (i.e., introduction, method, results and discussion) but allows students to see the important role argument plays in science. We have included sentence starters in the "Draft Report" section of the Investigation Handout to help facilitate the writing process (see Figure 16). The sentence starters are intended to act a guide for students as they learn to write in the context of science and should make the assignment less intimidating for students. We recommend that you use these sentence starters to encourage students to write in a clear and concise manner. Over time, you may decide that you no longer need to use them.

Stage 6 of ADI is important because it allows students to learn how to construct an explanation (SEP 6), argue from evidence (SEP 7), and communicate information (SEP 8). It also enables students to master the disciplinary-based writing skills outlined in the *Common Core*

State Standards in English language arts (*CCSS ELA;* NGAC and CCSSO 2010). As discussed in "Stage 3: Create a Draft Argument," allowing emerging bilingual students to write and communicate what they have learned in their first language, English, or a combination of the two will help support these students in learning science and English at the same time. It is important to scaffold this process for the emerging bilingual students based on their current level of English language proficiency because (a) students' development of academic language and academic content knowledge are interrelated processes (Echevarría, Short, and Powers 2006; Gottlieb, Katz, and Ernst-Slavit 2009) and (b) students' access to instructional tasks that require complex thinking is enhanced when linguistic complexity and instructional support match their levels of language proficiency (Gottlieb, Katz, and Ernst-Slavit 2009).

FIGURE 16

Students writing a draft report to share what they figured out during the investigation

Keep the following points in mind during stage 6 of ADI:

- There are many writing supports (i.e., sentence starters, pre-writing, graphic organizers) already embedded into this stage. You might not need to provide much extra support for your students.
- Don't worry if students do not produce a perfect report at this point. All the reports will be reviewed and revised before they are turned in to you. The reports should be viewed as a starting point. Encourage students to "just get something down on paper" so "we can work on it together."
- Writing is an important component of this model. Do not skip this stage. When a student writes the report on his or her own, it not only helps each student *learn to write* in the context of science (an important literacy skill) but also gives each student opportunity to *write to learn* (develop a better understanding of the core ideas of science by writing about them).
- Do not overscaffold the writing progress. Mistakes are opportunities to learn. Do not take these opportunities away from your students.

Stage 7: Peer Review

The students have an opportunity to review the reports in pairs using the peer-review guide and teacher scoring rubric (PRG/TSR; see Appendix 4) during the seventh stage of ADI. The PRG/TSR contains specific criteria that are to be used by a pair of students as they evaluate the quality of each section of the investigation report as well as quality of the writing. There is also space for the reviewers to provide the author with feedback about how to improve the report. Once a pair of students finishes reviewing a report as a team, they are given another report to review. When students are grouped together in pairs, they only need to review two different reports. Be sure to give students only 15 minutes to review each report (we recommend setting a timer to help manage time). When students are grouped into pairs and given 15 minutes to complete each review, the entire peer-review process can be completed in 30 minutes (2 different reports × 15 minutes = 30 minutes).

Reviewing each report as a pair using the PRG/TSR is an important component of the peer-review process because it provides students with a forum to discuss what counts as high quality or acceptable and, in so doing, forces them to reach a consensus during the review process. This method also helps prevent students from checking off "yes" for each criterion on the PRG/TSR without thorough consideration of the merits of the report. It is also important for students to provide constructive and specific feedback to the author when areas of the paper are found to not meet the standards established by the PRG/TSR. The peer-review process provides students with an opportunity to read good and bad examples of the reports. This helps the students learn new ways to organize and present information, which in turn will help them write better on subsequent reports. It also provides an opportunity and a mechanism for all students in the classroom, including emerging bilingual students, to develop shared norms for what counts as high-quality writing in the context of science.

This stage of the model is intended to give students opportunities to ask questions (SEP 1) and obtain and evaluate information (SEP 8). This stage also gives students an opportunity to develop the reading skills that they need to be successful in science. Students must be able to determine the central ideas or conclusions of a text and determine the meaning of symbols, key terms, and other domain-specific words. In addition, students must be able to assess the reasoning and evidence that an author includes in a text to support his or her claim and compare or contrast findings presented in a text with those from other sources when they read a scientific text. Students can develop all these skills, as well as the other discipline-based reading skills found in the *CCSS ELA*, when they are required to read and critically review reports written by their classmates. This stage is also beneficial for emerging bilingual students because students learn language through meaningful use, such as reading and reviewing a report, and interaction with others (Brown 2007; García 2005; García and Hamayan 2006).

Keep the following points in mind during stage 7 of ADI:

- The goal of the peer-review process is not to assign grades. Rather, it gives students an opportunity to give and receive feedback about their writing.
- The peer-review process is one of the best ways for students to learn how to write. When students read and review two different reports, they have an opportunity to (1) see examples of texts written by others for the same purpose, (2) discuss what counts as quality and why some reports are stronger that others, (3) discuss ways to strengthen a report, and (4) pick up things that they can do to improve their own reports.
- You can review the writing of the emerging bilingual students in your class depending on their English language proficiency, but make sure that your emerging bilingual students participate in the peer-review process.
- Don't skip this step—you will see tremendous growth in your students' writing skills the more they participate in the peer-review process.

Stage 8: Revise the Report

The final stage in the ADI instructional model is to revise the report. Each student is required to rewrite his or her report using the reviewers' comments and suggestions as a guideline. The author is also required to explain what he or she did to improve each section of the report in response to the reviewers' suggestions in the author response section of the PRG/TSR.

Once the report is revised, it is turned in to the teacher for evaluation with the original draft report and the PRG/TSR attached. The teacher can then provide a score on the PRG/TSR in the column labeled "Teacher Score" and use these ratings to assign an overall grade for the report. This approach provides all students with a chance to improve their writing mechanics and develop their reasoning and understanding of the content. This process also offers students the added benefit of reducing academic pressure by providing support in obtaining the highest possible grade for their final product.

The PRG/TSR is designed to be used with any ADI investigation, thus allowing teachers to use the same scoring rubric throughout the entire year. This is beneficial for several reasons. First, the criteria for what counts as a high-quality report do not change from investigation to investigation. Students therefore quickly learn what is expected from them when they write a report, and teachers do not have to spend valuable class time explaining the various components of the PRG/TSR each time they assign a report. Second, the PRG/TSR makes it clear which components of a report need to be improved next time, because the grade is not based on a holistic evaluation of the report. Students, as a result, can see

Keep the following points in mind during stage 8 of ADI:

- There are many writing supports (i.e., sentence starters, pre-writing, graphic organizers) already embedded in this stage. You might not need to provide much extra support for your students as they rewrite their reports.
- Students don't always use the feedback they receive from others to improve the quality of their report. Make sure you move from student to student and encourage them to make changes based on the feedback.
- If students get stuck as they are writing their final report, model how to use feedback. For example, you might say, "Okay, based on the peer-review guide, I need to do more to describe what I did to collect data. I bet if I add information about ____, readers will have a better idea of what I did."

which aspects of their writing are strong and which aspects need improvement. Finally, and perhaps most important, the PRG/TSR provides teachers with a standardized measure of student performance that can be compared over multiple reports across semesters, thus allowing teachers to track improvement over time.

The Role of the Teacher During Argument-Driven Inquiry

If the ADI instructional model is to be successful and student learning is to be optimized, the role of the teacher during an investigation that was designed using this model must be different than the teacher's role during a more traditional science lesson. The teacher *must* act as a resource for the students, rather than as a director, as students work through each stage of the activity; the teacher must encourage students to think about *what they are doing* and *why they made that decision* throughout the process. This encouragement should take the form of probing questions that teachers ask as they walk around the classroom, such as "Why do you want to set up your equipment that way?" or "What type of data will you need to collect to be able to answer that question?" Teachers must also restrain themselves from telling or showing students how to "properly" conduct the investigation. However, teachers must emphasize the need to maintain high standards for a scientific investigation by requiring students to use rigorous standards for what counts as a good method or a strong argument in the context of science.

Finally, and perhaps most important for the success of these investigations, teachers must be willing to let students try and fail, and then help them learn from their mistakes. Teachers should not try to make the investigations included in this book "student-proof" by providing additional directions to ensure that students do everything right the first time. We have found that students often learn more from an ADI investigation when they design a flawed method to collect data during stage 2 or analyze their results in an inappropriate manner during stage 3, because their classmates quickly point out these mistakes during the argumentation session (stage 4) and it leads to more teachable moments.

Because the teacher's role during an ADI investigation is different from what typically happens in a classroom, we've provided a chart describing teacher behaviors that are consistent and inconsistent with each stage of the instructional model (see Table 1). This table is organized by stage because what the students and the teacher need to accomplish during each stage is different. It might be helpful to keep this table handy as a guide when you are first attempting to implement the investigations found in the book.

TABLE 1

Teacher behaviors during the stages of the ADI instructional model

Stage	What the teacher does that is...	
	Consistent with ADI model	**Inconsistent with ADI model**
1: Introduce the task and the guiding question	• "Creates a need" for students to design and carry out an investigation by introducing a phenomenon • Reads the introduction aloud to the students • Supplies students with the materials and equipment they will need • Holds a "tool talk" to show students how to use the materials and equipment • Reviews relevant safety precautions and protocols • Allows students to tinker with the equipment they will be using later	• Provides a list of vocabulary terms • Tells students what they will figure out • Tells students that there is one correct answer • Makes students do everything on their own • Skips going over the safety precautions • Skips introducing the phenomenon to save time or does it as a demonstration
2: Design a method and collect data	• Encourages students to ask questions as they design their investigations • Encourages students to use the crosscutting concept (CC) to decide what data is important to collect • Asks groups probing questions about their investigation plan (e.g., "Why did you do it this way?") and the type of data they expect from that design • Checks over the investigation proposal and offers feedback as needed • Assists students as they get stuck • Ensures that all students are safe	• Gives students a procedure to follow • Does not question students about the method they design or the type of data they expect to collect • Approves vague or incomplete investigation proposals • Makes everyone in the class collect data the same way

Continued

TABLE 1 (*continued*)

Stage	What the teacher does that is... Consistent with ADI model	Inconsistent with ADI model
3: Create a draft argument	• Reminds students of the guiding question and what counts as appropriate evidence in science • Requires students to generate an argument that provides and supports a claim with genuine evidence • Encourages students to use the disciplinary core idea (DCI) and the CC in the justification of the evidence • Encourages all students in the group to make equal contributions	• Requires only one student to be prepared to discuss the argument • Moves to groups to check on progress without asking students questions about why they are doing what they are doing • Tells students the right answer • Has the class create a single argument together • Allows one student to do all the work for a group
4: Argumentation session	• Establishes and maintains classroom norms for discussions • Encourages student-to-student talk • Keeps the discussion focused on the elements of the argument • Encourages students to use appropriate criteria for determining what does and does not count	• Allows students to negatively respond to others • Asks questions about students' claims before other students can ask • Allows students to discuss ideas that are not supported by evidence • Allows students to use inappropriate criteria for determining what does and does not count
5: Reflective discussion	• Encourages students to discuss what they learned about the DCI and the CC • Encourages students to discuss what they learned about a nature of scientific knowledge (NOSK) or nature of scientific inquiry (NOSI) concept • Encourages students to think of ways to design better investigations in future • Asks students what they think	• Provides a lecture on the content • Skips over the discussion about a NOSK or NOSI concept to save time • Tells students what they "should have learned" or what they "should have figured out"
6: Write a draft report	• Reminds students about the audience, topic, and purpose of the report • Provides an example of a good report and an example of a bad report • "Chunks" the writing process into manageable pieces • Provides just-in-time instruction as students get stuck	• Has students write only a portion of the report • Moves on to the next activity/topic without providing feedback • Expects students to complete the entire report with little or no assistance

Continued

TABLE 1 *(continued)*

Stage	What the teacher does that is... Consistent with ADI model	Inconsistent with ADI model
7: Peer review	• Establishes and maintains classroom norms for the review process • Encourages students to remember that while grammar and punctuation are important, the main goal is an acceptable scientific claim with supporting evidence and justification • Reminds students of what counts as specific and useful feedback • Holds the reviewers accountable	• Allows students to make critical comments about the author (e.g., "This person is stupid") rather than their work (e.g., "This claim needs to be supported by evidence") • Allows students to just check off "Yes" on each item • Allows students to skip giving feedback during the peer-review process • Has students review reports on their own
8: Revise the report	• Requires students to edit their reports based on the reviewers' comments • Requires students to respond to the reviewers' feedback • Has students complete the Checkout Questions after they have turned in their report	• Allows students to turn in a report without a completed peer-review guide • Allows students to turn in a report without revising it first

How to Keep Students Safe During ADI Investigations

It is important for all of us to do what we can to make school science investigations safer for everyone in the classroom. We recommend four important guidelines to follow. First, we need to have proper safety equipment such as, but not limited to, fire extinguishers and an eye wash in our classrooms. Second, we need to ensure that students use appropriate personal protective equipment (PPE; e.g., sanitized indirectly vented chemical-splash goggles and nonlatex gloves) during all parts of the investigations (i.e., setup, the hands-on investigation, and cleanup) when students are using potentially harmful supplies, equipment, or chemicals. At a minimum, the PPE we provide for students to use must meet the ANSI/ISEA Z87.1D3 standard. Third, we must review and comply with all safety policies and procedures, including but not limited to appropriate chemical management, that have been established by our place of employment. Finally, and perhaps most important, we all need to adopt legal safety standards and better professional safety practices and enforce them inside the classroom.

We provide safety precautions for each investigation and recommend that all teachers follow these safety precautions to provide a safer learning experience inside the classroom. The safety precautions associated with each investigation are based, in part, on the use of the recommended materials and instructions, legal safety standards, and better professional safety practices. Selection of alternative materials or procedures for these investigations may jeopardize the level of safety and therefore is at the user's own risk. Remember that an investigation includes three parts: (1) setup, which is what you do to prepare the materials for students to use; (2) the actual investigation, which involves students using the materials and equipment; and (3) the cleanup, which includes cleaning the materials and putting them away for later use. The safety procedures and PPE we recommend for each investigation apply to all three parts.

We also recommend that you go over the 11 safety rules that are included as part of the safety acknowledgment form with your students before beginning the first investigation. Once you are done going over these rules with your students, have them sign the safety acknowledgment form. You should also send the form home for a parent or guardian to read and sign to acknowledge that they understand the safety procedures that must be followed by their child. The safety acknowledgment form can be found in Appendix 5. Another elementary science safety acknowledgment form can be found on the NSTA Safety Portal at *http://static.nsta.org/pdfs/SafetyAcknowledgmentForm-ElementarySchool.pdf.*

References

Abd-El-Khalick, F., and N. G. Lederman. 2000. Improving science teachers' conceptions of nature of science: A critical review of the literature. *International Journal of Science Education* 22: 665–701.

Blanchard, M., and V. Sampson. 2018. Fostering impactful research experiences for teachers (RETs). *Eurasia Journal of Mathematics, Science and Technology Education* 14 (1): 447–465.

Brown, D. H. 2007. *Principles of language learning and teaching.* 5th ed. White Plains, NY: Pearson.

Duschl, R. A., H. A. Schweingruber, and A. W. Shouse, eds. 2007. *Taking science to school: Learning and teaching science in grades K–8*. Washington, DC: National Academies Press.

Echevarría, J., D. Short, and K. Powers. 2006. School reform and standards-based education: A model for English-language learners. *Journal of Educational Research* 99 (4): 195–210.

Escamilla, K., and S. Hopewell. 2010. Transitions to biliteracy: Creating positive academic trajectories for emerging bilinguals in the United States. In *International perspectives on bilingual education: Policy, practice, controversy,* ed. J. E. Petrovic, 69–94. Charlotte, NC: Information Age Publishing.

García, E. E. 2005. *Teaching and learning in two languages: Bilingualism and schooling in the United States.* New York: Teachers College Press.

García, E. E., and E. Hamayan. 2006. What is the role of culture in language learning? In *English language learners at school: A guide for administrators,* eds. E. Hamayan and R. Freeman, 61–64. Philadelphia, PA: Caslon Publishing.

García, O., and J. Kleifgen. 2010. *Educating emergent bilinguals: Policies, programs, and practices for English language learners.* New York: Teachers College Press

Goldenberg, C., and R. Coleman. 2010. *Promoting academic achievement among English learners: A guide to the research.* Thousand Oaks, CA: Corwin Press.

González, N., L. Moll, and C. Amanti. 2005. *Funds of knowledge: Theorizing practices in households, communities and classrooms.* Mahwah, NJ: Erlbaum.

Gottlieb, M., A. Katz, and G. Ernst-Slavit. 2009. *Paper to practice: Using the English language proficiency standards in PreK–12 classrooms.* Alexandria, VA: Teachers of English to Speakers of Other Languages.

Grooms, J., P. Enderle, and V. Sampson. 2015. Coordinating scientific argumentation and the *Next Generation Science Standards* through argument driven inquiry. *Science Educator* 24 (1): 45–50.

Lederman, N. G., and J. S. Lederman. 2004. Revising instruction to teach the nature of science. *The Science Teacher* 71 (9): 36–39.

National Governors Association Center for Best Practices and Council of Chief State School Officers (NGAC and CCSSO). 2010. *Common core state standards.* Washington, DC: NGAC and CCSSO.

National Research Council (NRC). 2012. *A framework for K–12 science education: Practices, crosscutting concepts, and core ideas.* Washington, DC: National Academies Press.

NGSS Lead States. 2013. *Next Generation Science Standards: For states, by states.* Washington, DC: National Academies Press. *www.nextgenscience.org/next-generation-science-standards.*

Sampson, V., and L. Gleim. 2009. Argument-driven inquiry to promote the understanding of important concepts and practices in biology. *American Biology Teacher* 71 (8): 471–477.

Sampson, V., J. Grooms, and J. Walker. 2009. Argument-driven inquiry: A way to promote learning during laboratory activities. *The Science Teacher* 76 (7): 42–47.

Sampson, V., J. Grooms, and J. Walker. 2011. Argument-driven inquiry as a way to help students learn how to participate in scientific argumentation and craft written arguments: An exploratory study. *Science Education* 95 (2): 217–257.

Schwartz, R. S., N. Lederman, and B. Crawford. 2004. Developing views of nature of science in an authentic context: An explicit approach to bridging the gap between nature of science and scientific inquiry. *Science Education* 88: 610–645.

Strimaitis, A., S. Southerland, V. Sampson, P. Enderle, and J. Grooms. 2017. Promoting equitable biology lab instruction by engaging all students in a broad range of science practices: An exploratory study. *School Science and Mathematics* 117 (3–4): 92–103.

Wallace, C., B. Hand, and V. Prain, eds. 2004. *Writing and learning in the science classroom.* Boston: Kluwer Academic Publishers.

Chapter 2
The Investigations

How to Use the Investigations

This book includes 14 argument-driven inquiry (ADI) investigations. These investigations are not designed to replace an existing science curriculum, but rather are designed to function as a tool that teachers can use to integrate three-dimensional instruction into their existing curriculum. These investigations are also intended to provide teachers with a way to help children develop fundamental literacy and mathematics skills in the context of science, because students have numerous opportunities to read, write, talk, and use mathematics throughout each investigation. Finally, teachers can use these investigations to turn the classroom into a language-learning environment that provides emerging bilingual students with the opportunities they need to use receptive (listening and reading) and productive (speaking and writing) language. These investigations can therefore provide a context for emerging bilingual students to learn English and science at the same time as they interact with other people, the available materials, and ideas to develop a new understanding about how the world works.

We do not expect teachers to use every investigation included in this book. We do, however, recommend that teachers try to incorporate as many of these investigations into their third-grade science curriculum as possible to give students more opportunities to learn how to use disciplinary core ideas (DCIs), crosscutting concepts (CCs), and scientific and engineering practices (SEPs) to figure things out (NGSS Lead States 2013; NRC 2012). The more ADI investigations that students complete, the more progress that they will make on each aspect of science proficiency (Grooms, Enderle, and Sampson 2015; Sampson et al. 2013; Strimaitis et al. 2017).

These investigations are designed to function as stand-alone lessons, which gives teachers the flexibility they need to decide which ones to use and when to use them during the academic year. The investigations are organized by topic into different sections in the book. Sections 2–6 can be taught in any order. The investigations within a section, however, should be taught in order because many of the investigations within a section use similar DCIs and are designed to build on what students figured out during a previous investigation. For example, Investigations 3 and 4 both require students to use the same DCI (PS2.A: Forces and Motion) to make sense of the world around them, but students can also use what they figured out during Investigation 3 (how to predict the motion of a ball rolling back and forth in a half-pipe) as they attempt to figure out how changing the strength and direction of the forces acting on a cart affects the motion of that cart during Investigation 4.

We have aligned the 14 investigations with the following sources to facilitate curriculum and lesson planning:

- *A Framework for K–12 Science Education* (NRC 2012; see Standards Matrix A in Appendix 1)
- The *Next Generation Science Standards,* or *NGSS,* performance expectations (NGSS Lead States 2013; see Standards Matrix B in Appendix 1)
- Aspects of the nature of scientific knowledge (NOSK) and the nature of scientific inquiry (NOSI) (see Standards Matrix C in Appendix 1; see also the discussion of NOSK and NOSI concepts in Chapter 1)

We wrote all the investigations included in this book to align with a specific performance expectation in the *NGSS* for grade 3. Teachers who teach in states that use science standards other than the *NGSS* will need to determine how well each investigation aligns with the specific state standards for grade 3. In states that have adopted standards based on the *Framework,* there is likely a great deal of overlap because the state standards will include many, if not all, of the DCIs, CCs, and SEPs that were used to write the *NGSS;* the standards will just be worded differently. In states that have standards that are not based on the *Framework,* there might be less overlap. However, given the fact that all 14 investigations are designed to give children an opportunity to learn science by doing science, it is likely that the investigations will still be useful for helping students learn the concepts and inquiry or process skills found in state standards that are based on something other than the *Framework* (such as the *Benchmarks for Science Literacy* [AAAS 1993] or the *National Science Education Standards* [NRC 1996]). It is important to note, however, that some of these investigations might be better aligned with content and inquiry or process skills standards that are addressed in grade 4 or 5 in states that have standards that are not the *NGSS* or based on the *Framework.* If these investigations align better with the content and inquiry or process skills for grade 4 or 5, they can be used in those grade levels instead of in grade 3 classrooms.

The investigations in this book, as noted earlier, create a context where students read, write, talk, and use mathematics to figure out how something works or why something happens. Teachers can therefore use these investigations to help students develop important literacy skills, understand and use mathematical concepts and practices, or acquire a second language. With this in mind, we have also aligned each investigation with the following sources:

- The *Common Core State Standards* in English language arts, or *CCSS ELA* (NGAC and CCSSO 2010; see Standards Matrix D in Appendix 1)
- The *Common Core State Standards* in mathematics, or *CCSS Mathematics* (NGAC and CCSSO 2010; see Standards Matrix E in Appendix 1)
- The *English Language Proficiency (ELP) Standards* (CCSSO 2014; see Standards Matrix F in Appendix 1)

Teacher Notes

We have included Teacher Notes for each investigation to help teachers decide when to use each investigation and how to help guide students through each stage of an ADI. These notes include information about the purpose of the investigation, information about the DCI and the CC that students use during the investigation, and what students figure out by the end of it. The notes also include information about the time needed to implement each stage of the model, the materials that students need, safety precautions, and a detailed lesson plan by stage.

The Teacher Notes also include a "Connections to Standards" section showing how each investigation is aligned with the *NGSS* performance expectations, the *CCSS ELA* standards, the *CSSS Mathematics* standards, and the *ELP* standards. In the following subsections, we will describe the information provided in each section of the Teacher Notes.

Purpose

This section describes what the students will do during the investigation. It also identifies the DCI and the CC that the students will use during the investigation to accomplish the task and the NOSK or NOSI concept that will be highlighted during the reflective discussion. Please note that because of the nature of the ADI approach, you do not need to "teach the vocabulary first" or make sure that your students "know the content" before the investigation begins. Students will learn more about the DCI and the CC along with the SEPs and any important vocabulary as they work through each stage of ADI to *investigate* the phenomenon; *make sense* of the phenomenon; and *evaluate and refine* their ideas, explanations, and arguments.

The Disciplinary Core Idea

This section of the Teacher Notes provides a basic overview of the DCI that students will use during the investigation. The overview is based on the *Framework* and describes what students should know about the DCI by the end of grade 2.

The Crosscutting Concept

This section of the Teacher Notes provides a basic overview of the CC that students will use during the investigation. The overview of the CC is based on the *Framework* and describes what students should know about the CC in the early grades.

What Students Figure Out

This section of the Teacher Notes describes what new understanding of the natural world the students are likely to develop as they work through the investigation. This information will also help teachers get a better sense of what students will learn as they *investigate* the phenomenon; *make sense* of the phenomenon; and *evaluate and refine* their ideas, explanations, and arguments.

Timeline

This section of the Teacher Notes provides information about how much time each stage of the investigation should take. Unlike typical science lessons, ADI investigations take between three and five hours to complete. In most cases, an investigation can be completed over eight days (one day for each stage) during the designated science time in the daily schedule, or it can be completed in one day. The amount of time it will take to complete each investigation will vary depending on how long it takes to collect data and how familiar students are with the stages of ADI. The time needed to complete each stage of ADI will take longer the first few times that students work through the process, but the time the students need will be reduced as they become familiar with using DCIs, CCs, and SEPs to figure things out.

Materials and Preparation

This section describes the consumables and equipment that the students will need to complete the investigation. The items are listed by group, by individual student, or by class. We have also included specific suggestions for some lab supplies, based on our findings that these supplies worked best during the field tests. However, if needed, substitutions can be made. Always be sure to test all materials before starting an investigation.

This section also describes any preparation that needs to be done *before* students can do the investigation. Please note that the preparation for some investigations may need to be done several days in advance.

Safety Precautions

This section provides an overview of potential safety hazards as well as safety protocols that should be followed to make the investigation safer for students. These are based on legal safety standards and current better professional safety practices. Teachers should also review and follow all local polices and protocols used within their school district and school (e.g., the district chemical hygiene plan, Board of Education safety policies).

Lesson Plan by Stage

This section provides a detailed lesson plan for each stage that includes information about what the teacher can do as he or she guides students through the investigation. The plan includes directions to give students and sample questions to ask students at different points in the lesson, but these directions and questions should not be followed like a script. As professionals, we believe that teachers know the unique needs of their students and how to best support them. The detailed lesson plans should be viewed as a starting point and should be modified and revised based on the unique needs of the students in a particular classroom and the professional judgment of the teacher.

It is important to keep in mind that the goal of ADI is give students more opportunities to learn how to use DCIs, CCs, and SEPs to figure out how or why something happens, so teachers should resist the urge to tell students exactly what to do or how to complete a task before they start. It is also important not to reduce the complexity of the tasks by providing too much scaffolding. We believe that it is far better to keep the complexity of the task high by giving students more voice and choice, allowing for productive struggle during the investigation, and then providing scaffolding only when it is needed.

Many different teachers have tested these investigations and have helped us refine them and make them run smoother. As a result, we have collected a lot of advice about how to support children as they investigate a phenomenon; make sense of that phenomenon; and evaluate and refine their ideas, explanations, and arguments. We have organized this advice around some common questions that teachers have as they are guiding students through each stage, such as the following:

- What are some things I can do to encourage productive talk among my students?
- What can I do to help my students comprehend more of what they read?
- What are some things I can do to support my emerging bilingual students?
- What are some things I need to remember during each stage of ADI?
- What should a student-designed investigation look like?
- What should a table or graph look like for this investigation?
- What are some things I can do if my students get stuck?

The "Lesson Plan by Stage" section also includes advice for supporting students as they work through each stage. Some of this advice can be found in "hint boxes" that describe, for example, what a student-designed investigation should look like and what a table or graph for this investigation should look like. In addition, teachers should refer to the hints for each stage of ADI found in Chapter 1 and to the frequently asked questions found in Appendix 3.

How to Use the Checkout Questions

This section describes how to use an optional assessment that we call Checkout Questions. It includes information about when to use this assessment and what types of answers to look for to determine if students understand the DCI and the CC from the investigation.

Connections to Standards

This section is designed to inform curriculum and lesson planning by highlighting how the investigation can be used to address specific performance expectations from the *NGSS*, the *CCSS ELA*, the *CCSS Mathematics*, and the *ELP* standards.

Instructional Materials

The instructional materials included in this book are reproducible copy masters that are designed to support students as they participate in an ADI investigation. The materials needed for each investigation include an Investigation Handout, the peer-review guide and teacher scoring rubric (PRG/TSR), and a set of Checkout Questions. Some investigations also require supplementary materials.

Investigation Handout

At the beginning of each ADI investigation, each student should be given a copy of the Investigation Handout. This handout provides information about the phenomenon that they will investigate and a guiding question for the students to answer. In addition, the handout provides space for students to design their investigation, record their observations and measurements, and create graphs to analyze the data they collect during the investigation. The handout also has space for the students to keep track of critiques, suggestions for improvement, and good ideas that arise during the argumentation session. To help support students as they learn how to write in this context, the handout includes sentence starters that they can use when writing the draft report.

Peer-Review Guide and Teacher Scoring Rubric

The PRG/TSR is designed to make the criteria that are used to judge the quality of an investigation report explicit. The PRG/TSR can be found in Appendix 4. We recommend that teachers make one copy of the PRG/TSR for each student. Then during the peer-review stage, have each pair of reviewers fill out one PRG/TSR for each report that they review. The reviewers should rate the report on each criterion and then provide advice to the author about ways to improve it based on their rating. Once the review is complete, the author needs to revise his or her report and respond to the reviewers' rating and comments in the appropriate sections in the PRG/TSR. The PRG/TSR should be turned in to the teacher along with the first draft and the final report for a final evaluation.

To score the report, the teacher can simply fill out the "Teacher Score" column of the PRG/TSR and then total the scores. There is also space at the bottom of the PRG/TSR for teacher feedback.

Checkout Questions

To facilitate formative assessment inside the classroom, we have included a set of Checkout Questions for each investigation. The questions target the DCI and the CC that the students used during the investigation. Students should complete the Checkout Questions one day after they turn in their final report. One handout is needed for each student. The students should complete these questions on their own. The teacher can use the students' responses to the Checkout Questions, along with what they write in their report, to determine if the students learned what they needed to learn during the lab, and then reteach as needed.

Supplementary Materials

Some investigations include supplementary materials such as full-color images and links to videos. Students will need to be able to use these materials during their investigation. These materials can be downloaded from the book's Extras page at *www.nsta.org/adi-3rd*.

References

American Association for the Advancement of Science (AAAS). 1993. *Benchmarks for science literacy.* Project 2061. New York: Oxford University Press.

Council of Chief State School Officers (CCSSO). 2014. *English language proficiency (ELP) standards.* Washington, DC: NGAC and CCSSO. *www.ccsso.org/resource-library/english-language-proficiency-elp-standards.*

Grooms, J., P. Enderle, and V. Sampson. 2015. Coordinating scientific argumentation and the *Next Generation Science Standards* through argument driven inquiry. *Science Educator* 24 (1): 45–50.

National Governors Association Center for Best Practices and Council of Chief State School Officers (NGAC and CCSSO). 2010. *Common core state standards.* Washington, DC: NGAC and CCSSO.

National Research Council (NRC). 1996. *National Science Education Standards.* Washington, DC: National Academies Press.

National Research Council (NRC). 2012. *A framework for K–12 science education: Practices, crosscutting concepts, and core ideas.* Washington, DC: National Academies Press.

NGSS Lead States. 2013. *Next Generation Science Standards: For states, by states.* Washington, DC: National Academies Press. *www.nextgenscience.org/next-generation-science-standards.*

Sampson, V., P. Enderle, J. Grooms, and S. Witte. 2013. Writing to learn and learning to write during the school science laboratory: Helping middle and high school students develop argumentative writing skills as they learn core ideas. *Science Education* 97 (5): 643–670.

Strimaitis, A., S. Southerland, V. Sampson, P. Enderle, and J. Grooms. 2017. Promoting equitable biology lab instruction by engaging all students in a broad range of science practices: An exploratory study. *School Science and Mathematics* 117 (3–4): 92–103.

Section 2

Motion and Stability: Forces and Interactions

Investigation 1

Magnetic Attraction: What Types of Objects Are Attracted to a Magnet?

Purpose

The purpose of this investigation is to give students an opportunity to use the disciplinary core idea (DCI) of PS1.A: Structure and Properties of Matter and the crosscutting concept (CC) of Patterns from *A Framework for K–12 Science Education* (NRC 2012) to figure out what types of materials are attracted to a magnet. Students will also learn about the difference between observations and inferences during the reflective discussion.

The Disciplinary Core Idea

Students in third grade should understand the following about Structure and Properties of Matter and should be able to use this DCI to figure out which materials are attracted to a magnet:

> *Different kinds of matter exist (e.g., wood, metal, water), and many of them can be either solid or liquid, depending on temperature. Matter can be described and classified by its observable properties ..., by its uses, and by whether it occurs naturally or is manufactured. (NRC 2012, p. 108)*

The Crosscutting Concept

Students in third grade should understand the following about the CC Patterns:

> *Noticing patterns is often a first step to organizing and asking scientific questions about why and how the patterns occur. One major use of pattern recognition is in classification, which depends on careful observation of similarities and differences; objects can be classified into groups on the basis of similarities of visible or microscopic features or on the basis of similarities of function. Such classification is useful in codifying relationships and organizing a multitude of objects or processes into a limited number of groups. (NRC 2012, p. 85)*

Students in third grade should also be given opportunities to "investigate the characteristics that allow classification of animal types (e.g., mammals, fish, insects), of plants (e.g., trees, shrubs, grasses), or of materials (e.g., wood, rock, metal, plastic)" (NRC 2012, p. 86).

Students should be encouraged to use their developing understanding of patterns as a tool or a way of thinking about a phenomenon during this investigation to help them figure out what types of materials are attracted to a magnet and which are not.

What Students Figure Out

A *magnet* is an object that gives off a magnetic field. Magnets apply a force on other magnets or materials that contain iron. Substances such as glass, plastic, cloth, and wood are not attracted to magnets. Most metals, such as copper, zinc, tin, and aluminum, are also not attracted to magnets. People can use magnets to determine if a material has iron in it or not.

Timeline

The time needed to complete this investigation is 250 minutes (4 hours and 10 minutes). The amount of instructional time needed for each stage of the investigation is as follows:

- *Stage 1.* Introduce the task and the guiding question: 35 minutes
- *Stage 2.* Design a method and collect data: 40 minutes
- *Stage 3.* Create a draft argument: 35 minutes
- *Stage 4.* Argumentation session: 30 minutes
- *Stage 5.* Reflective discussion: 15 minutes
- *Stage 6.* Write a draft report: 35 minutes
- *Stage 7.* Peer review: 30 minutes
- *Stage 8.* Revise the report: 30 minutes

This investigation can be completed in one day or over eight days (one day for each stage) during your designated science time in the daily schedule.

Materials and Preparation

The materials needed for this investigation are listed in Table 1.1 (p. 42). The items can be purchased from a big-box retail store such as Wal-Mart or Target or through an online retailer such as Amazon. The materials for this investigation can also be purchased as a complete kit (which includes enough materials for 24 students, or six groups) at *www.argumentdriveninquiry.com*.

TABLE 1.1

Materials for Investigation 1

Item	Quantity
Safety glasses or goggles	1 per student
Ceramic magnet	1 per student
Aluminum foil, 4" × 4" square	1 per group
Copper wire, 3" piece	1 per group
Penny	1 per group
Nickel	1 per group
Nail	1 per group
Paper clips	3 per group
Steel BBs	3 per group
Plastic BBs	3 per group
Washer	1 per group
Lead fishing sinker	1 per group
Tin split shot fishing sinker	1 per group
String, 3" piece	1 per group
Index card	1 per group
Acrylic tile	1 per group
Glass tile	1 per group
Marble	1 per group
Wood block	1 per group
Whiteboard, 2' × 3'*	1 per group
Investigation Handout	1 per student
Peer-review guide and teacher scoring rubric	1 per student
Checkout Questions (optional)	1 per student

*As an alternative, students can use computer and presentation software such as Microsoft PowerPoint or Apple Keynote to create their arguments.

We recommend putting all the materials except for the ceramic magnet and paper clips into a bag, a bowl, or some other type of container that can be given to each group. This will make it easier to pass out and collect all these different materials during stage 2.

Be sure to use a set routine for distributing and collecting the materials. One option is to set up the materials for each group in a kit that you can deliver to each group. A second option is to have all the materials on a table or cart at a central location. You can then assign a member of each group to be the "materials manager." This individual is responsible for collecting all the materials his or her group needs from the table or cart during class and for returning all the materials at the end of the class.

Safety Precautions

Remind students to follow all normal safety rules. In addition, tell the students to take the following safety precautions:

- Wear sanitized safety glasses or goggles during setup, investigation activity, and cleanup.
- Do not throw objects or put any objects in their mouth.
- Use caution when handling sharp objects (e.g., wire, nails, paper clips). They can cut or puncture skin.
- Immediately pick up any slip or fall hazards (e.g., marbles) from the floor.
- Wash their hands with soap and water when done collecting the data.

Lesson Plan by Stage

Stage 1: Introduce the Task and the Guiding Question (35 minutes)

1. Ask the students to sit in six groups, with three or four students in each group.
2. Ask students to clear off their desks except for a pencil (and their *Student Workbook for Argument-Driven Inquiry in Third-Grade Science* if they have one).
3. Pass out an Investigation Handout to each student (or ask students to turn to Investigation Handout 1 in their workbook).
4. Read the first paragraph of the "Introduction" aloud to the class. Ask the students to follow along as you read.
5. Pass out a magnet and three paper clips to each group.
6. Remind students of the safety rules and explain the safety precautions for this investigation.
7. Tell students to play with the magnet and paper clips and then record their observations and questions in the "OBSERVED/WONDER" chart in the "Introduction" section of their Investigation Handout (or the investigation log in their workbook).
8. Ask students to share *what they observed* as they played with the magnet and paper clips.
9. Ask students to share *what questions they have* about the behavior of the magnet and paper clips.
10. Tell the students, "Some of your questions might be answered by reading the rest of the 'Introduction.'"
11. Ask the students to read the rest of the "Introduction" on their own *or* ask them to follow along as you read aloud.

12. Once the students have read the rest of the "Introduction," ask them to fill out the "KNOW/NEED" chart on their Investigation Handout (or in their investigation log) as a group.
13. Ask students to share what they learned from the reading. Add these ideas to a class "know / need to figure out" chart.
14. Ask students to share what they think they will need to figure out based on what they read. Add these ideas to the class "know / need to figure out" chart.
15. Tell the students, "It looks like we have something to figure out. Let's see what we will need to do during our investigation."
16. Read the task and the guiding question aloud.
17. Tell the students, "I have lots of materials here that you can use."
18. Introduce the students to the materials available for them to use during the investigation by either (a) holding each one up and then asking what it might be used for or (b) giving them a kit with all the materials in it and giving them three or four minutes to play with them. If you give the students an opportunity to play with the materials, be sure to collect them from each group before moving on to stage 2.

Stage 2: Design a Method and Collect Data (40 minutes)

1. Tell the students, "I am now going to give you and the other members of your group about 15 minutes to plan your investigation. Before you begin, I want you all to take a couple of minutes to discuss the following questions with the rest of your group."
2. Show the following questions on the screen or board:
   - What information should we collect so we can *compare and contrast* different materials?
   - What types of *patterns* might we look for to help answer the guiding question?
3. Tell the students, "Please take a few minutes to come up with an answer to these questions." Give the students two or three minutes to discuss these two questions.
4. Ask two or three different groups to share their answers. Be sure to highlight or write down any important ideas on the board so students can refer to them later.
5. If possible, use a document camera to project an image of the graphic organizer for this investigation on a screen or board (or take a picture of it and project the picture on a screen or board). Tell the students, "I now want you all to plan out your investigation. To do that, you will need to create an investigation proposal by filling out this graphic organizer."

6. Point to the box labeled "Our guiding question:" and tell the students, "You can put the question we are trying to answer in this box." Then ask, "Where can we find the guiding question?"
7. Wait for a student to answer where to find the guiding question (the answer is "in the handout").
8. Point to the box labeled "We will collect the following data:" and tell the students, "You can list the measurements or observations that you will need to collect during the investigation in this box."
9. Point to the box labeled "These are the steps we will follow to collect data:" and tell the students, "You can list what you are going to do to collect the data you need and what you will do with your data once you have it. Be sure to give enough detail that I could do your investigation for you."
10. Ask the students, "Do you have any questions about what you need to do?"
11. Answer any questions that come up.
12. Tell the students, "Once you are done, raise your hand and let me know. I'll then come by and look over your proposal and give you some feedback. You may not begin collecting data until I have approved your proposal by signing it. You need to have your proposal done in the next 15 minutes."
13. Give the students 15 minutes to work in their groups on their investigation proposal. As they work, move from group to group to check in, ask probing questions, and offer a suggestion if a group gets stuck.

What should a student-designed investigation look like?

The students' investigation proposal should include the following information:

- The guiding question is "What types of objects are attracted to a magnet?"
- The data that the students should collect are (1) name of the object, (2) what materials make up the object, (3) if the object sticks to a magnet or not.
- The steps that the students will follow to collect the data might include (1) lay the objects on a table, (2) place a magnet on each object, (3) see if each object moves or sticks to the magnet, (4) create two groups of objects, ones that stick to a magnet and ones that do not, and (5) look for things that objects that stick to a magnet have in common.

14. As each group finishes its investigation proposal, be sure to read it over and determine if it will be productive or not. If you feel the investigation will be productive (not necessarily what you would do or what the other groups are doing), sign your name on the proposal and let the group start collecting data. If the plan needs to be changed, offer some suggestions or ask some probing questions, and have the group make the changes before you approve it.
15. Pass out the materials or have one student from each group collect the materials they need from a central supply table or cart for the groups that have an approved proposal.
16. Remind students of the safety rules and precautions for this investigation.
17. Tell the students to collect their data and record their observations or measurements in the "Collect Your Data" box in their Investigation Handout (or the investigation log in their workbook).
18. Give the students 10 minutes to collect their data.
19. Be sure to collect the materials from each group before asking them to analyze their data.

Stage 3: Create a Draft Argument (35 minutes)

1. Tell the students, "Now that we have all this data, we need to analyze the data so we can figure out an answer to the guiding question."
2. If possible, project an image of the "Analyze Your Data" text and box for this investigation on a screen or board using a document camera (or take a picture of it and project the picture on a screen or board). Point to the box and tell the students, "You can create a table or a graph as a way to analyze your data. You can make your table or graph in this box."
3. Ask the students, "What information do we need to include in a table or graph?"
4. Tell the students, "Please take a few minutes to discuss this question with your group, and be ready to share."
5. Give the students five minutes to discuss.
6. Ask two or three different groups to share their answers. Be sure to highlight or write down any important ideas on the board so students can refer to them later.
7. Tell the students, "I am now going to give you and the other members of your group about 10 minutes to create your table or graph." If the students are having trouble making a table or graph, you can take a few minutes to provide a mini-lesson about how to create a table or graph from a bunch of observations or measurements (this strategy is called just-in-time instruction because it is offered only when students get stuck).

What should a table or graph look like for this investigation?

There are a number of different ways that students can analyze the observations or measurements they collect during this investigation. One of the most straightforward ways is to create a table that has three columns. The first column includes the names of the materials that make up the objects (iron, aluminum, copper, plastic, wood), the second column includes the number of objects that were made up of each material, and the third column includes the number of objects that were attracted to a magnet. An example of this type of table can be seen in Figure 1.1 (p. 48). This information can also be displayed as a bar graph with the names of the different materials on the horizontal axis, or *x*-axis, and the number of objects that stuck to a magnet on the vertical axis, or *y*-axis. There are other options for analyzing the data they collected. Students often come up with some unique ways of analyzing their data, so be sure to give them some voice and choice during this stage.

8. Give the students 10 minutes to analyze their data. As they work, move from group to group to check in, ask probing questions, and offer suggestions.
9. Tell the students, "I am now going to give you and the other members of your group 15 minutes to create an argument to share what you have learned and convince others that they should believe you. Before you do that, we need to take a few minutes to discuss what you need to include in your argument."
10. If possible, use a document camera to project the "Argument Presentation on a Whiteboard" image from the Investigation Handout (or the investigation log in their workbook) on a screen or board (or take a picture of it and project the picture on a screen or board).
11. Point to the box labeled "The Guiding Question:" and tell the students, "You can put the question we are trying to answer here on your whiteboard."
12. Point to the box labeled "Our Claim:" and tell the students, "You can put your claim here on your whiteboard. The claim is your answer to the guiding question."
13. Point to the box labeled "Our Evidence:" and tell the students, "You can put the evidence that you are using to support your claim here on your whiteboard. Your evidence will need to include the analysis you just did and an explanation of what your analysis means or shows. Scientists always need to support their claims with evidence."

14. Point to the box labeled "Our Justification of the Evidence:" and tell the students, "You can put your justification of your evidence here on your whiteboard. Your justification needs to explain why your evidence is important. Scientists often use core ideas to explain why the evidence they are using matters. Core ideas are important concepts that scientists use to help them make sense of what happens during an investigation."
15. Ask the students, "What are some core ideas that we read about earlier that might help us explain why the evidence we are using is important?"
16. Ask students to share some of the core ideas from the "Introduction" section of the Investigation Handout (or the investigation log in the workbook). List these core ideas on the board.
17. Tell the students, "That is great. I would like to see everyone try to include these core ideas in your justification of the evidence. Your goal is to use these core ideas to help explain why your evidence matters and why the rest of us should pay attention to it."
18. Ask the students, "Do you have any questions about what you need to do?"
19. Answer any questions that come up.
20. Tell the students, "Okay, go ahead and start working on your arguments. You need to have your argument done in the next 15 minutes. It doesn't need to be perfect. We just need something down on the whiteboards so we can share our ideas."
21. Give the students 15 minutes to work in their groups on their arguments. As they work, move from group to group to check in, ask probing questions, and offer a suggestion if a group gets stuck. Figure 1.1 shows an example of an argument created by students for this investigation.

FIGURE 1.1

Example of an argument

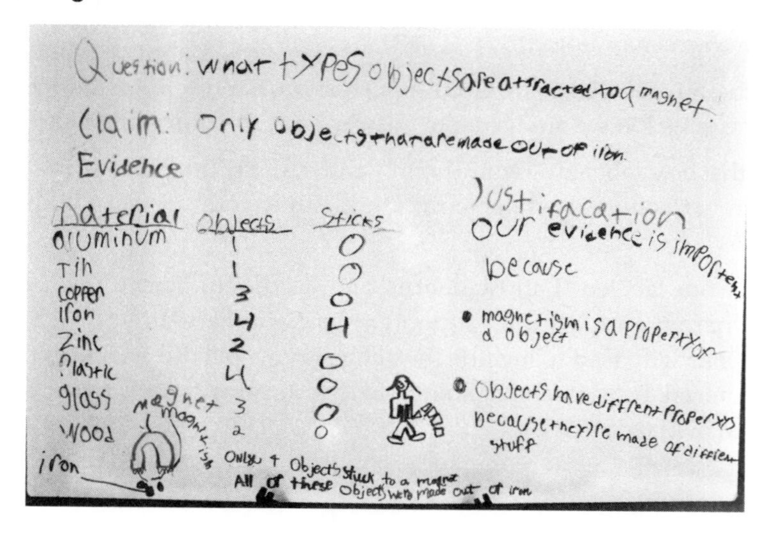

Stage 4: Argumentation Session (30 minutes)

The argumentation session can be conducted in a whole-class presentation format, a gallery walk format, or a modified gallery walk format. We recommend using a whole-class presentation format for the first investigation, but try to transition to either the gallery walk or modified gallery walk format as soon as possible because that will maximize student voice and choice inside the classroom. The following list shows the steps for the three formats; unless otherwise noted, the steps are the same for all three formats.

1. Begin by introducing the use of the whiteboard.
   - *If using the whole-class presentation format,* tell the students, "We are now going to share our arguments. Please set up your whiteboard so everyone can see them."
   - *If using the gallery walk or modified gallery walk format,* tell the students, "We are now going to share our arguments. Please set up your whiteboard so they are facing the walls."
2. Allow the students to set up their whiteboards.
   - *If using the whole-class presentation format,* the whiteboards should be set up on stands or chairs so they are facing toward the center of the room.
   - *If using the gallery walk or modified gallery walk format,* the whiteboards should be set up on stands or chairs so they are facing toward the outside of the room.
3. Give the following instructions to the students:
   - *If using the whole-class presentation format or the modified gallery walk format,* tell the students, "Okay, before we get started I want to explain what we are going to do next. I'm going to ask some of you to present your arguments to your classmates. If you are presenting your argument, your job is to share your group's claim, evidence, and justification of the evidence. The rest of you will be reviewers. If you are a reviewer, your job is to listen to the presenters, ask the presenters questions if you do not understand something, and then offer them some suggestions about ways to make their argument better. After we have a chance to learn from each other, I'm going to give you some time to revise your arguments and make them better."
   - *If using the gallery walk format,* tell the students, "Okay, before we get started I want to explain what we are going to do next. You are going to have an opportunity to read the arguments that were created by other groups. Your group will go to a different group's argument. I'll give you a few minutes to read it and review it. Your job is to offer them some suggestions about ways to make their argument better. You can use sticky notes to give them suggestions. Please be specific about what you want to change and be specific about how you think they should change it. After we have a chance to learn from each other, I'm going to give you some time to revise your arguments and make them better."

4. Use a document camera to project the "Ways to IMPROVE our argument …" box from the Investigation Handout (or the investigation log in their workbook) on a screen or board (or take a picture of it and project the picture on a screen or board).
   - *If using the whole-class presentation format or the modified gallery walk format,* point to the box and tell the students, "If you are a presenter, you can write down the suggestions you get from the reviewers here. If you are a reviewer, and you see a good idea from another group, you can write down that idea here. Once we are done with the presentations, I will give you a chance to use these suggestions or ideas to improve your arguments.
   - *If using the gallery walk format,* point to the box and tell the students, "If you see good ideas from another group, you can write them down here. Once we are done reviewing the different arguments, I will give you a chance to use these ideas to improve your own arguments. It is important to share ideas like this."

   Ask the students, "Do you have any questions about what you need to do?"
5. Answer any questions that come up.
6. Give the following instructions:
   - *If using the whole-class presentation format,* tell the students, "Okay. Let's get started."
   - *If using the gallery walk format,* tell the students, "Okay, I'm now going to tell you which argument to go to and review.
   - *If using the modified gallery walk format,* tell the students, "Okay, I'm now going to assign you to be a presenter or a reviewer." Assign one or two students from each group to be presenters and one or two students from each group to be reviewers.
7. Begin the review of the arguments.
   - *If using the whole-class presentation format,* have four or five groups present their argument one at a time. Give each group only two to three minutes to present their argument. Then give the class two to three minutes to ask them questions and offer suggestions. Be sure to encourage as much participation from the students as possible.
   - *If using the gallery walk format,* tell the students, "Okay. Let's get started. Each group, move one argument to the left. Don't move to the next argument until I tell you to move. Once you get there, read the argument and then offer suggestions about how to make it better. I will put some sticky notes next to each argument. You can use the sticky notes to leave your suggestions." Give each group about three to four minutes to read the arguments, talk, and offer suggestions.

     a. Tell the students, "Okay. Let's rotate. Move one group to the left."

b. Again, give each group three or four minutes to read, talk, and offer suggestions.

c. Repeat this process for two more rotations.

- *If using the modified gallery walk format,* tell the students, "Okay. Let's get started. Reviewers, move one group to the left. Don't move to the next group until I tell you to move. Presenters, go ahead and share your argument with the reviewers when they get there." Give each group of presenters and reviewers about three to four minutes to talk.

  a. Tell the students, "Okay. Let's rotate. Reviewers, move one group to the left."

  b. Again, give each group of presenters and reviewers about three or four minutes to talk.

  c. Repeat this process for two more rotations.

8. Tell the students to return to their workstations.
9. Give the following instructions about revising the argument:
   - *If using the whole-class presentation format,* tell the students, "I'm now going to give you about 10 minutes to revise your argument. Take a few minutes to talk in your groups and determine what you want to change to make your argument better. Once you have decided what to change, go ahead and make the changes to your whiteboard."
   - *If using the gallery walk format,* tell the students, "I'm now going to give you about 10 minutes to revise your argument. Take a few minutes to read the suggestions that were left at your argument. Then talk in your groups and determine what you want to change to make your argument better. Once you have decided what to change, go ahead and make the changes to your whiteboard."
   - *If using the modified gallery walk format,* "I'm now going to give you about 10 minutes to revise your argument. Please return to your original groups." Wait for the students to move back into their original groups and then tell the students, "Okay, take a few minutes to talk in your groups and determine what you want to change to make your argument better. Once you have decided what to change, go ahead and make the changes to your whiteboard."

   Ask the students, "Do you have any questions about what you need to do?"
10. Answer any questions that come up.
11. Tell the students, "Okay. Let's get started."
12. Give the students 10 minutes to work in their groups on their arguments. As they work, move from group to group to check in, ask probing questions, and offer a suggestion if a group gets stuck.

Stage 5: Reflective Discussion (15 minutes)

1. Tell the students, "We are now going to take a minute to talk about what we did and what we have learned."
2. Show an image of several different objects on the screen. These objects should be made from different materials such as iron, copper, plastic, glass, and wood.
3. Ask the students, "What do you all see going on here?"
4. Allow students to share their ideas.
5. Ask the students, "How can we tell what these different objects are made from?"
6. Allow students to share their ideas.
7. Ask the students, "What makes materials different from each other?"
8. Allow students to share their ideas. Keep probing until someone mentions that materials have different properties.
9. Ask the students, "What are some examples of different properties that we can use to tell one material from a different material?"
10. Allow students to share their ideas.
11. Ask the students, "How can we tell if any of these objects are made from iron?"
12. Allow students to share their ideas. A student should mention that they could use a magnet.
13. Ask the students, "Why do objects that are made of iron stick to magnets?"
14. Allow students to share their ideas. Keep probing students until someone mentions that magnetism is a physical property of magnets.
15. Tell the students, "Okay, let's make sure we are on the same page. Objects are made out of different types of material. We can identify different types of materials because materials have specific physical properties. Physical properties are things like color, how hard it is, if it bends or not, and if it sticks to a magnet. The fact that different types of materials have different physical properties is a really important core idea in science."
16. Ask the students, "Does anyone have any questions about this core idea?"
17. Answer any questions that come up.
18. Tell the students, "We also looked for patterns during our investigation." Then ask, "Can anyone tell me why we needed to look for patterns?"
19. Allow students to share their ideas.
20. Tell the students, "Patterns are really important in science. Scientists look for patterns all the time. In fact, they even use patterns to help classify different things just like we did."

21. Tell the students, "We are now going take a minute to talk about what went well and what didn't go so well during our investigation. We need to talk about this because you are going to be planning and carrying out your own investigations like this a lot this year, and I want to help you all get better at it."
22. Show an image of the question "What made your investigation scientific?" on the screen. Tell the students, "Take a few minutes to talk about how you would answer this question with the other people in your group. Be ready to share with the rest of the class." Give the students two to three minutes to talk in their group.
23. Ask the students, "What do you all think? Who would like to share an idea?"
24. Allow three or four students to share their ideas. Be sure to expand on their ideas about what makes an investigation scientific.
25. Show an image of the question "What made your investigation not so scientific?" on the screen. Tell the students, "Take a few minutes to talk about how you would answer this question with the other people in your group. Be ready to share with the rest of the class." Give the students two to three minutes to talk in their group.
26. Ask the students, "What do you all think? Who would like to share an idea?"
27. Allow students to share their ideas. Be sure to expand on their ideas about what makes an investigation less scientific.
28. Show an image of the question "What rules can we put into place to help us make sure our next investigation is more scientific?" on the screen. Tell the students, "Take a few minutes to talk about how you would answer this question with the other people in your group. Be ready to share with the rest of the class." Give the students two to three minutes to talk in their group.
29. Ask the students, "What do you all think? Who would like to share an idea?"
30. Allow students to share their ideas. Once they have shared their ideas, offer a suggestion for a possible class rule.
31. Ask the students, "What do you all think? Should we make this a rule?"
32. If the students agree, write the rule on the board or make a class "Rules for Scientific Investigation" chart so you can refer to it during the next investigation.
33. Tell the students, "We are now going take a minute to talk about what makes scientific knowledge different from other types of knowledge."
34. Show an image of the question "What is the difference between an observation and an inference?" on the screen. Tell the students, "Take a few minutes to talk about how you would answer this question with the other people in your group. Be ready to share with the rest of the class." Give the students two to three minutes to talk in their group.
35. Ask the students, "What do you all think? Who would like to share an idea?"

36. Allow students to share their ideas.
37. Tell the students, "Okay, let's make sure we are all using the same definition. An *observation* is a descriptive statement about something. An *inference* is an interpretation of an observation."
38. Show an image of a piece of metal along with the statement "This object is a piece of metal" on the screen.
39. Ask the students, "Is this statement an observation or an inference and why?"
40. Allow students to share their ideas.
41. Tell the students, "That statement is an inference because it is an interpretation of some observations."
42. Ask the students, "What are some observations we can make about this object?"
43. Allow students to share their ideas.
44. Ask the students, "Does anyone have any questions about the difference between an observation and an inference?"
45. Answer any questions that come up.

Stage 6: Write a Draft Report (35 minutes)

Your students will use either the Investigation Handout or the investigation log in the student workbook when writing the draft report. When you give the directions shown in quotes in the following steps, substitute "investigation log" (as shown in brackets) for "handout" if they are using the workbook.

1. Tell the students, "You are now going to write an investigation report to share what you have learned. Please take out a pencil and turn to the 'Draft Report' section of your handout [investigation log]."
2. If possible, use a document camera to project the "Introduction" section of the draft report from the Investigation Handout (or the investigation log in their workbook) on a screen or board (or take a picture of it and project the picture on a screen or board).
3. Tell the students, "The first part of the report is called the 'Introduction.' In this section of the report you want to explain to the reader what you were investigating, why you were investigating it, and what question you were trying to answer. All of this information can be found in the text at the beginning of your handout [investigation log]." Point to the image and say, "There are some sentence starters here to help you begin writing the report." Ask the students, "Do you have any questions about what you need to do?"
4. Answer any questions that come up.
5. Tell the students, "Okay. Let's write."

6. Give the students 10 minutes to write the "Introduction" section of the report. As they work, move from student to student to check in, ask probing questions, and offer a suggestion if a student gets stuck.
7. If possible, use a document camera to project the "Method" section of the draft report from the Investigation Handout (or the investigation log in their workbook) on a screen or board (or take a picture of it and project the picture on a screen or board).
8. Tell the students, "The second part of the report is called the 'Method.' In this section of the report you want to explain to the reader what you did during the investigation, what data you collected and why, and how you went about analyzing your data. All of this information can be found in the 'Plan Your Investigation' section of your handout [investigation log]. Remember that you all planned and carried out different investigations, so do not assume that the reader will know what you did." Point to the image and say, "There are some sentence starters here to help you begin writing this part of the report." Ask the students, "Do you have any questions about what you need to do?"
9. Answer any questions that come up.
10. Tell the students, "Okay. Let's write."
11. Give the students 10 minutes to write the "Method" section of the report. As they work, move from student to student to check in, ask probing questions, and offer a suggestion if a student gets stuck.
12. If possible, use a document camera to project the "Argument" section of the draft report from the Investigation Handout (or the investigation log in their workbook) on a screen or board (or take a picture of it and project the picture on a screen or board).
13. Tell the students, "The last part of the report is called the 'Argument.' In this section of the report you want to share your claim, evidence, and justification of the evidence with the reader. All of this information can be found on your whiteboard." Point to the image and say, "There are some sentence starters here to help you begin writing this part of the report." Ask the students, "Do you have any questions about what you need to do?"
14. Answer any questions that come up.
15. Tell the students, "Okay. Let's write."
16. Give the students 10 minutes to write the "Argument" section of the report. As they work, move from student to student to check in, ask probing questions, and offer a suggestion if a student gets stuck.

Stage 7: Peer Review (30 minutes)

Your students will use either the Investigation Handout or the investigation log in the student workbook when doing the peer review. When you give the directions shown in quotes in the following steps, substitute "workbook" (as shown in brackets) for "Investigation Handout" if they are using the workbook.

1. Tell the students, "We are now going to review our reports to find ways to make them better. I'm going to come around and collect your Investigation Handout [workbook]. While I do that, please take out a pencil."
2. Collect the Investigation Handouts or workbooks from the students.
3. If possible, use a document camera to project the peer-review guide (PRG; see Appendix 4) on a screen or board (or take a picture of it and project the picture on a screen or board).
4. Tell the students, "We are going to use this peer-review guide to give each other feedback." Point to the image.
5. Give the following instructions:
   - *If using the Investigation Handout,* tell the students, "I'm going to ask you to work with a partner to do this. I'm going to give you and your partner a draft report to read and a peer-review guide to fill out. You two will then read the report together. Once you are done reading the report, I want you to answer each of the questions on the peer-review guide." Point to the review questions on the image of the PRG.
   - *If using the workbook,* tell the students, "I'm going to ask you to work with a partner to do this. I'm going to give you and your partner a draft report to read. You two will then read the report together. Once you are done reading the report, I want you to answer each of the questions on the peer-review guide that is right after the report in the investigation log." Point to the review questions on the image of the PRG.
6. Tell the students, "You can check 'yes,' 'almost,' or 'no' after each question." Point to the checkboxes on the image of the PRG.
7. Tell the students, "This will be your rating for this part of the report. Make sure you agree on the rating you give the author. If you mark 'almost' or 'no,' then you need to tell the author what he or she needs to do to get a 'yes.'" Point to the space for the reviewer feedback on the image of the PRG.
8. Tell the students, "It is really important for you to give the authors feedback that is helpful. That means you need to tell them exactly what they need to do to make their reports better." Ask the students, "Do you have any questions about what you need to do?"
9. Answer any questions that come up.

10. Tell the students, "Please sit with a partner who is not in your current group." Allow the students time to sit with a partner.
11. Give the following instructions:
    - *If using the Investigation Handout,* tell the students, "Okay, I am now going to give you one report to read and one peer-review guide to fill out." Pass out one report to each pair. Make sure that the report you give a pair was not written by one of the students in that pair. Give each pair one PRG to fill out as a team.
    - *If using the workbook,* tell the students, "Okay, I am now going to give you one report to read." Pass out a workbook to each pair. Make sure that the workbook you give a pair is not from one of the students in that pair.
12. Tell the students, "Okay, I'm going to give you 15 minutes to read the report I gave you and to fill out the peer-review guide. Go ahead and get started."
13. Give the students 15 minutes to work. As they work, move around from pair to pair to check in and see how things are going, answer questions, and offer advice.
14. After 15 minutes pass, tell the students, "Okay, time is up." *If using the Investigation Handout,* say, "Please give me the report and the peer-review guide that you filled out." *If using the workbook,* say, "Please give me the workbook that you have."
15. Collect the Investigation Handouts and the PRGs, or collect the workbooks if they are being used. Be sure you keep the handout and the PRG together.
16. Give the following instructions:
    - *If using the Investigation Handout,* tell the students, "Okay, I am now going to give you a different report to read and a new peer-review guide to fill out." Pass out another report to each pair. Make sure that this report was not written by one of the students in that pair. Give each pair a new PRG to fill out as a team.
    - *If using the workbook,* tell the students, "Okay, I am now going to give you a different report to read." Pass out a different workbook to each pair. Make sure that the workbook you give a pair is not from one of the students in that pair.
17. Tell the students, "Okay, I'm going to give you 15 minutes to read this new report and to fill out the peer-review guide. Go ahead and get started."
18. Give the students 15 minutes to work. As they work, move around from pair to pair to check in and see how things are going, answer questions, and offer advice.
19. After 15 minutes pass, tell the students, "Okay, time is up." *If using the Investigation Handout,* say, "Please give me the report and the peer-review guide that you filled out." *If using the workbook,* say, "Please give me the workbook that you have."
20. Collect the Investigation Handouts and the PRGs, or collect the workbooks if they are being used. Be sure you keep the handout and the PRG together.

Stage 8: Revise the Report (30 minutes)

Your students will use either the Investigation Handout or the investigation log in the student workbook when revising the report. Except where noted below, the directions are the same whether using the handout or the log.

1. Give the following instructions:
   - *If using the Investigation Handout,* tell the students, "You are now going to revise your investigation report based on the feedback you get from your classmates. Please take out a pencil while I hand back your draft report and the peer-review guide."
   - *If using the investigation log in the student workbook,* tell the students, "You are now going to revise your investigation report based on the feedback you get from your classmates. Please take out a pencil while I hand back your investigation logs."
2. *If using the Investigation Handout,* pass back the handout and the PRG to each student. *If using the investigation log,* pass back the log to each student.
3. Tell the students, "Please take a few minutes to read over the peer-review guide. You should use it to figure out what you need to change in your report and how you will change the report."
4. Allow the students time to read the PRG.
5. *If using the investigation log,* if possible use a document camera to project the "Write Your Final Report" section from the investigation log on a screen or board (or take a picture of it and project the picture on a screen or board).
6. Give the following instructions:
   - *If using the Investigation Handout,* tell the students, "Okay. Let's revise our reports. Please take out a piece of paper. I would like you to rewrite your report. You can use your draft report as a starting point, but use the feedback on the peer-review guide to help make it better."
   - *If using the investigation log,* tell the students, "Okay. Let's revise our reports. I would like you to rewrite your report in the section of the investigation log that says 'Write Your Final Report.'" Point to the image on the screen and tell the students, "You can use your draft report as a starting point, but use the feedback on the peer-review guide to help make your report better."

   Ask the students, "Do you have any questions about what you need to do?"
7. Answer any questions that come up.
8. Tell the students, "Okay. Let's write."
9. Give the students 30 minutes to rewrite their report. As they work, move from student to student to check in, ask probing questions, and offer a suggestion if a student gets stuck.

10. Give the following instructions:
    - *If using the Investigation Handout,* tell the students, "Okay. Time's up. I will now come around and collect your Investigation Handout, the peer-review guide, and your final report."
    - *If using the investigation log,* tell the students, "Okay. Time's up. I will now come around and collect your workbooks."
11. *If using the Investigation Handout,* collect all the Investigation Handouts, PRGs, and final reports. *If using the investigation log,* collect all the workbooks.
12. *If using the Investigation Handout,* use the "Teacher Score" columns in the PRG to grade the final report. *If using the investigation log,* use the "ADI Investigation Report Grading Rubric" in the investigation log to grade the final report. Whether you are using the handout or the log, you can give the students feedback about their writing in the "Teacher Comments" section.

How to Use the Checkout Questions

The Checkout Questions are an optional assessment. We recommend giving them to students one day after they finish stage 8 of the ADI investigation. The Checkout Questions can be used as a formative or summative assessment of student thinking. If you plan to use them as a formative assessment, we recommend that you look over the student answers to determine if you need to reteach the core idea and/or crosscutting concept from the investigation, but do not grade them. If you plan to use them as a summative assessment, we have included a "Teacher Scoring Rubric" at the end of the Checkout Questions that you can use to score a student's ability to apply the core idea in a new scenario and explain their use of a crosscutting concept. The rubric includes a 4-point scale that ranges from 0 (the student cannot apply the core idea correctly in all cases and cannot explain the [crosscutting concept]) to 3 (the student can apply the core idea correctly in all cases and can fully explain the [crosscutting concept]). The Checkout Questions, regardless of how you decide to use them, are a great way to make student thinking visible so you can determine if the students have learned the core idea and the crosscutting concept.

A student who can apply the core idea correctly in all cases and can explain the pattern would indicate that a nail and a paper clip will stick to a magnet and then explain that objects that are made out of iron stick to magnets but objects made out of other materials do not.

Connections to Standards

Table 1.2 (p. 60) highlights how the investigation can be used to address specific performance expectations from the *NGSS, Common Core State Standards (CCSS)* in English language arts (ELA) and in mathematics, and *English Language Proficiency (ELP) Standards.*

TABLE 1.2

Investigation 1 alignment with standards

***NGSS* performance expectations**	Strong alignment • 2-PS1-1: Plan and conduct an investigation to describe and classify different kinds of materials by their observable properties. Moderate alignment (this investigation can be used to build toward these performance expectations) • 3-PS2-3: Ask questions to determine cause and effect relationships of electric or magnetic interactions between two objects not in contact with each other. • 3-PS2-4: Define a simple design problem that can be solved by applying scientific ideas about magnets.
***CCSS ELA*—Reading: Informational Text**	Key ideas and details • CCSS.ELA-LITERACY.RI.3.1: Ask and answer questions to demonstrate understanding of a text, referring explicitly to the text as the basis for the answers. • CCSS.ELA-LITERACY.RI.3.2: Determine the main idea of a text; recount the key details and explain how they support the main idea. • CCSS.ELA-LITERACY.RI.3.3: Describe the relationship between a series of historical events, scientific ideas or concepts, or steps in technical procedures in a text, using language that pertains to time, sequence, and cause/effect. Craft and structure • CCSS.ELA-LITERACY.RI.3.4: Determine the meaning of general academic and domain-specific words and phrases in a text relevant to a *grade 3 topic or subject area*. • CCSS.ELA-LITERACY.RI.3.5: Use text features and search tools (e.g., key words, sidebars, hyperlinks) to locate information relevant to a given topic efficiently. • CCSS.ELA-LITERACY.RI.3.6: Distinguish their own point of view from that of the author of a text. Integration of knowledge and ideas • CCSS.ELA-LITERACY.RI.3.7: Use information gained from illustrations (e.g., maps, photographs) and the words in a text to demonstrate understanding of the text (e.g., where, when, why, and how key events occur). • CCSS.ELA-LITERACY.RI.3.8: Describe the logical connection between particular sentences and paragraphs in a text (e.g., comparison, cause/effect, first/second/third in a sequence). • CCSS.ELA-LITERACY.RI.3.9: Compare and contrast the most important points and key details presented in two texts on the same topic. Range of reading and level of text complexity • CCSS.ELA-LITERACY.RI.3.10: By the end of the year, read and comprehend informational texts, including history/social studies, science, and technical texts, at the high end of the grades 2–3 text complexity band independently and proficiently.

Continued

Table 1.2 (*continued*)

***CCSS ELA*—Writing**	Text types and purposes • CCSS.ELA-LITERACY.W.3.1: Write opinion pieces on topics or texts, supporting a point of view with reasons. o CCSS.ELA-LITERACY.W.3.1.A: Introduce the topic or text they are writing about, state an opinion, and create an organizational structure that lists reasons. o CCSS.ELA-LITERACY.W.3.1.B: Provide reasons that support the opinion. o CCSS.ELA-LITERACY.W.3.1.C: Use linking words and phrases (e.g., *because*, *therefore*, *since, for example*) to connect opinion and reasons. o CCSS.ELA-LITERACY.W.3.1.D: Provide a concluding statement or section. • CCSS.ELA-LITERACY.W.3.2: Write informative/explanatory texts to examine a topic and convey ideas and information clearly. o CCSS.ELA-LITERACY.W.3.2.A: Introduce a topic and group related information together; include illustrations when useful to aiding comprehension. o CCSS.ELA-LITERACY.W.3.2.B: Develop the topic with facts, definitions, and details. o CCSS.ELA-LITERACY.W.3.2.C: Use linking words and phrases (e.g., *also*, *another*, *and*, *more*, *but*) to connect ideas within categories of information. o CCSS.ELA-LITERACY.W.3.2.D: Provide a concluding statement or section. Production and distribution of writing • CCSS.ELA-LITERACY.W.3.4: With guidance and support from adults, produce writing in which the development and organization are appropriate to task and purpose. • CCSS.ELA-LITERACY.W.3.5: With guidance and support from peers and adults, develop and strengthen writing as needed by planning, revising, and editing. • CCSS.ELA-LITERACY.W.3.6: With guidance and support from adults, use technology to produce and publish writing (using keyboarding skills) as well as to interact and collaborate with others. Research to build and present knowledge • CCSS.ELA-LITERACY.W.3.8: Recall information from experiences or gather information from print and digital sources; take brief notes on sources and sort evidence into provided categories. Range of writing • CCSS.ELA-LITERACY.W.3.10: Write routinely over extended time frames (time for research, reflection, and revision) and shorter time frames (a single sitting or a day or two) for a range of discipline-specific tasks, purposes, and audiences.

Continued

Table 1.2 *(continued)*

***CCSS ELA*— Speaking and Listening**	Comprehension and collaboration • CCSS.ELA-LITERACY.SL.3.1: Engage effectively in a range of collaborative discussions (one-on-one, in groups, and teacher-led) with diverse partners on *grade 3 topics and texts,* building on others' ideas and expressing their own clearly. o CCSS.ELA-LITERACY.SL.3.1.A: Come to discussions prepared, having read or studied required material; explicitly draw on that preparation and other information known about the topic to explore ideas under discussion. o CCSS.ELA-LITERACY.SL.3.1.B: Follow agreed-upon rules for discussions (e.g., gaining the floor in respectful ways, listening to others with care, speaking one at a time about the topics and texts under discussion). o CCSS.ELA-LITERACY.SL.3.1.C: Ask questions to check understanding of information presented, stay on topic, and link their comments to the remarks of others. o CCSS.ELA-LITERACY.SL.3.1.D: Explain their own ideas and understanding in light of the discussion. • CCSS.ELA-LITERACY.SL.3.2: Determine the main ideas and supporting details of a text read aloud or information presented in diverse media and formats, including visually, quantitatively, and orally. • CCSS.ELA-LITERACY.SL.3.3: Ask and answer questions about information from a speaker, offering appropriate elaboration and detail. Presentation of knowledge and ideas • CCSS.ELA-LITERACY.SL.3.4: Report on a topic or text, tell a story, or recount an experience with appropriate facts and relevant, descriptive details, speaking clearly at an understandable pace. • CCSS.ELA-LITERACY.SL.3.6: Speak in complete sentences when appropriate to task and situation in order to provide requested detail or clarification.
***CCSS Mathematics*— Measurement and Data**	Represent and interpret data. • CCSS.MATH.CONTENT.3.MD.B.3: Draw a scaled picture graph and a scaled bar graph to represent a data set with several categories. Solve one- and two-step "how many more" and "how many less" problems using information presented in scaled bar graphs.

Continued

Table 1.2 (*continued*)

ELP Standards	Receptive modalities • ELP 1: Construct meaning from oral presentations and literary and informational text through grade-appropriate listening, reading, and viewing. • ELP 8: Determine the meaning of words and phrases in oral presentations and literary and informational text. Productive modalities • ELP 3: Speak and write about grade-appropriate complex literary and informational texts and topics. • ELP 4: Construct grade-appropriate oral and written claims and support them with reasoning and evidence. • ELP 7: Adapt language choices to purpose, task, and audience when speaking and writing. Interactive modalities • ELP 2: Participate in grade-appropriate oral and written exchanges of information, ideas, and analyses, responding to peer, audience, or reader comments and questions. • ELP 5: Conduct research and evaluate and communicate findings to answer questions or solve problems. • ELP 6: Analyze and critique the arguments of others orally and in writing. Linguistic structures of English • ELP 9: Create clear and coherent grade-appropriate speech and text. • ELP 10: Make accurate use of standard English to communicate in grade-appropriate speech and writing.

Investigation 1

Magnetic Attraction: What Types of Objects Are Attracted to a Magnet?

Introduction

We use magnets every day. We hold pictures up on the refrigerator with magnets. We also use magnets to keep the refrigerator door shut. There are even magnets in many different kinds of toys. Take a few minutes to explore what happens when you bring a magnet near a bunch of paper clips. Keep track of what you observe and what you are wondering about in the boxes below.

Things I OBSERVED ...	Things I WONDER about ...

Magnets come in different shapes and sizes. Most magnets are made of iron. Magnets are amazing because they can stick to an object. Magnets can also make an object move without touching it. Scientists call the force produced by a magnet that can make things stick to it or that makes an object move without touching it *magnetism*.

You have probably seen magnets stick to some objects but not others. Magnets do not stick to all objects because objects are made of different materials. For example, beverage cans are made out of aluminum. We use wood and graphite to make pencils. Wires are made out of copper and plastic. These different materials have different *physical properties*. An example of a physical property is color. The ability to stick to a magnet is another example of a physical property. Materials are different from each other because they have different physical properties.

Your goal in this investigation is to *figure out* what types of objects are attracted to a magnet and what types of objects are not attracted to a magnet. To complete this task, you will need to test different objects to see if they stick to a magnet or not. You will then need to look for patterns in the objects that are attracted to magnets and the objects that are not attracted to magnets. Scientists often look for patterns in nature like this. They then use these patterns to classify objects or sort objects into groups. You can use the patterns you find to figure out what is similar about objects that are attracted to a magnet and what makes them different from objects that are not attracted to a magnet.

Things we KNOW from what we read …	What we will NEED to figure out …

Your Task

Use what you know about magnetism, properties of matter, and patterns to design and carry out an investigation to determine what types of materials are attracted to a magnet and what types of materials are not.

The *guiding question* of this investigation is, ***What types of objects are attracted to a magnet?***

Materials

You may use any of the following materials during your investigation:

- Safety glasses or goggles (required)
- Ceramic magnet (iron)
- Foil (aluminum)
- Wire (copper)
- Penny (zinc and copper)
- Nickel (copper and nickel)
- Nail (iron)
- Paper clips (iron)
- BBs (iron)
- BBs (plastic)
- Washer (zinc and iron)
- Fishing sinker (lead)
- String (cotton)
- Index card (paper)
- Tile (plastic)
- Tile (glass)
- Marble (glass)
- Block (wood)

Safety Rules

Follow all normal safety rules. In addition, be sure to follow these rules:

- Wear sanitized safety glasses or goggles during setup, investigation activity, and cleanup.
- Do not throw objects or put any objects in your mouth.
- Use caution when handling sharp objects (e.g., wire, nails, paper clips). They can cut or puncture skin.
- Immediately pick up any slip or fall hazards (e.g., marbles) from the floor.
- Wash your hands with soap and water when you are done collecting the data.

Plan Your Investigation

Prepare a plan for your investigation by filling out the chart that follows; this plan is called an *investigation proposal*. Before you start developing your plan, be sure to discuss the following questions with the other members of your group:

- What information should we collect so we can **compare and contrast** different materials?
- What types of **patterns** might we look for to help answer the guiding question?

Our guiding question:

We will collect the following data:

These are the steps we will follow to collect data:

I approve of this investigation proposal.

Teacher's signature

Date

Collect Your Data

Keep a record of what you measure or observe during your investigation in the space below.

Analyze Your Data

You will need to analyze the data you collected before you can develop an answer to the guiding question. In the space below, create a table or figure that shows how objects that are attracted to a magnet are similar to each other and different from objects that are not attracted to a magnet.

Draft Argument

Develop an argument on a whiteboard. It should include the following parts:

1. A *claim:* Your answer to the guiding question.
2. *Evidence:* An analysis of the data and an explanation of what the analysis means.
3. A *justification of the evidence:* Why your group thinks the evidence is important.

The Guiding Question:	
Our Claim:	
Our Evidence:	Our Justification of the Evidence:

Argumentation Session

Share your argument with your classmates. Be sure to ask them how to make your draft argument better. Keep track of their suggestions in the space below.

Ways to IMPROVE our argument …

Draft Report

Prepare an *investigation report* to share what you have learned. Use the information in this handout and your group's final argument to write a *draft* of your investigation report.

Introduction

We have been studying ______________________________ in class.

Before we started this investigation, we explored ______________________________

____

____

____

____

We noticed ______________________________

____

____

____

____

My goal for this investigation was to figure out ______________________________

____

____

____

____

The guiding question was ______________________________

____

____

____

____

Method

To gather the data I needed to answer this question, I ______________________________

____

____

____

____

____

____

____

I then analyzed the data I collected by ______________________________

____

____

____

____

____

____

____

____

Argument

My claim is ____

____

____

____

____

____

The ____________________ below shows ____

____

____

This evidence is important because ____

____

____

____

____

Review

Your friends need your help! Review the draft of their investigation reports and give them ideas about how to improve. Use the peer-review guide when doing your review.

Submit Your Final Report

Once you have received feedback from your friends about your draft report, create your final investigation report and hand it in to your teacher.

Checkout Questions

Investigation 1.
Magnetic Attraction: What Types of Objects Are Attracted to a Magnet?

1. Listed below are some objects. Place an *X* next to the objects that will stick to a magnet.

☐ Book	☐ Paper clip	☐ Piece of tin
☐ Coffee cup	☐ Piece of aluminum foil	☐ Plastic toy
☐ Nail	☐ Piece of copper wire	☐ Seed
☐ Nickel	☐ Piece of paper	☐ Wood block

2. Explain your thinking. What *pattern* from your investigation did you use to decide if an object will or will not stick to a magnet?

____

____

____

____

____

____

____

____

____

Teacher Scoring Rubric for the Checkout Questions

Level	Description
3	The student can apply the core idea correctly in all cases and can fully explain the pattern.
2	The student can apply the core idea correctly in all cases but cannot fully explain the pattern.
1	The student cannot apply the core idea correctly in all cases but can fully explain the pattern.
0	The student cannot apply the core idea correctly in all cases and cannot explain the pattern.

Investigation 2

Magnetic Force: How Does Changing the Distance Between Two Magnets Affect Magnetic Force Strength?

Purpose

The purpose of this investigation is to give students an opportunity to use the disciplinary core idea (DCI) of PS2.B: Types of Interactions and the crosscutting concept (CC) of Cause and Effect from *A Framework for K–12 Science Education* (NRC 2012) to figure out what types of materials are attracted to a magnet. Students will also learn about the difference between data and evidence during the reflective discussion.

The Disciplinary Core Idea

Students in third grade should understand the following about Types of Interactions and be able to use this DCI to figure out how changing the distance between two magnets affects the magnetic force strength between the two magnets:

> *When objects touch or collide, they push on one another and can change motion or shape. (NRC 2012, p. 117)*

The Crosscutting Concept

Students in third grade should understand the following about the CC Cause and Effect:

> *Repeating patterns in nature, or events that occur together with regularity, are clues that scientists can use to start exploring causal, or cause-and-effect, relationships. … Any application of science, or any engineered solution to a problem, is dependent on understanding the cause-and-effect relationships between events; the quality of the application or solution often can be improved as knowledge of the relevant relationships is improved. (NRC 2012, p. 87)*

Students in third grade should also be given opportunities to "begin to look for and analyze patterns—whether in their observations of the world or in the relationships between different quantities in data (e.g., the sizes of plants over time)"; "they can also begin to consider what might be causing these patterns and relationships and design tests that gather more evidence to support or refute their ideas" (NRC 2012, pp. 88–89).

Students should be encouraged to use their developing understanding of cause and effect as a tool or a way of thinking about a phenomenon during this investigation to help them figure out what causes a change in the strength of the magnetic force that exists between two magnets.

What Students Figure Out

Magnetism is a non-contact force. Two magnets do not have to be touching to apply a force on each other. The size or strength of the magnetic force between any two magnets depends on how far apart the two magnets are from each other. The strength of the magnet force that exists between the two magnets weakens as the distance between the magnets increases.

Timeline

The time needed to complete this investigation is 250 minutes (4 hours and 10 minutes). The amount of instructional time needed for each stage of the investigation is as follows:

- *Stage 1.* Introduce the task and the guiding question: 35 minutes
- *Stage 2.* Design a method and collect data: 40 minutes
- *Stage 3.* Create a draft argument: 35 minutes
- *Stage 4.* Argumentation session: 30 minutes
- *Stage 5.* Reflective discussion: 15 minutes
- *Stage 6.* Write a draft report: 35 minutes
- *Stage 7.* Peer review: 30 minutes
- *Stage 8.* Revise the report: 30 minutes

This investigation can be completed in one day or over eight days (one day for each stage) during your designated science time in the daily schedule.

Materials and Preparation

The materials needed for this investigation are listed in Table 2.1 (p. 76). Except for the Alnico cylinder magnets, most of the items can be purchased from a big-box retail store such as Wal-Mart or Target or through an online retailer such as Amazon. The Alnico cylinder magnets can be purchased online from a specialized magnet retailer such as AMF Magnets (see *www.amfmagnets.com*). The materials for this investigation can also be purchased as a complete kit (which includes enough materials for 24 students, or six groups) at *www.argumentdriveninquiry.com.*

TABLE 2.1
Materials for Investigation 2

Item	Quantity
Safety glasses or goggles	1 per student
Alnico cylinder magnets	2 per group
Plastic ruler (with groove), 12"/30 cm	1 per group
Ceramic ring magnets	3 per class
Whiteboard, 2' × 3'*	1 per group
Investigation Handout	1 per student
Peer-review guide and teacher scoring rubric	1 per student
Checkout Questions (optional)	1 per student

*As an alternative, students can use computer and presentation software such as Microsoft PowerPoint or Apple Keynote to create their arguments.

Be sure to use a set routine for distributing and collecting the materials. One option is to set up the materials for each group in a kit that you can deliver to each group. A second option is to have all the materials on a table or cart at a central location. You can then assign a member of each group to be the "materials manager." This individual is responsible for collecting all the materials his or her group needs from the table or cart during class and for returning all the materials at the end of the class.

Safety Precautions

Remind students to follow all normal safety rules. In addition, tell the students to take the following safety precautions:

- Wear sanitized safety glasses or goggles during setup, investigation activity, and cleanup.
- Do not throw objects or put any objects in their mouth.
- Wash their hands with soap and water when done collecting the data.

Lesson Plan by Stage

Stage 1: Introduce the Task and the Guiding Question (35 minutes)

1. Ask the students to sit in six groups, with three or four students in each group.
2. Ask students to clear off their desks except for a pencil (and their *Student Workbook for Argument-Driven Inquiry in Third-Grade Science* if they have one).
3. Pass out an Investigation Handout to each student (or ask students to turn to Investigation Log 2 in their workbook).

4. Read the first paragraph of the "Introduction" aloud to the class. Ask the students to follow along as you read.
5. Pass out two Alnico cylinder magnets to each group.
6. Remind students of the safety rules and explain the safety precautions for this investigation.
7. Tell students to play with the magnets and then record their observations and questions in the "OBSERVED/WONDER" chart in the "Introduction" section of their Investigation Handout (or the investigation log in their workbook).
8. Ask students to share *what they observed* as they played with the magnets.
9. Ask students to share *what questions they have* about the behavior of the two magnets.
10. Tell the students, "Some of your questions might be answered by reading the rest of the 'Introduction.'"
11. Ask the students to read the rest of the "Introduction" on their own *or* ask them to follow along as you read aloud.
12. Once the students have read the rest of the "Introduction," ask them to fill out the "KNOW/NEED" chart on their Investigation Handout (or in their investigation log) as a group.
13. Ask students to share what they learned from the reading. Add these ideas to a class "know / need to figure out" chart.
14. Ask students to share what they think they will need to figure out based on what they read. Add these ideas to the class "know / need to figure out" chart.
15. Tell the students, "It looks like we have something to figure out. Let's see what we will need to do during our investigation."
16. Read the task and the guiding question aloud.
17. Tell the students, "I have lots of materials here that you can use."
18. Introduce the students to the materials available for them to use during the investigation by showing them how to use a ruler to measure the magnetic force strength between two magnets. Magnetic force strength can be measured by (1) putting two Alnico cylinder magnets in the groove of a ruler with like poles facing each other (see Figure 2.1, p. 78), (2) holding the first magnet in place and then bringing the second magnet closer to first magnet without letting go of either magnet, (3) letting go of the second magnet, and (4) measuring how far the second magnet moves away from the first magnet. This method can be used to measure the strength of the magnetic force between two magnets because magnetism, which is a force, can cause an object to move and the amount the object moves is directly related to how much force acts on it (more force means more movement, less force means less movement). Therefore, the more force that acts on the second magnet, the farther it will move away from the first magnet.

FIGURE 2.1

How to measure the magnetic force strength between two magnets

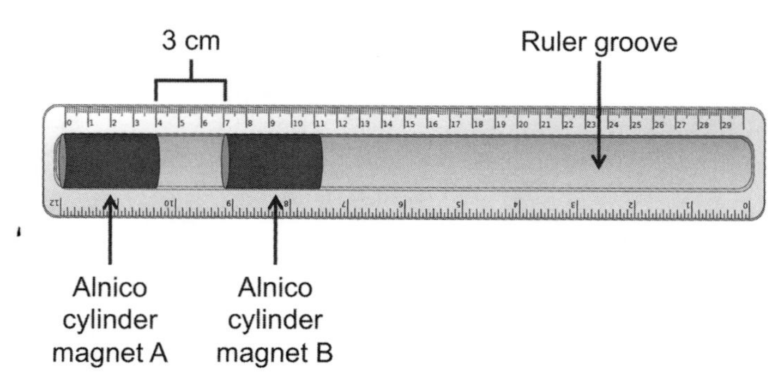

Stage 2: Design a Method and Collect Data (40 minutes)

1. Tell the students, "I am now going to give you and the other members of your group about 15 minutes to plan your investigation. Before you begin, I want you all to take a couple of minutes to discuss the following questions with the rest of your group."
2. Show the following questions on the screen or board:
   - What types of *patterns* might we look for to help answer the guiding question?
   - What information do we need to find a *cause-and-effect relationship?*
3. Tell the students, "Please take a few minutes to come up with an answer to these questions." Give the students two or three minutes to discuss these two questions.
4. Ask two or three different groups to share their answers. Be sure to highlight or write down any important ideas on the board so students can refer to them later.
5. If possible, use a document camera to project an image of the graphic organizer for this investigation on a screen or board (or take a picture of it and project the picture on a screen or board). Tell the students, "I now want you all to plan out your investigation. To do that, you will need to create an investigation proposal by filling out this graphic organizer."
6. Point to the box labeled "Our guiding question:" and tell the students, "You can put the question we are trying to answer in this box." Then ask, "Where can we find the guiding question?"
7. Wait for a student to answer where to find the guiding question (the answer is "in the handout").
8. Point to the box labeled "We will collect the following data:" and tell the students, "You can list the measurements or observations that you will need to collect during the investigation in this box."

9. Point to the box labeled "This is a picture of how we will set up the equipment:" and tell the students, "You can draw a picture of how you plan to set up the equipment during the investigation so you can collect the data you need in this box."
10. Point to the box labeled "These are the steps we will follow to collect data:" and tell the students, "You can list what you are going to do to collect the data you need and what you will do with your data once you have it. Be sure to give enough detail that I could do your investigation for you."
11. Ask the students, "Do you have any questions about what you need to do?"
12. Answer any questions that come up.
13. Tell the students, "Once you are done, raise your hand and let me know. I'll then come by and look over your proposal and give you some feedback. You may not begin collecting data until I have approved your proposal by signing it. You need to have your proposal done in the next 15 minutes."
14. Give the students 15 minutes to work in their groups on their investigation proposal. As they work, move from group to group to check in, ask probing questions, and offer a suggestion if a group gets stuck.

What should a student-designed investigation look like?

The students' investigation proposal should include the following information: The guiding question is "How does changing the distance between two magnets affect magnetic force strength?"

- The data that the students should collect are (1) initial distance between the two magnets and (2) distance the magnet moves.
- To collect the data, the students might conduct a series of tests in which they hold two magnets a set distance apart on a ruler (starting with 1 cm and increasing 1 cm on each test, up to 4 cm), let go of one magnet, and measure how far the magnet moves. This is just one example of how they can collect the data, and there should be a lot of variation in the student-designed investigations.

15. As each group finishes its investigation proposal, be sure to read it over and determine if it will be productive or not. If you feel the investigation will be productive (not necessarily what you would do or what the other groups are doing), sign your name on the proposal and let the group start collecting data. If the plan needs to be changed, offer some suggestions or ask some probing questions, and have the group make the changes before you approve it.

16. Pass out the materials or have one student from each group collect the materials they need from a central supply table or cart for the groups that have an approved proposal.
17. Remind students of the safety rules and precautions for this investigation.
18. Tell the students to collect their data and record their observations or measurements in the "Collect Your Data" box in their Investigation Handout (or the investigation log in their workbook).
19. Give the students 10 minutes to collect their data.
20. Be sure to collect the materials from each group before asking them to analyze their data.

Stage 3: Create a Draft Argument (35 minutes)

1. Tell the students, "Now that we have all this data, we need to analyze the data so we can figure out an answer to the guiding question."
2. If possible, project an image of the "Analyze Your Data" section for this investigation on a screen or board using a document camera (or take a picture of it and project the picture on a screen or board). Point to the section and tell the students, "You can create a graph as a way to analyze your data. You can make your graph in this section."
3. Ask the students, "What information do we need to include in a graph?"
4. Tell the students, "Please take a few minutes to discuss this question with your group, and be ready to share."
5. Give the students five minutes to discuss.
6. Ask two or three different groups to share their answers. Be sure to highlight or write down any important ideas on the board so students can refer to them later.
7. Tell the students, "I am now going to give you and the other members of your group about 10 minutes to create your graph." If the students are having trouble making a graph, you can take a few minutes to provide a mini-lesson about how to create a graph from a bunch of observations or measurements (this strategy is called just-in-time instruction because it is offered only when students get stuck).

What should a graph look like for this investigation?

There are a number of different ways that students can analyze the observations or measurements they collect during this investigation. One of the most straightforward ways is to create a bar graph. This bar graph should have the starting distance between the magnets on the horizontal or *x*-axis and the distance the magnets moved on the vertical or *y*-axis. An example of a graph can be seen in Figure 2.2 (p. 82). There are other options for analyzing the data they collected. Students often come up with some unique ways of analyzing their data, so be sure to give them some voice and choice during this stage.

8. Give the students 10 minutes to analyze their data. As they work, move from group to group to check in, ask probing questions, and offer suggestions.
9. Tell the students, "I am now going to give you and the other members of your group 15 minutes to create an argument to share what you have learned and convince others that they should believe you. Before you do that, we need to take a few minutes to discuss what you need to include in your argument."
10. If possible, use a document camera to project the "Argument Presentation on a Whiteboard" image from the Investigation Handout (or the investigation log in their workbook) on a screen or board (or take a picture of it and project the picture on a screen or board).
11. Point to the box labeled "The Guiding Question:" and tell the students, "You can put the question we are trying to answer here on your whiteboard."
12. Point to the box labeled "Our Claim:" and tell the students, "You can put your claim here on your whiteboard. The claim is your answer to the guiding question."
13. Point to the box labeled "Our Evidence:" and tell the students, "You can put the evidence that you are using to support your claim here on your whiteboard. Your evidence will need to include the analysis you just did and an explanation of what your analysis means or shows. Scientists always need to support their claims with evidence."
14. Point to the box labeled "Our Justification of the Evidence:" and tell the students, "You can put your justification of your evidence here on your whiteboard. Your justification needs to explain why your evidence is important. Scientists often use core ideas to explain why the evidence they are using matters. Core ideas are important concepts that scientists use to help them make sense of what happens during an investigation."

15. Ask the students, "What are some core ideas that we read about earlier that might help us explain why the evidence we are using is important?"
16. Ask students to share some of the core ideas from the "Introduction" section of the Investigation Handout (or the investigation log in the workbook). List these core ideas on the board.
17. Tell the students, "That is great. I would like to see everyone try to include these core ideas in your justification of the evidence. Your goal is to use these core ideas to help explain why your evidence matters and why the rest of us should pay attention to it."
18. Ask the students, "Do you have any questions about what you need to do?"
19. Answer any questions that come up.
20. Tell the students, "Okay, go ahead and start working on your arguments. You need to have your argument done in the next 15 minutes. It doesn't need to be perfect. We just need something down on the whiteboards so we can share our ideas."
21. Give the students 15 minutes to work in their groups on their arguments. As they work, move from group to group to check in, ask probing questions, and offer a suggestion if a group gets stuck. Figure 2.2 shows an example of an argument created by students for this investigation.

FIGURE 2.2

Example of an argument

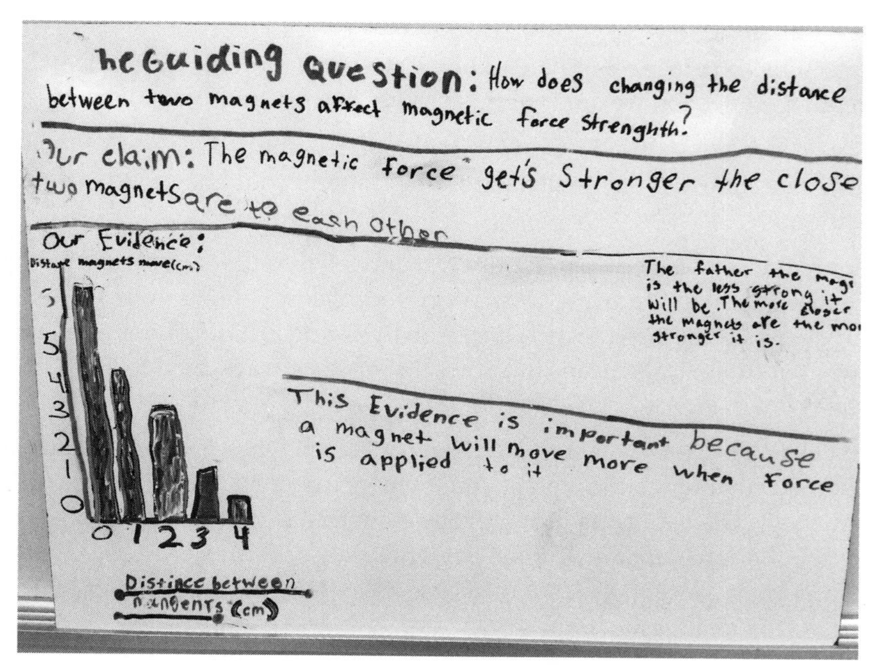

Stage 4: Argumentation Session (30 minutes)

The argumentation session can be conducted in a whole-class presentation format, a gallery walk format, or a modified gallery walk format. We recommend using a whole-class presentation format for the first investigation, but try to transition to either the gallery walk or modified gallery walk format as soon as possible because that will maximize student voice and choice inside the classroom. The following list shows the steps for the three formats; unless otherwise noted, the steps are the same for all three formats.

1. Begin by introducing the use of the whiteboard.
   - *If using the whole-class presentation format,* tell the students, "We are now going to share our arguments. Please set up your whiteboards so everyone can see them."
   - *If using the gallery walk or modified gallery walk format,* tell the students, "We are now going to share our arguments. Please set up your whiteboard so they are facing the walls."
2. Allow the students to set up their whiteboards.
   - *If using the whole-class presentation format,* the whiteboards should be set up on stands or chairs so they are facing toward the center of the room.
   - *If using the gallery walk or modified gallery walk format,* the whiteboards should be set up on stands or chairs so they are facing toward the outside of the room.
3. Give the following instructions to the students:
   - *If using the whole-class presentation format or the modified gallery walk format,* tell the students, "Okay, before we get started I want to explain what we are going to do next. I'm going to ask some of you to present your arguments to your classmates. If you are presenting your argument, your job is to share your group's claim, evidence, and justification of the evidence. The rest of you will be reviewers. If you are a reviewer, your job is to listen to the presenters, ask the presenters questions if you do not understand something, and then offer them some suggestions about ways to make their argument better. After we have a chance to learn from each other, I'm going to give you some time to revise your arguments and make them better."
   - *If using the gallery walk format,* tell the students, "Okay, before we get started I want to explain what we are going to do next. You are going to have an opportunity to read the arguments that were created by other groups. Your group will go to a different group's argument. I'll give you a few minutes to read it and review it. Your job is to offer them some suggestions about ways to make their argument better. You can use sticky notes to give them suggestions. Please be specific about what you want to change and be specific about how you think they should change it. After we have a chance to learn from each

other, I'm going to give you some time to revise your arguments and make them better."

4. Use a document camera to project the "Ways to IMPROVE our argument ..." box from the Investigation Handout (or the investigation log in their workbook) on a screen or board (or take a picture of it and project the picture on a screen or board).
   - *If using the whole-class presentation format or the modified gallery walk format,* point to the box and tell the students, "If you are a presenter, you can write down the suggestions you get from the reviewers here. If you are a reviewer, and you see a good idea from another group, you can write down that idea here. Once we are done with the presentations, I will give you a chance to use these suggestions or ideas to improve your arguments.
   - *If using the gallery walk format,* point to the box and tell the students, "If you see good ideas from another group, you can write them down here. Once we are done reviewing the different arguments, I will give you a chance to use these ideas to improve your own arguments. It is important to share ideas like this."

   Ask the students, "Do you have any questions about what you need to do?"
5. Answer any questions that come up.
6. Give the following instructions:
   - *If using the whole-class presentation format,* tell the students, "Okay. Let's get started."
   - *If using the gallery walk format,* tell the students, "Okay, I'm now going to tell you which argument to go to and review.
   - *If using the modified gallery walk format,* tell the students, "Okay, I'm now going to assign you to be a presenter or a reviewer." Assign one or two students from each group to be presenters and one or two students from each group to be reviewers.
7. Begin the review of the arguments.
   - *If using the whole-class presentation format,* have four or five groups present their argument one at a time. Give each group only two to three minutes to present their argument. Then give the class two to three minutes to ask them questions and offer suggestions. Be sure to encourage as much participation from the students as possible.
   - *If using the gallery walk format,* tell the students, "Okay. Let's get started. Each group, move one argument to the left. Don't move to the next argument until I tell you to move. Once you get there, read the argument and then offer suggestions about how to make it better. I will put some sticky notes next to each argument. You can use the sticky notes to leave your suggestions." Give

each group about three to four minutes to read the arguments, talk, and offer suggestions.

a. Tell the students, "Okay. Let's rotate. Move one group to the left."

b. Again, give each group three or four minutes to read, talk, and offer suggestions.

c. Repeat this process for two more rotations.

- *If using the modified gallery walk format,* tell the students, "Okay. Let's get started. Reviewers, move one group to the left. Don't move to the next group until I tell you to move. Presenters, go ahead and share your argument with the reviewers when they get there." Give each group of presenters and reviewers about three to four minutes to talk.

  a. Tell the students, "Okay. Let's rotate. Reviewers, move one group to the left."

  b. Again, give each group of presenters and reviewers about three or four minutes to talk.

  c. Repeat this process for two more rotations.

8. Tell the students to return to their workstations.
9. Give the following instructions about revising the argument:
   - *If using the whole-class presentation format,* tell the students, "I'm now going to give you about 10 minutes to revise your argument. Take a few minutes to talk in your groups and determine what you want to change to make your argument better. Once you have decided what to change, go ahead and make the changes to your whiteboard."
   - *If using the gallery walk format,* tell the students, "I'm now going to give you about 10 minutes to revise your argument. Take a few minutes to read the suggestions that were left at your argument. Then talk in your groups and determine what you want to change to make your argument better. Once you have decided what to change, go ahead and make the changes to your whiteboard."
   - *If using the modified gallery walk format,* "I'm now going to give you about 10 minutes to revise your argument. Please return to your original groups." Wait for the students to move back into their original groups and then tell the students, "Okay, take a few minutes to talk in your groups and determine what you want to change to make your argument better. Once you have decided what to change, go ahead and make the changes to your whiteboard."

   Ask the students, "Do you have any questions about what you need to do?"
10. Answer any questions that come up.
11. Tell the students, "Okay. Let's get started."

12. Give the students 10 minutes to work in their groups on their arguments. As they work, move from group to group to check in, ask probing questions, and offer a suggestion if a group gets stuck.

Stage 5: Reflective Discussion (15 minutes)

1. Tell the students, "We are now going to take a minute to talk about what we did and what we have learned."
2. Hold up a pencil. Put a ceramic ring magnet on the pencil. Next, put a second ceramic ring magnet on the pencil (be sure to put the second ceramic ring magnet on the pencil so the like poles of the two magnets are facing each other). The second ceramic ring magnet should float above the first one.
3. Ask the students, "What do you all see going on here?"
4. Allow students to share their ideas.
5. Ask the students, "Why don't these two magnets stick together?"
6. Allow students to share their ideas. Keep probing until someone mentions that magnets have different poles and that like poles repel.
7. Ask the students, "What do you think will happen if I add another magnet to the pencil?"
8. Allow students to share their ideas.
9. Add a third ceramic ring magnet to the pencil (be sure to put the third ceramic ring magnet on the pencil so that the second and third magnets have opposite poles facing each other). The third magnet should stick to the second magnet.
10. Ask the students, "Why did these two magnets stick together?"
11. Allow students to share their ideas. Keep probing until someone mentions that magnets have different poles and that different poles attract.
12. Ask the students, "What causes these magnets to move when I bring them together?"
13. Allow students to share their ideas. Keep probing until someone mentions magnetic force.
14. Ask the students, "How could we make this magnet move more?"
15. Allow students to share their ideas. Keep probing until someone mentions adding more force or making the magnetic force stronger.
16. Ask the students, "Why do objects that are made of iron stick to magnets?"
17. Allow students to share their ideas. Keep probing students until one mentions that magnetism is a physical property of magnets.

18. Tell the students, "Okay, let's make sure we are on the same page. Magnetism is a type of force. Two magnets do not have to be touching to apply a force on each other. Two magnets with like poles facing each other will repel. You can make magnets move farther apart by increasing the strength of the magnetic force between them. The way that non-contact forces can cause an object to move is a really important core idea in science."
19. Ask the students, "Does anyone have any questions about this core idea?"
20. Answer any questions that come up.
21. Tell the students, "We also looked for a cause-and-effect relationship during our investigation." Then ask, "Can anyone tell me why it is useful to look for a cause-and-effect relationship?"
22. Allow students to share their ideas.
23. Tell the students, "Cause-and-effect relationships are important because they allow us to predict what will happen in the future."
24. Ask the students, "What was the cause and what was the effect that we uncovered today?"
25. Allow students to share their ideas. Keep probing until someone mentions that decreasing the distance between two magnets will increase the strength of the magnetic force between them.
26. Tell the students, "That is great, and if we know that we can predict what will happen the next time we move two magnets closer to each other or farther apart."
27. Tell the students, "We are now going take a minute to talk about what went well and what didn't go so well during our investigation. We need to talk about this because you are going to be planning and carrying out your own investigations like this a lot this year, and I want to help you all get better at it."
28. Show an image of the question "What made your investigation scientific?" on the screen. Tell the students, "Take a few minutes to talk about how you would answer this question with the other people in your group. Be ready to share with the rest of the class." Give the students two to three minutes to talk in their group.
29. Ask the students, "What do you all think? Who would like to share an idea?"
30. Allow students to share their ideas. Be sure to expand on their ideas about what makes an investigation scientific.
31. Show an image of the question "What made your investigation not so scientific?" on the screen. Tell the students, "Take a few minutes to talk about how you would answer this question with the other people in your group. Be ready to share with the rest of the class." Give the students two to three minutes to talk in their group.

32. Ask the students, "What do you all think? Who would like to share an idea?"
33. Allow students to share their ideas. Be sure to expand on their ideas about what makes an investigation less scientific.
34. Show an image of the question "What rules can we put into place to help us make sure our next investigation is more scientific?" on the screen. Tell the students, "Take a few minutes to talk about how you would answer this question with the other people in your group. Be ready to share with the rest of the class." Give the students two to three minutes to talk in their group.
35. Ask the students, "What do you all think? Who would like to share an idea?"
36. Allow students to share their ideas. Once they have shared their ideas, offer a suggestion for a possible class rule.
37. Ask the students, "What do you all think? Should we make this a rule?"
38. If the students agree, write the rule on the board or make a class "Rules for Scientific Investigation" chart so you can refer to it during the next investigation.
39. Tell the students, "We are now going take a minute to talk about what makes scientific knowledge different from other types of knowledge."
40. Show an image of the question "What is the difference between data and evidence?" on the screen. Tell the students, "Take a few minutes to talk about how you would answer this question with the other people in your group. Be ready to share with the rest of the class." Give the students two to three minutes to talk in their group.
41. Ask the students, "What do you all think? Who would like to share an idea?"
42. Allow students to share their ideas.
43. Tell the students, "Okay, let's make sure we are all using the same definition. Data is a bunch of observations or measurements. Evidence is an analysis of the data and an interpretation of an analysis."
44. Show an image of a magnet next to another magnet along with the statement "The second magnet moved 2 cm away from the first magnet" on the screen.
45. Ask the students, "Is this statement data or evidence and why?"
46. Allow students to share their ideas.
47. Tell the students, "That statement is data because it is a measurement."
48. Show the same image of a magnet next to another magnet along with an image of a graph of distance between two magnets versus distance the magnet moved on the screen.
49. Ask the students, "Is this graph data or evidence and why?"
50. Allow students to share their ideas.

51. Tell the students, "That graph is an analysis of data but there is no interpretation of that analysis, so it is still not evidence."
52. Show the same two images shown previously in step 48 (a magnet next to another magnet and a graph of distance between two magnets versus distance the magnet moved) along with the statement, "This graph suggests that magnets move farther away from each other as they get closer together" on the screen.
53. Ask the students, "Is this information data or evidence and why?"
54. Allow students to share their ideas.
55. Tell the students, "That is an example of evidence because it includes an analysis of the data and an interpretation of an analysis."
56. Ask the students, "Does anyone have any questions about the difference between data and evidence?"
57. Answer any questions that come up.

Stage 6: Write a Draft Report (35 minutes)

Your students will use either the Investigation Handout or the investigation log in the student workbook when writing the draft report. When you give the directions shown in quotes in the following steps, substitute "investigation log" (as shown in brackets) for "handout" if they are using the workbook.

1. Tell the students, "You are now going to write an investigation report to share what you have learned. Please take out a pencil and turn to the 'Draft Report' section of your handout [investigation log]."
2. If possible, use a document camera to project the "Introduction" section of the draft report from the Investigation Handout (or the investigation log in their workbook) on a screen or board (or take a picture of it and project the picture on a screen or board).
3. Tell the students, "The first part of the report is called the 'Introduction.' In this section of the report you want to explain to the reader what you were investigating, why you were investigating it, and what question you were trying to answer. All of this information can be found in the text at the beginning of your handout [investigation log]." Point to the image and say, "There are some sentence starters here to help you begin writing the report." Ask the students, "Do you have any questions about what you need to do?"
4. Answer any questions that come up.
5. Tell the students, "Okay. Let's write."
6. Give the students 10 minutes to write the "Introduction" section of the report. As they work, move from student to student to check in, ask probing questions, and offer a suggestion if a student gets stuck.

7. If possible, use a document camera to project the "Method" section of the draft report from the Investigation Handout (or the investigation log in their workbook) on a screen or board (or take a picture of it and project the picture on a screen or board).
8. Tell the students, "The second part of the report is called the 'Method.' In this section of the report you want to explain to the reader what you did during the investigation, what data you collected and why, and how you went about analyzing your data. All of this information can be found in the 'Plan Your Investigation' section of your handout [investigation log]. Remember that you all planned and carried out different investigations, so do not assume that the reader will know what you did." Point to the image and say, "There are some sentence starters here to help you begin writing this part of the report." Ask the students, "Do you have any questions about what you need to do?"
9. Answer any questions that come up.
10. Tell the students, "Okay. Let's write."
11. Give the students 10 minutes to write the "Method" section of the report. As they work, move from student to student to check in, ask probing questions, and offer a suggestion if a student gets stuck.
12. If possible, use a document camera to project the "Argument" section of the draft report from the Investigation Handout (or the investigation log in their workbook) on a screen or board (or take a picture of it and project the picture on a screen or board).
13. Tell the students, "The last part of the report is called the 'Argument.' In this section of the report you want to share your claim, evidence, and justification of the evidence with the reader. All of this information can be found on your whiteboard." Point to the image and say, "There are some sentence starters here to help you begin writing this part of the report." Ask the students, "Do you have any questions about what you need to do?"
14. Answer any questions that come up.
15. Tell the students, "Okay. Let's write."
16. Give the students 10 minutes to write the "Argument" section of the report. As they work, move from student to student to check in, ask probing questions, and offer a suggestion if a student gets stuck.

Stage 7: Peer Review (30 minutes)

Your students will use either the Investigation Handout or the investigation log in the student workbook when doing the peer review. When you give the directions shown in quotes in the following steps, substitute "workbook" (as shown in brackets) for "Investigation Handout" if they are using the workbook.

1. Tell the students, "We are now going to review our reports to find ways to make them better. I'm going to come around and collect your Investigation Handout [workbook]. While I do that, please take out a pencil."
2. Collect the Investigation Handouts or workbooks from the students.
3. If possible, use a document camera to project the peer-review guide (PRG; see Appendix 4) on a screen or board (or take a picture of it and project the picture on a screen or board).
4. Tell the students, "We are going to use this peer-review guide to give each other feedback." Point to the image.
5. Give the following instructions:
   - *If using the Investigation Handout,* tell the students, "I'm going to ask you to work with a partner to do this. I'm going to give you and your partner a draft report to read and a peer-review guide to fill out. You two will then read the report together. Once you are done reading the report, I want you to answer each of the questions on the peer-review guide." Point to the review questions on the image of the PRG.
   - *If using the workbook,* tell the students, "I'm going to ask you to work with a partner to do this. I'm going to give you and your partner a draft report to read. You two will then read the report together. Once you are done reading the report, I want you to answer each of the questions on the peer-review guide that is right after the report in the investigation log." Point to the review questions on the image of the PRG.
6. Tell the students, "You can check 'yes,' 'almost,' or 'no' after each question." Point to the checkboxes on the image of the PRG.
7. Tell the students, "This will be your rating for this part of the report. Make sure you agree on the rating you give the author. If you mark 'almost' or 'no,' then you need to tell the author what he or she needs to do to get a 'yes.'" Point to the space for the reviewer feedback on the image of the PRG.
8. Tell the students, "It is really important for you to give the authors feedback that is helpful. That means you need to tell them exactly what they need to do to make their reports better." Ask the students, "Do you have any questions about what you need to do?"
9. Answer any questions that come up.
10. Tell the students, "Please sit with a partner who is not in your current group." Allow the students time to sit with a partner.
11. Give the following instructions:
    - *If using the Investigation Handout,* tell the students, "Okay, I am now going to give you one report to read and one peer-review guide to fill out." Pass out one

report to each pair. Make sure that the report you give a pair was not written by one of the students in that pair. Give each pair one PRG to fill out as a team.

- *If using the workbook,* tell the students, "Okay, I am now going to give you one report to read." Pass out a workbook to each pair. Make sure that the workbook you give a pair is not from one of the students in that pair.

12. Tell the students, "Okay, I'm going to give you 15 minutes to read the report I gave you and to fill out the peer-review guide. Go ahead and get started."
13. Give the students 15 minutes to work. As they work, move around from pair to pair to check in and see how things are going, answer questions, and offer advice.
14. After 15 minutes pass, tell the students, "Okay, time is up." *If using the Investigation Handout,* say, "Please give me the report and the peer-review guide that you filled out." *If using the workbook,* say, "Please give me the workbook that you have."
15. Collect the Investigation Handouts and the PRGs, or collect the workbooks if they are being used. Be sure you keep the handout and the PRG together.
16. Give the following instructions:
    - *If using the Investigation Handout,* tell the students, "Okay, I am now going to give you a different report to read and a new peer-review guide to fill out." Pass out another report to each pair. Make sure that this report was not written by one of the students in that pair. Give each pair a new PRG to fill out as a team.
    - *If using the workbook,* tell the students, "Okay, I am now going to give you a different report to read." Pass out a different workbook to each pair. Make sure that the workbook you give a pair is not from one of the students in that pair.
17. Tell the students, "Okay, I'm going to give you 15 minutes to read this new report and to fill out the peer-review guide. Go ahead and get started."
18. Give the students 15 minutes to work. As they work, move around from pair to pair to check in and see how things are going, answer questions, and offer advice.
19. After 15 minutes pass, tell the students, "Okay, time is up." *If using the Investigation Handout,* say, "Please give me the report and the peer-review guide that you filled out." *If using the workbook,* say, "Please give me the workbook that you have."
20. Collect the Investigation Handouts and the PRGs, or collect the workbooks if they are being used. Be sure you keep the handout and the PRG together.

Stage 8: Revise the Report (30 minutes)

Your students will use either the Investigation Handout or the investigation log in the student workbook when revising the report. Except where noted below, the directions are the same whether using the handout or the log.

1. Give the following instructions:
   - *If using the Investigation Handout,* tell the students, "You are now going to revise your investigation report based on the feedback you get from your classmates. Please take out a pencil while I hand back your draft report and the peer-review guide."
   - *If using the investigation log in the student workbook,* tell the students, "You are now going to revise your investigation report based on the feedback you get from your classmates. Please take out a pencil while I hand back your investigation logs."
2. *If using the Investigation Handout,* pass back the handout and the PRG to each student. *If using the investigation log,* pass back the log to each student.
3. Tell the students, "Please take a few minutes to read over the peer-review guide. You should use it to figure out what you need to change in your report and how you will change the report."
4. Allow the students time to read the PRG.
5. *If using the investigation log,* if possible use a document camera to project the "Write Your Final Report" section from the investigation log on a screen or board (or take a picture of it and project the picture on a screen or board).
6. Give the following instructions:
   - *If using the Investigation Handout,* tell the students, "Okay. Let's revise our reports. Please take out a piece of paper. I would like you to rewrite your report. You can use your draft report as a starting point, but use the feedback on the peer-review guide to help make it better."
   - *If using the investigation log,* tell the students, "Okay. Let's revise our reports. I would like you to rewrite your report in the section of the investigation log that says 'Write Your Final Report.'" Point to the image on the screen and tell the students, "You can use your draft report as a starting point, but use the feedback on the peer-review guide to help make your report better."

   Ask the students, "Do you have any questions about what you need to do?"
7. Answer any questions that come up.
8. Tell the students, "Okay. Let's write."
9. Give the students 30 minutes to rewrite their report. As they work, move from student to student to check in, ask probing questions, and offer a suggestion if a student gets stuck.

10. Give the following instructions:
    - *If using the Investigation Handout,* tell the students, "Okay. Time's up. I will now come around and collect your Investigation Handout, the peer-review guide, and your final report."
    - *If using the investigation log,* tell the students, "Okay. Time's up. I will now come around and collect your workbooks."
11. *If using the Investigation Handout,* collect all the Investigation Handouts, PRGs, and final reports. *If using the investigation log,* collect all the workbooks.
12. *If using the Investigation Handout,* use the "Teacher Score" columns in the PRG to grade the final report. *If using the investigation log,* use the "ADI Investigation Report Grading Rubric" in the investigation log to grade the final report. Whether you are using the handout or the log, you can give the students feedback about their writing in the "Teacher Comments" section.

How to Use the Checkout Questions

The Checkout Questions are an optional assessment. We recommend giving them to students one day after they finish stage 8 of the ADI investigation. The Checkout Questions can be used as a formative or summative assessment of student thinking. If you plan to use them as a formative assessment, we recommend that you look over the student answers to determine if you need to reteach the core idea and/or crosscutting concept from the investigation, but do not grade them. If you plan to use them as a summative assessment, we have included a "Teacher Scoring Rubric" at the end of the Checkout Questions that you can use to score a student's ability to apply the core idea in a new scenario and explain their use of a crosscutting concept. The rubric includes a 4-point scale that ranges from 0 (the student cannot apply the core idea correctly in all cases and cannot explain the [crosscutting concept]) to 3 (the student can apply the core idea correctly in all cases and can fully explain the [crosscutting concept]). The Checkout Questions, regardless of how you decide to use them, are a great way to make student thinking visible so you can determine if the students have learned the core idea and the crosscutting concept.

A student who can apply the core idea correctly in all cases and can explain the cause-and-effect relationship would rank the images as follows: A-4, B-1, C-3, and D-2. He or she should then be able to explain that the strength of the magnetic force between any two magnets gets bigger (increases) as the distance between the two magnets gets smaller (decreases).

Connections to Standards

Table 2.2 highlights how the investigation can be used to address specific performance expectations from the *NGSS, Common Core State Standards (CCSS)* in English language arts (ELA) and in mathematics, and *English Language Proficiency (ELP) Standards.*

TABLE 2.2

Investigation 2 alignment with standards

***NGSS* performance expectations**	Strong alignment • 3-PS2-3: Ask questions to determine cause and effect relationships of electric or magnetic interactions between two objects not in contact with each other. Moderate alignment (this investigation can be used to build toward this performance expectation) • 3-PS2-4: Define a simple design problem that can be solved by applying scientific ideas about magnets.
***CCSS ELA*—Reading: Informational Text**	Key ideas and details • CCSS.ELA-LITERACY.RI.3.1: Ask and answer questions to demonstrate understanding of a text, referring explicitly to the text as the basis for the answers. • CCSS.ELA-LITERACY.RI.3.2: Determine the main idea of a text; recount the key details and explain how they support the main idea. • CCSS.ELA-LITERACY.RI.3.3: Describe the relationship between a series of historical events, scientific ideas or concepts, or steps in technical procedures in a text, using language that pertains to time, sequence, and cause/effect. Craft and structure • CCSS.ELA-LITERACY.RI.3.4: Determine the meaning of general academic and domain-specific words and phrases in a text relevant to a *grade 3 topic or subject area*. • CCSS.ELA-LITERACY.RI.3.5: Use text features and search tools (e.g., key words, sidebars, hyperlinks) to locate information relevant to a given topic efficiently. • CCSS.ELA-LITERACY.RI.3.6: Distinguish their own point of view from that of the author of a text. Integration of knowledge and ideas • CCSS.ELA-LITERACY.RI.3.7: Use information gained from illustrations (e.g., maps, photographs) and the words in a text to demonstrate understanding of the text (e.g., where, when, why, and how key events occur). • CCSS.ELA-LITERACY.RI.3.8: Describe the logical connection between particular sentences and paragraphs in a text (e.g., comparison, cause/effect, first/second/third in a sequence). • CCSS.ELA-LITERACY.RI.3.9: Compare and contrast the most important points and key details presented in two texts on the same topic. Range of reading and level of text complexity • CCSS.ELA-LITERACY.RI.3.10: By the end of the year, read and comprehend informational texts, including history/social studies, science, and technical texts, at the high end of the grades 2–3 text complexity band independently and proficiently.

Continued

TABLE 2.2 (*continued*)

***CCSS ELA*—Writing**	Text types and purposes • CCSS.ELA-LITERACY.W.3.1: Write opinion pieces on topics or texts, supporting a point of view with reasons. o CCSS.ELA-LITERACY.W.3.1.A: Introduce the topic or text they are writing about, state an opinion, and create an organizational structure that lists reasons. o CCSS.ELA-LITERACY.W.3.1.B: Provide reasons that support the opinion. o CCSS.ELA-LITERACY.W.3.1.C: Use linking words and phrases (e.g., *because*, *therefore*, *since, for example*) to connect opinion and reasons. o CCSS.ELA-LITERACY.W.3.1.D: Provide a concluding statement or section. • CCSS.ELA-LITERACY.W.3.2: Write informative or explanatory texts to examine a topic and convey ideas and information clearly. o CCSS.ELA-LITERACY.W.3.2.A: Introduce a topic and group related information together; include illustrations when useful to aiding comprehension. o CCSS.ELA-LITERACY.W.3.2.B: Develop the topic with facts, definitions, and details. o CCSS.ELA-LITERACY.W.3.2.C: Use linking words and phrases (e.g., *also*, *another*, *and*, *more*, *but*) to connect ideas within categories of information. o CCSS.ELA-LITERACY.W.3.2.D: Provide a concluding statement or section. Production and distribution of writing • CCSS.ELA-LITERACY.W.3.4: With guidance and support from adults, produce writing in which the development and organization are appropriate to task and purpose. • CCSS.ELA-LITERACY.W.3.5: With guidance and support from peers and adults, develop and strengthen writing as needed by planning, revising, and editing. • CCSS.ELA-LITERACY.W.3.6: With guidance and support from adults, use technology to produce and publish writing (using keyboarding skills) as well as to interact and collaborate with others. Research to build and present knowledge • CCSS.ELA-LITERACY.W.3.8: Recall information from experiences or gather information from print and digital sources; take brief notes on sources and sort evidence into provided categories. Range of writing • CCSS.ELA-LITERACY.W.3.10: Write routinely over extended time frames (time for research, reflection, and revision) and shorter time frames (a single sitting or a day or two) for a range of discipline-specific tasks, purposes, and audiences.

Continued

TABLE 2.2 (*continued*)

***CCSS ELA*—Speaking and Listening**	Comprehension and collaboration • CCSS.ELA-LITERACY.SL.3.1: Engage effectively in a range of collaborative discussions (one-on-one, in groups, and teacher-led) with diverse partners on *grade 3 topics and texts*, building on others' ideas and expressing their own clearly. o CCSS.ELA-LITERACY.SL.3.1.A: Come to discussions prepared, having read or studied required material; explicitly draw on that preparation and other information known about the topic to explore ideas under discussion. o CCSS.ELA-LITERACY.SL.3.1.B: Follow agreed-upon rules for discussions (e.g., gaining the floor in respectful ways, listening to others with care, speaking one at a time about the topics and texts under discussion). o CCSS.ELA-LITERACY.SL.3.1.C: Ask questions to check understanding of information presented, stay on topic, and link their comments to the remarks of others. o CCSS.ELA-LITERACY.SL.3.1.D: Explain their own ideas and understanding in light of the discussion. • CCSS.ELA-LITERACY.SL.3.2: Determine the main ideas and supporting details of a text read aloud or information presented in diverse media and formats, including visually, quantitatively, and orally. • CCSS.ELA-LITERACY.SL.3.3: Ask and answer questions about information from a speaker, offering appropriate elaboration and detail. Presentation of knowledge and ideas • CCSS.ELA-LITERACY.SL.3.4: Report on a topic or text, tell a story, or recount an experience with appropriate facts and relevant, descriptive details, speaking clearly at an understandable pace. • CCSS.ELA-LITERACY.SL.3.6: Speak in complete sentences when appropriate to task and situation in order to provide requested detail or clarification.
***CCSS Mathematics*—Number and Operations in Base Ten**	Use place value understanding and properties of operations to perform multi-digit arithmetic. • CCSS.MATH.CONTENT.3.NBT.A.1: Use place value understanding to round whole numbers to the nearest 10 or 100. • CCSS.MATH.CONTENT.3.NBT.A.2: Fluently add and subtract within 1,000 using strategies and algorithms based on place value, properties of operations, and/or the relationship between addition and subtraction.

Continued

TABLE 2.2 (*continued*)

CCSS Mathematics— **Measurement and Data**	Represent and interpret data. • CCSS.MATH.CONTENT.3.MD.B.3: Draw a scaled picture graph and a scaled bar graph to represent a data set with several categories. Solve one- and two-step "how many more" and "how many less" problems using information presented in scaled bar graphs. • CCSS.MATH.CONTENT.3.MD.B.4: Generate measurement data by measuring lengths using rulers marked with halves and fourths of an inch. Show the data by making a line plot, where the horizontal scale is marked off in appropriate units—whole numbers, halves, or quarters.
ELP Standards	Receptive modalities • ELP 1: Construct meaning from oral presentations and literary and informational text through grade-appropriate listening, reading, and viewing. • ELP 8: Determine the meaning of words and phrases in oral presentations and literary and informational text. Productive modalities • ELP 3: Speak and write about grade-appropriate complex literary and informational texts and topics. • ELP 4: Construct grade-appropriate oral and written claims and support them with reasoning and evidence. • ELP 7: Adapt language choices to purpose, task, and audience when speaking and writing. Interactive modalities • ELP 2: Participate in grade-appropriate oral and written exchanges of information, ideas, and analyses, responding to peer, audience, or reader comments and questions. • ELP 5: Conduct research and evaluate and communicate findings to answer questions or solve problems. • ELP 6: Analyze and critique the arguments of others orally and in writing. Linguistic structures of English • ELP 9: Create clear and coherent grade-appropriate speech and text. • ELP 10: Make accurate use of standard English to communicate in grade-appropriate speech and writing.

Investigation 2

Magnetic Force: How Does Changing the Distance Between Two Magnets Affect Magnetic Force Strength?

Introduction

Magnets are very useful. We can use a magnet to pull objects that are made of iron or contain iron closer to us without touching them. We can also use magnets to hold things in place because the force of a magnet can travel through objects. For example, we can use a magnet to hold a piece of paper on a refrigerator. Magnets can also push or pull on other magnets. Take a moment to explore what happens when you bring one magnet near another magnet. Keep track of what you observe and what you are wondering about in the boxes below.

Things I OBSERVED …	Things I WONDER about …

There is a *magnetic field* around every magnet. We cannot see the magnetic field, but it can apply a push or pull on another magnet. Scientists call this push or pull a *non-contact* force because it does not require the two magnets to be touching each other. The magnetic field is strongest around the two ends of a magnet. The ends of a magnet are called poles. One end of the magnet is called the *north pole,* and the other end of the magnet is called the *south pole.*

The two poles of a magnet may look the same, but they do not act the same. If you put the south pole of a magnet near the north pole of a second magnet, the two magnets will pull together (attract). If you put the south pole of a magnet near the south pole of a second magnet, or if you put the north pole of one magnet near the north pole of a different magnet, the two magnets will push away from each other (repel). In all magnets, identical poles repel each other and different poles attract.

The magnetic field surrounding a magnet is not very large. That is why you must bring two magnets close together for the two magnets to attract or repel. Strong magnets have larger magnetic fields than weak magnets. The magnetic field around a strong magnet will also cause a greater push or pull than the push or pull that is caused by a weak magnet.

Your goal in this investigation is to figure out how changing the distance between two magnets (a cause) changes or does not change the strength of the magnetic force between the two magnets (the effect). You can figure out the strength of the magnetic force between the two magnets by measuring how far one magnet moves away from the other magnet when the two magnets are placed near each other. You can figure out how strong the magnetic force is between two magnets by measuring how far one magnet moves away from another magnet for two reasons. The first reason is that a non-contact force such as magnetism can cause an object to move. The second reason is that an object will travel a greater distance when more force is applied to it.

Things we KNOW from what we read …	What we will NEED to figure out …

Your Task

Use what you know about magnets and cause and effect to design and carry out an investigation to determine how the strength of the magnetic force between two magnets changes as the distance between them changes.

The *guiding question* of this investigation is, ***How does changing the distance between two magnets affect magnetic force strength?***

Materials

You may use any of the following materials during your investigation:

- Safety glasses or goggles (required)
- 2 Alnico cylinder magnets
- Ruler with groove

Safety Rules

Follow all normal safety rules. In addition, be sure to follow these rules:

- Wear sanitized safety glasses or goggles during setup, investigation activity, and cleanup.
- Do not throw objects or put any objects in your mouth.
- Wash your hands with soap and water when you are done collecting the data.

Plan Your Investigation

Prepare a plan for your investigation by filling out the chart that follows; this plan is called an *investigation proposal.* Before you start developing your plan, be sure to discuss the following questions with the other members of your group:

- What types of **patterns** might we look for to help answer the guiding question?
- What information do we need to find a **cause-and-effect relationship?**

Our guiding question:

This is a picture of how we will set up the equipment:

We will collect the following data:

These are the steps we will follow to collect data:

I approve of this investigation proposal.

Teacher's signature

Date

Collect Your Data

Keep a record of what you measure or observe during your investigation in the space below.

Analyze Your Data

You will need to analyze the data you collected before you can develop an answer to the guiding question. To do this, create a graph that shows the relationship between the cause and the effect.

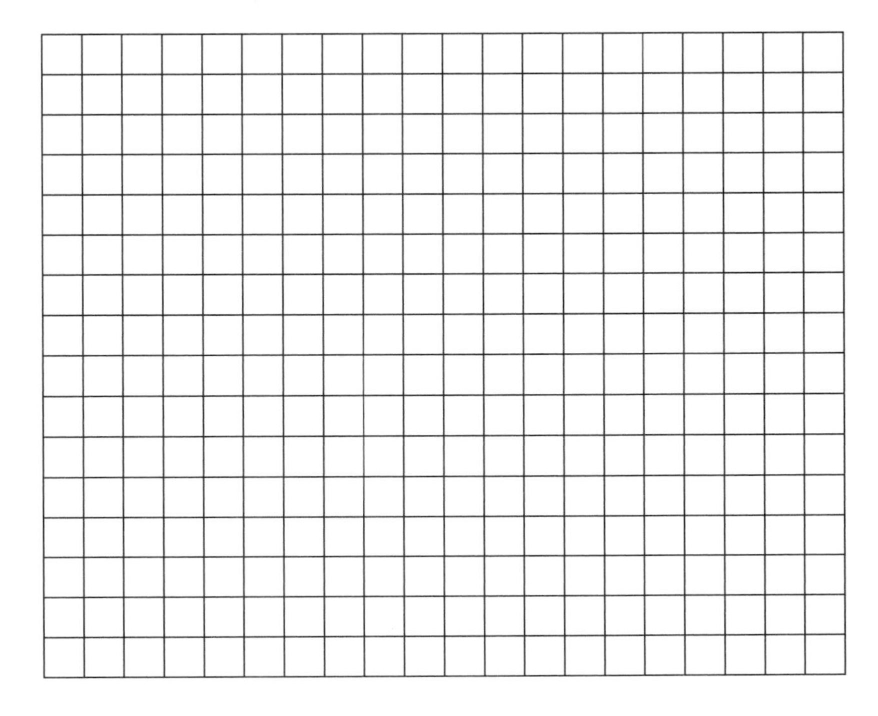

Draft Argument

Develop an argument on a whiteboard. It should include the following parts:

1. A *claim:* Your answer to the guiding question.
2. *Evidence:* An analysis of the data and an explanation of what the analysis means.
3. A *justification of the evidence:* Why your group thinks the evidence is important.

The Guiding Question:	
Our Claim:	
Our Evidence:	Our Justification of the Evidence:

Argumentation Session

Share your argument with your classmates. Be sure to ask them how to make your draft argument better. Keep track of their suggestions in the space below.

Ways to IMPROVE our argument …

Draft Report

Prepare an *investigation report* to share what you have learned. Use the information in this handout and your group's final argument to write a *draft* of your investigation report.

Introduction

We have been studying ______________________________ in class.

Before we started this investigation, we explored ______________________________

____

____

We noticed ______________________________

____

____

My goal for this investigation was to figure out ______________________________

____

____

The guiding question was ______________________________

____

____

Method

To gather the data I needed to answer this question, I ______________________________

____

____

____

____

____

I then analyzed the data I collected by ______________________________

____

____

____

Argument

My claim is ______________________________

____

____

____

The graph below shows ______________________________

____

____

____

____

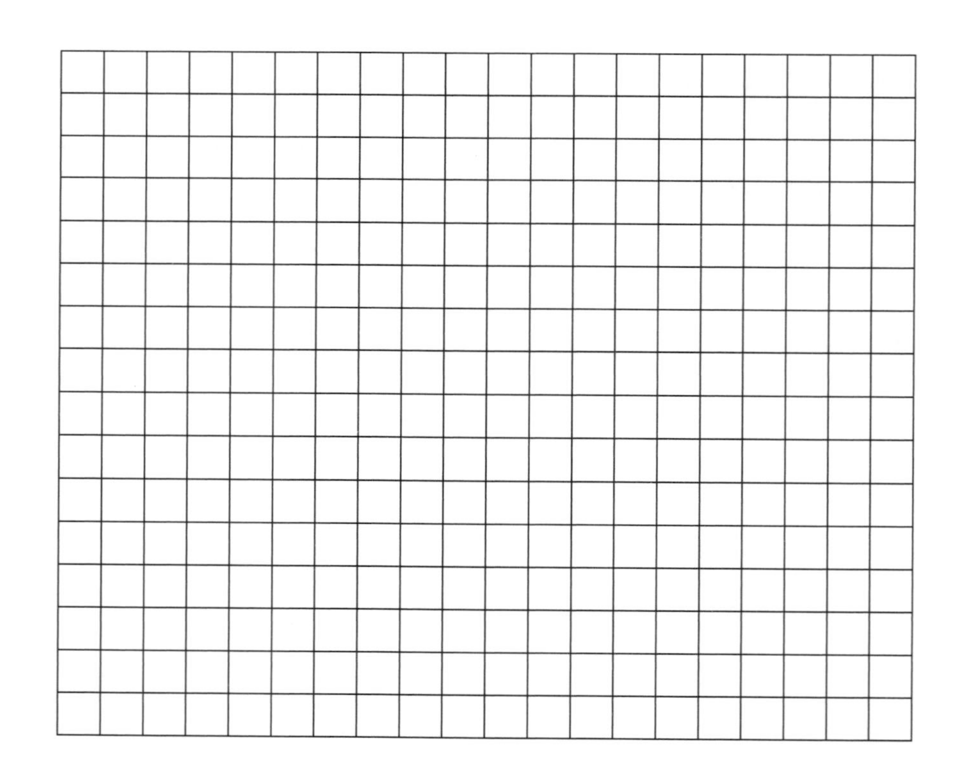

This evidence is important because ____

____

____

____

____

____

____

____

____

____

Review

Your friends need your help! Review the draft of their investigation reports and give them ideas about how to improve. Use the *peer-review guide* when doing your review.

Submit Your Final Report

Once you have received feedback from your friends about your draft report, create your final investigation report and hand it in to your teacher.

Checkout Questions

Investigation 2. Magnetic Force: How Does Changing the Distance Between Two Magnets Affect Magnetic Force Strength?

1. Listed below are four sets of magnets. Rank the sets of magnets by the strength of the *repelling* force that exists between them. Use a 1 for the *strongest* force and use a 4 for the *weakest* force.

	Magnets	Rank
A.	[N S] [S N] (far apart)	______
B.	[N S] [S N] (close together)	______
C.	[S N] [N S] (far apart)	______
D.	[S N] [N S] (closer)	______

2. Explain your thinking. What *cause-and-effect relationship* did you use to rank the strength of the repelling force between the magnets?

____

____

____

____

____

____

Teacher Scoring Rubric for the Checkout Questions

Level	Description
3	The student can apply the core idea correctly in all cases and can fully explain the cause-and-effect relationship.
2	The student can apply the core idea correctly in all cases but cannot fully explain the cause-and-effect relationship.
1	The student cannot apply the core idea correctly in all cases but can fully explain the cause-and-effect relationship.
0	The student cannot apply the core idea correctly in all cases and cannot explain the cause-and-effect relationship.

Investigation 3

Changes in Motion: Where Will the Marble Be Located Each Time It Changes Direction in a Half-Pipe?

Purpose

The purpose of this investigation is to give students an opportunity to use the disciplinary core idea (DCI) of PS2.A: Forces and Motion and the crosscutting concept (CC) of Patterns from *A Framework for K–12 Science Education* (NRC 2012) to figure out how to predict the motion of a marble as it rolls back and forth in a half-pipe. Students will also learn about what makes science different from other ways of knowing during the reflective discussion.

The Disciplinary Core Idea

Students in third grade should understand the following about Forces and Motion and be able to use this DCI to figure how to predict the motion of a marble rolling back and forth in a half-pipe:

> *Objects pull or push each other when they collide or are connected. Pushes and pulls can have different strengths and directions. Pushing or pulling on an object can change the speed or direction of its motion and can start or stop it. An object sliding on a surface or sitting on a slope experiences a pull due to friction on the object due to the surface that opposes the object's motion. (NRC 2012, p. 115)*

The Crosscutting Concept

Students in third grade should understand the following about the CC Patterns:

> *Noticing patterns is often a first step to organizing and asking scientific questions about why and how the patterns occur. One major use of pattern recognition is in classification, which depends on careful observation of similarities and differences; objects can be classified into groups on the basis of similarities of visible or microscopic features or on the basis of similarities of function. Such classification is useful in codifying relationships and organizing a multitude of objects or processes into a limited number of groups. (NRC 2012, p. 85)*

Students in third grade should also be given opportunities to "investigate the characteristics that allow classification of animal types (e.g., mammals, fish, insects), of plants (e.g., trees, shrubs, grasses), or of materials (e.g., wood, rock, metal, plastic)" (NRC 2012, p. 86).

Students should be encouraged to use their developing understanding of patterns as a tool or a way of thinking about a phenomenon during this investigation to help them figure out how to predict the motion of a marble rolling back and forth in a half-pipe.

What Students Figure Out

When the past motion of an object exhibits a regular pattern, future motion can be predicted from the past motion using the same pattern. People can use measurements to describe the motion of an object in terms of the direction an object moved (left, right, up, or down), how far it moved (its displacement), how long it took for the object to reach the new position (time), or its speed (how fast it moves).

Timeline

The time needed to complete this investigation is 260 minutes (4 hours and 20 minutes). The amount of instructional time needed for each stage of the investigation is as follows:

- *Stage 1.* Introduce the task and the guiding question: 35 minutes
- *Stage 2.* Design a method and collect data: 50 minutes
- *Stage 3.* Create a draft argument: 35 minutes
- *Stage 4.* Argumentation session: 30 minutes
- *Stage 5.* Reflective discussion: 15 minutes
- *Stage 6.* Write a draft report: 35 minutes
- *Stage 7.* Peer review: 30 minutes
- *Stage 8.* Revise the report: 30 minutes

This investigation can be completed in one day or over eight days (one day for each stage) during your designated science time in the daily schedule.

Materials and Preparation

The materials needed for this investigation are listed in Table 3.1. The items can be purchased from a big-box retail store such as Wal-Mart or Target or through an online retailer such as Amazon. The materials for this investigation can also be purchased as a complete kit (which includes enough materials for 24 students, or six groups) at *www.argumentdriveninquiry.com*.

TABLE 3.1

Materials for Investigation 3

Item	Quantity
Safety glasses or goggles	1 per student
Marble	1 per group
Plastic flexible track (used with toy cars)	3 per group
Stopwatch	1 per group
Roll of ruler tape	1 per class
Washer	1 per class
String, 12" in length	1 per group
Whiteboard, 2' × 3'*	1 per group
Investigation Handout	1 per student
Peer-review guide and teacher scoring rubric	1 per student
Checkout Questions (optional)	1 per student

*As an alternative, students can use computer and presentation software such as Microsoft PowerPoint or Apple Keynote to create their arguments.

You will need to create a simple half-pipe for students to use during this investigation before class begins. To create the half-pipe, you will need three pieces of flexible plastic track, which is used for toy cars, and some books. Attach the three pieces of track together. Add a piece of ruler tape that covers the length of the track (see Figure 3.1). Draw a circle at the midpoint of the three pieces of track using a permanent marker (see Figure 3.1). This spot will serve as the reference point in the half-pipe during the investigation. Student can then create a half-pipe by simply placing several books under each end of the track as shown in Figure 3.2 (p. 112).

FIGURE 3.1

How to place the ruler tape on the plastic track and mark the reference point with a permanent marker

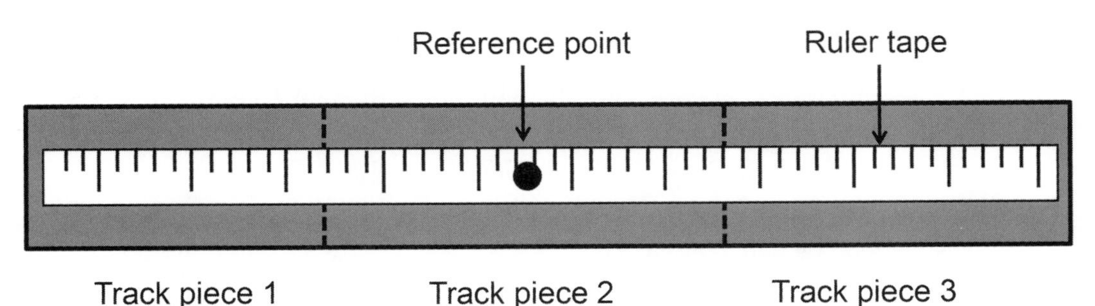

FIGURE 3.2

How to use the track to create a half-pipe

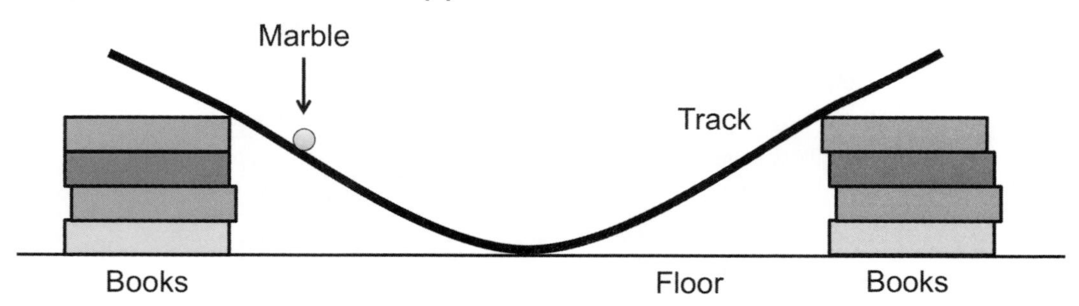

Be sure to use a set routine for distributing and collecting the materials. One option is to set up the materials for each group in a kit that you can deliver to each group. A second option is to have all the materials on a table or cart at a central location. You can then assign a member of each group to be the "materials manager." This individual is responsible for collecting all the materials his or her group needs from the table or cart during class and for returning all the materials at the end of the class.

Safety Precautions

Remind students to follow all normal safety rules. In addition, tell the students to take the following safety precautions:

- Wear sanitized safety glasses or goggles during setup, investigation activity, and cleanup.
- Do not throw objects or put any objects in their mouth.
- Immediately pick up any slip or fall hazards (e.g., marbles) from the floor.
- Wash their hands with soap and water when done collecting the data.

Lesson Plan by Stage

Stage 1: Introduce the Task and the Guiding Question (35 minutes)

1. Ask the students to sit in six groups, with three or four students in each group).
2. Ask students to clear off their desks except for a pencil (and their *Student Workbook for Argument-Driven Inquiry in Third-Grade Science* if they have one).
3. Pass out an Investigation Handout to each student (or ask students to turn to Investigation Log 3 in their workbook).
4. Read the first paragraph of the "Introduction" aloud to the class. Ask the students to follow along as you read.

5. Pass out the assembled pieces of track, a few books, and a marble to each group.
6. Remind students of the safety rules and explain the safety precautions for this investigation.
7. Tell students to place a few books under each end of the track to make a simple half-pipe. Once they have set up the track, tell them to place the marble on the track in different places and let go and then record their observations and questions in the "OBSERVED/WONDER" chart in the "Introduction" section of their Investigation Handout (or the investigation log in their workbook).
8. Ask students to share *what they observed* as they played with the marbles.
9. Ask students to share *what questions they have* about the behavior of the marble.
10. Tell the students, "Some of your questions might be answered by reading the rest of the 'Introduction.'"
11. Ask the students to read the rest of the "Introduction" on their own *or* ask them to follow along as you read aloud.
12. Once the students have read the rest of the "Introduction," ask them to fill out the "KNOW/NEED" chart on their Investigation Handout (or in their investigation log) as a group.
13. Ask students to share what they learned from the reading. Add these ideas to a class "know / need to figure out" chart.
14. Ask students to share what they think they will need to figure out based on what they read. Add these ideas to the class "know / need to figure out" chart.
15. Tell the students, "It looks like we have something to figure out. Let's see what we will need to do during our investigation."
16. Read the task and the guiding question aloud.
17. Tell the students, "I have lots of materials here that you can use."
18. Introduce the students to the materials available for them to use during the investigation by showing them how to use a half-pipe and some different ways to describe the motion of an object using numbers. The first way to describe the motion of an object using numbers is to measure how far an object moves from a reference point in a given direction. Scientists call this measurement the *displacement* of an object. The second way is to record how long it takes for an object to move from one position to a different position. Scientists call this measurement the *velocity* of an object. Velocity is how fast an object moves in a specific direction over a period of time. With this information, the students can describe the motion of a marble by reporting the direction it moved (left or right), how far it moved (its displacement), how long it took for the marble to reach a new position (time), and how fast the marble moved in a specific direction (its velocity).

Stage 2: Design a Method and Collect Data (50 minutes)

1. Tell the students, "I am now going to give you and the other members of your group about 15 minutes to plan your investigation. Before you begin, I want you all to take a couple of minutes to discuss the following questions with the rest of your group."
2. Show the following questions on the screen or board:
   - What information should we collect so we can *describe* the motion of the marble?
   - What types of *patterns* might we look for to help answer the guiding question?
3. Tell the students, "Please take a few minutes to come up with an answer to these questions." Give the students two or three minutes to discuss these two questions.
4. Ask two or three different groups to share their answers. Be sure to highlight or write down any important ideas on the board so students can refer to them later.
5. If possible, use a document camera to project an image of the graphic organizer for this investigation on a screen or board (or take a picture of it and project the picture on a screen or board). Tell the students, "I now want you all to plan out your investigation. To do that, you will need to create an investigation proposal by filling out this graphic organizer."
6. Point to the box labeled "Our guiding question:" and tell the students, "You can put the question we are trying to answer in this box." Then ask, "Where can we find the guiding question?"
7. Wait for a student to answer where to find the guiding question (the answer is "in the handout").
8. Point to the box labeled "We will collect the following data:" and tell the students, "You can list the measurements or observations that you will need to collect during the investigation in this box."
9. Point to the box labeled "This is a picture of how we will set up the equipment:" and tell the students, "You can draw a picture of how you plan to set up the equipment during the investigation so you can collect the data you need in this box."
10. Point to the box labeled "These are the steps we will follow to collect data:" and tell the students, "You can list what you are going to do to collect the data you need and what you will do with your data once you have it. Be sure to give enough detail that I could do your investigation for you."
11. Ask the students, "Do you have any questions about what you need to do?"
12. Answer any questions that come up.

13. Tell the students, "Once you are done, raise your hand and let me know. I'll then come by and look over your proposal and give you some feedback. You may not begin collecting data until I have approved your proposal by signing it. You need to have your proposal done in the next 15 minutes."
14. Give the students 15 minutes to work in their groups on their investigation proposal. As they work, move from group to group to check in, ask probing questions, and offer a suggestion if a group gets stuck.

What should a student-designed investigation look like?

The students' investigation proposal should include the following information:

- The guiding question is "Where will the marble be located each time it changes direction in a half-pipe?"
- Students might decide to collect the following data:
  - Initial position of the marble on the half-pipe (in terms of the reference point)
  - Where on the ramp the marble changes direction and how long it takes to change directions
  - How long it takes for the marble to stop rolling back and forth
  - How many times it crosses the reference point
- To collect the data, students could follow these steps (but this is just an example, and there should be a lot of variation in the student-designed investigations):
  1. Place a marble 8 inches up from the reference point.
  2. Let go of the marble.
  3. Measure how far the marble moves up the far side of the ramp and how long it takes.
  4. Measure how far the marble moves up the near side of the ramp and how long it takes.
  5. Measure how far the marble moves up the far side of the ramp for a second time.

Repeat steps 1–5 but change the starting point of the marble to 6 inches, 4 inches, and 2 inches.

15. As each group finishes its investigation proposal, be sure to read it over and determine if it will be productive or not. If you feel the investigation will be productive (not necessarily what you would do or what the other groups are doing), sign your name on the proposal and let the group start collecting data. If the plan needs to be changed, offer some suggestions or ask some probing questions, and have the group make the changes before you approve it.
16. Pass out the materials or have one student from each group collect the materials they need from a central supply table or cart for the groups that have an approved proposal.
17. Remind students of the safety rules and precautions for this investigation.
18. Tell the students to collect their data and record their observations or measurements in the "Collect Your Data" box in their Investigation Handout (or the investigation log in their workbook).
19. Give the students 10 minutes to collect their data.
20. Be sure to collect the materials from each group before asking them to analyze their data.

Stage 3: Create a Draft Argument (35 minutes)

1. Tell the students, "Now that we have all this data, we need to analyze the data so we can figure out an answer to the guiding question."
2. If possible, project an image of the "Analyze Your Data" section for this investigation on a screen or board using a document camera (or take a picture of it and project the picture on a screen or board). Point to the section and tell the students, "You can create a graph as a way to analyze your data. You can make your graph in this section."
3. Ask the students, "What information do we need to include in a graph?"
4. Tell the students, "Please take a few minutes to discuss this question with your group, and be ready to share."
5. Give the students five minutes to discuss.
6. Ask two or three different groups to share their answers. Be sure to highlight or write down any important ideas on the board so students can refer to them later.
7. Tell the students, "I am now going to give you and the other members of your group about 10 minutes to create your graph." If the students are having trouble making a graph, you can take a few minutes to provide a mini-lesson about how to create a graph from a bunch of observations or measurements (this strategy is called just-in-time instruction because it is offered only when students get stuck).

What should a graph look like for this investigation?

There are a number of different ways that students can analyze the observations or measurements they collect during this investigation. One of the most interesting and informative ways to identify a pattern is to create a line graph to show how the marble changes position relative to the reference point (displacement) over time when it starts at a specific point on the half-pipe. To create this type of graph, *time* (or the number of times it crosses the reference point) should be on the horizontal or *x*-axis and the *distance* of the marble from the reference point should be on the vertical or *y*-axis. Students will need to create a graph that includes positive and negative values on the vertical or *y*-axis for this graph because the marble moves in different directions (left and right). Another option is to create a bar graph that shows (a) the starting point of the marble on the ramp versus the number of times the marble moved back and forth or (b) the change in direction number versus the displacement from the reference point. An example of this type of bar graph can be seen in Figure 3.3 on page 118 (in this example, a ball was used instead of a marble in the investigation). There are many different options for graphing the data. Students often come up with some unique ways of analyzing their data, so be sure to give them some voice and choice during this stage.

8. Give the students 10 minutes to analyze their data. As they work, be sure to move from group to group to check in, ask probing questions, and offer suggestions.
9. Tell the students, "I am now going to give you and the other members of your group 15 minutes to create an argument to share what you have learned and convince others that they should believe you. Before you do that, we need to take a few minutes to discuss what you need to include in your argument."
10. If possible, use a document camera to project the "Argument Presentation on a Whiteboard" image from the Investigation Handout (or the investigation log in their workbook) on a screen or board (or take a picture of it and project the picture on a screen or board).
11. Point to the box labeled "The Guiding Question:" and tell the students, "You can put the question we are trying to answer here on your whiteboard."
12. Point to the box labeled "Our Claim:" and tell the students, "You can put your claim here on your whiteboard. The claim is your answer to the guiding question."
13. Point to the box labeled "Our Evidence:" and tell the students, "You can put the evidence that you are using to support your claim here on your whiteboard.

Your evidence will need to include the analysis you just did and an explanation of what your analysis means or shows. Scientists always need to support their claims with evidence."

14. Point to the box labeled "Our Justification of the Evidence:" and tell the students, "You can put your justification of your evidence here on your whiteboard. Your justification needs to explain why your evidence is important. Scientists often use core ideas to explain why the evidence they are using matters. Core ideas are important concepts that scientists use to help them make sense of what happens during an investigation."
15. Ask the students, "What are some core ideas that we read about earlier that might help us explain why the evidence we are using is important?"
16. Ask students to share some of the core ideas from the "Introduction" section of the Investigation Handout (or the investigation log in the workbook). List these core ideas on the board.
17. Tell the students, "That is great. I would like to see everyone try to include these core ideas in your justification of the evidence. Your goal is to use these core ideas to help explain why your evidence matters and why the rest of us should pay attention to it."

FIGURE 3.3

Example of an argument

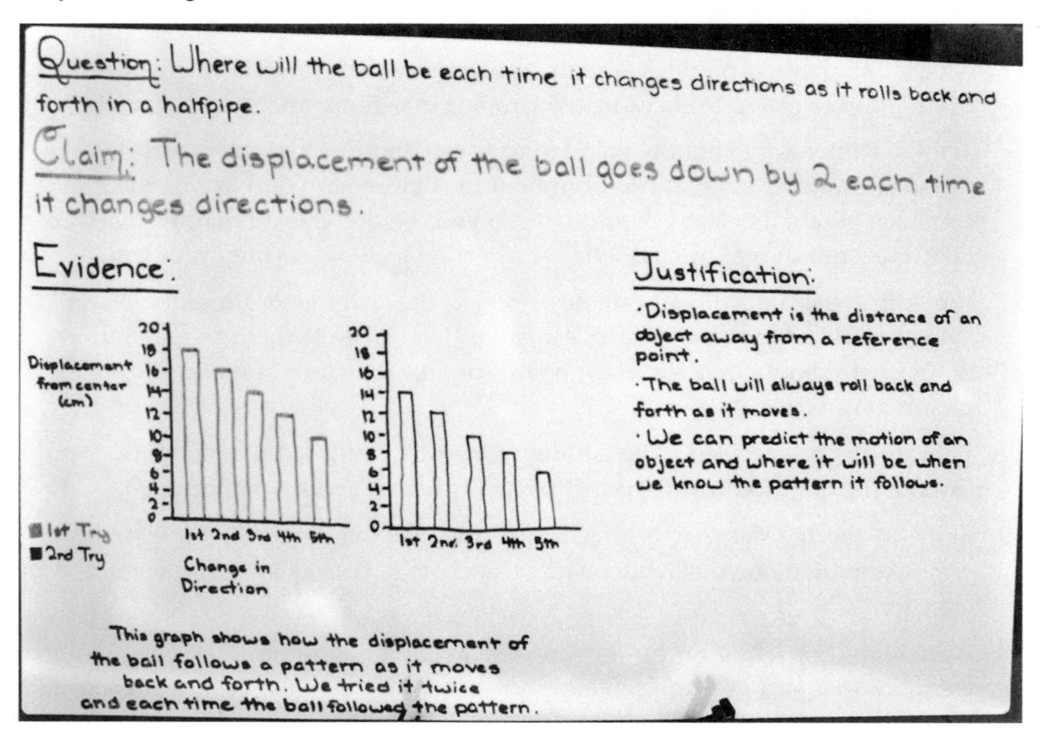

18. Ask the students, "Do you have any questions about what you need to do?"
19. Answer any questions that come up.
20. Tell the students, "Okay, go ahead and start working on your arguments. You need to have your argument done in the next 15 minutes. It doesn't need to be perfect. We just need something down on the whiteboards so we can share our ideas."
21. Give the students 15 minutes to work in their groups on their arguments. As they work, move from group to group to check in, ask probing questions, and offer a suggestion if a group gets stuck. Figure 3.3 shows an example of an argument created by students for this investigation.

Stage 4: Argumentation Session (30 minutes)

The argumentation session can be conducted in a whole-class presentation format, a gallery walk format, or a modified gallery walk format. We recommend using a whole-class presentation format for the first investigation, but try to transition to either the gallery walk or modified gallery walk format as soon as possible because that will maximize student voice and choice inside the classroom. The following list shows the steps for the three formats; unless otherwise noted, the steps are the same for all three formats.

1. Begin by introducing the use of the whiteboard.
    - *If using the whole-class presentation format,* tell the students, "We are now going to share our arguments. Please set up your whiteboard so everyone can see them."
    - *If using the gallery walk or modified gallery walk format,* tell the students, "We are now going to share our arguments. Please set up your whiteboard so they are facing the walls."
2. Allow the students to set up their whiteboards.
    - *If using the whole-class presentation format,* the whiteboards should be set up on stands or chairs so they are facing toward the center of the room.
    - *If using the gallery walk or modified gallery walk format,* the whiteboards should be set up on stands or chairs so they are facing toward the outside of the room.
3. Give the following instructions to the students:
    - *If using the whole-class presentation format or the modified gallery walk format,* tell the students, "Okay, before we get started I want to explain what we are going to do next. I'm going to ask some of you to present your arguments to your classmates. If you are presenting your argument, your job is to share your group's claim, evidence, and justification of the evidence. The rest of you will be reviewers. If you are a reviewer, your job is to listen to the presenters, ask the presenters questions if you do not understand something, and then offer them some suggestions about ways to make their argument better. After we have a

chance to learn from each other, I'm going to give you some time to revise your arguments and make them better."

- *If using the gallery walk format,* tell the students, "Okay, before we get started I want to explain what we are going to do next. You are going to have an opportunity to read the arguments that were created by other groups. Your group will go to a different group's argument. I'll give you a few minutes to read it and review it. Your job is to offer them some suggestions about ways to make their argument better. You can use sticky notes to give them suggestions. Please be specific about what you want to change and be specific about how you think they should change it. After we have a chance to learn from each other, I'm going to give you some time to revise your arguments and make them better."

4. Use a document camera to project the "Ways to IMPROVE our argument…" box from the Investigation Handout (or the investigation log in their workbook) on a screen or board (or take a picture of it and project the picture on a screen or board).
    - *If using the whole-class presentation format or the modified gallery walk format,* point to the box and tell the students, "If you are a presenter, you can write down the suggestions you get from the reviewers here. If you are a reviewer, and you see a good idea from another group, you can write down that idea here. Once we are done with the presentations, I will give you a chance to use these suggestions or ideas to improve your arguments.
    - *If using the gallery walk format,* point to the box and tell the students, "If you see good ideas from another group, you can write them down here. Once we are done reviewing the different arguments, I will give you a chance to use these ideas to improve your own arguments. It is important to share ideas like this."

    Ask the students, "Do you have any questions about what you need to do?"
5. Answer any questions that come up.
6. Give the following instructions:
    - *If using the whole-class presentation format,* tell the students, "Okay. Let's get started."
    - *If using the gallery walk format,* tell the students, "Okay, I'm now going to tell you which argument to go to and review.
    - *If using the modified gallery walk format,* tell the students, "Okay, I'm now going to assign you to be a presenter or a reviewer." Assign one or two students from each group to be presenters and one or two students from each group to be reviewers.
7. Begin the review of the arguments.
    - *If using the whole-class presentation format,* have four or five groups present their argument one at a time. Give each group only two to three minutes to present

their argument. Then give the class two to three minutes to ask them questions and offer suggestions. Be sure to encourage as much participation from the students as possible.

- *If using the gallery walk format,* tell the students, "Okay. Let's get started. Each group, move one argument to the left. Don't move to the next argument until I tell you to move. Once you get there, read the argument and then offer suggestions about how to make it better. I will put some sticky notes next to each argument. You can use the sticky notes to leave your suggestions." Give each group about three to four minutes to read the arguments, talk, and offer suggestions.

    a. Tell the students, "Okay. Let's rotate. Move one group to the left."

    b. Again, give each group three or four minutes to read, talk, and offer suggestions.

    c. Repeat this process for two more rotations.

- *If using the modified gallery walk format,* tell the students, "Okay. Let's get started. Reviewers, move one group to the left. Don't move to the next group until I tell you to move. Presenters, go ahead and share your argument with the reviewers when they get there." Give each group of presenters and reviewers about three to four minutes to talk.

    a. Tell the students, "Okay. Let's rotate. Reviewers, move one group to the left."

    b. Again, give each group of presenters and reviewers about three or four minutes to talk.

    c. Repeat this process for two more rotations.

8. Tell the students to return to their workstations.
9. Give the following instructions about revising the argument:
    - *If using the whole-class presentation format,* tell the students, "I'm now going to give you about 10 minutes to revise your argument. Take a few minutes to talk in your groups and determine what you want to change to make your argument better. Once you have decided what to change, go ahead and make the changes to your whiteboard."
    - *If using the gallery walk format,* tell the students, "I'm now going to give you about 10 minutes to revise your argument. Take a few minutes to read the suggestions that were left at your argument. Then talk in your groups and determine what you want to change to make your argument better. Once you have decided what to change, go ahead and make the changes to your whiteboard."
    - *If using the modified gallery walk format,* "I'm now going to give you about 10 minutes to revise your argument. Please return to your original groups." Wait for the students to move back into their original groups and then tell the students, "Okay, take a few minutes to talk in your groups and determine what

you want to change to make your argument better. Once you have decided what to change, go ahead and make the changes to your whiteboard."

Ask the students, "Do you have any questions about what you need to do?"

10. Answer any questions that come up.
11. Tell the students, "Okay. Let's get started."
12. Give the students 10 minutes to work in their groups on their arguments. As they work, move from group to group to check in, ask probing questions, and offer a suggestion if a group gets stuck.

Stage 5: Reflective Discussion (15 minutes)

1. Tell the students, "We are now going to take a minute to talk about what we did and what we have learned."
2. Hold up a washer attached to a string. Pull the washer to one side and let go. Let it swing back and forth.
3. Ask the students, "What do you all see going on here?"
4. Allow students to share their ideas.
5. Ask the students, "How can we use numbers to describe the motion of the washer?"
6. Allow students to share their ideas. Keep probing until someone mentions ideas like the direction it moves, how far it moves, how fast it moves, or how long it moves.
7. Ask the students, "What do you think will happen if I pull back the washer farther than I did before?"
8. Allow students to share their ideas.
9. Hold up the washer attached to a string for a second time. Pull the washer to one side farther than you did before and then let go. Let it swing back and forth.
10. Ask the students, "How accurate were the predictions you made?"
11. Allow students to share their ideas.
12. Ask the students, "Why can we predict the motion of an object like this?"
13. Allow students to share their ideas. Keep probing until someone mentions that it follows a pattern.
14. Tell the students, "Okay, let's make sure we are on the same page. People can use measurements to find a pattern in the motion of an object. We can describe the motion of object in terms of the direction it moves, how far it moves, how long it takes for an object to reach the new position, or how fast it moves. When the motion of an object exhibits a regular pattern, we can predict the future

motion of the object based on what it did before using the same pattern. The fact that we can use patterns to describe and predict the motion of an object is a really important core idea in science."

15. Ask the students, "Does anyone have any questions about this core idea?"
16. Answer any questions that come up.
17. Tell the students, "We are now going take a minute to talk about what went well and what didn't go so well during our investigation. We need to talk about this because you are going to be planning and carrying out your own investigations like this a lot this year, and I want to help you all get better at it."
18. Show an image of the question "What made your investigation scientific?" on the screen. Tell the students, "Take a few minutes to talk about how you would answer this question with the other people in your group. Be ready to share with the rest of the class." Give the students two to three minutes to talk in their group.
19. Ask the students, "What do you all think? Who would like to share an idea?"
20. Allow students to share their ideas. Be sure to expand on their ideas about what makes an investigation scientific.
21. Show an image of the question "What made your investigation not so scientific?" on the screen. Tell the students, "Take a few minutes to talk about how you would answer this question with the other people in your group. Be ready to share with the rest of the class." Give the students two to three minutes to talk in their group.
22. Ask the students, "What do you all think? Who would like to share an idea?"
23. Allow students to share their ideas. Be sure to expand on their ideas about what makes an investigation less scientific.
24. Show an image of the question "What rules can we put into place to help us make sure our next investigation is more scientific?" on the screen. Tell the students, "Take a few minutes to talk about how you would answer this question with the other people in your group. Be ready to share with the rest of the class." Give the students two to three minutes to talk in their group.
25. Ask the students, "What do you all think? Who would like to share an idea?"
26. Allow students to share their ideas. Once they have shared their ideas, offer a suggestion for a possible class rule.
27. Ask the students, "What do you all think? Should we make this a rule?"
28. If the students agree, write the rule on the board or make a class "Rules for Scientific Investigation" chart so you can refer to it during the next investigation.
29. Tell the students, "We are now going take a minute to talk about what makes scientific knowledge different from other types of knowledge."

30. Show an image of the question "What kinds of things do scientists do to figure things out?" on the screen. Tell the students, "Take a few minutes to talk about how you would answer this question with the other people in your group. Be ready to share with the rest of the class." Give the students two to three minutes to talk in their group.
31. Ask the students, "What do you all think? Who would like to share an idea?"
32. Allow students to share their ideas.
33. Tell the students, "Okay, these are all great ideas. Always remember that science is both a set of core ideas and a way to figure things out. Many different people use science to figure things out. Just like we did today."
34. Ask the students, "Does anyone have any questions about what makes science both a set of core ideas and a way to figure things out?"
35. Answer any questions that come up.

Stage 6: Write a Draft Report (35 minutes)

Your students will use either the Investigation Handout or the investigation log in the student workbook when writing the draft report. When you give the directions shown in quotes in the following steps, substitute "investigation log" (as shown in brackets) for "handout" if they are using the workbook.

1. Tell the students, "You are now going to write an investigation report to share what you have learned. Please take out a pencil and turn to the 'Draft Report' section of your handout [investigation log]."
2. If possible, use a document camera to project the "Introduction" section of the draft report from the Investigation Handout (or the investigation log in their workbook) on a screen or board (or take a picture of it and project the picture on a screen or board).
3. Tell the students, "The first part of the report is called the 'Introduction.' In this section of the report you want to explain to the reader what you were investigating, why you were investigating it, and what question you were trying to answer. All of this information can be found in the text at the beginning of your handout [investigation log]." Point to the image and say, "There are some sentence starters here to help you begin writing the report." Ask the students, "Do you have any questions about what you need to do?"
4. Answer any questions that come up.
5. Tell the students, "Okay. Let's write."
6. Give the students 10 minutes to write the "Introduction" section of the report. As they work, move from student to student to check in, ask probing questions, and offer a suggestion if a student gets stuck.

7. If possible, use a document camera to project the "Method" section of the draft report from the Investigation Handout (or the investigation log in their workbook) on a screen or board (or take a picture of it and project the picture on a screen or board).
8. Tell the students, "The second part of the report is called the 'Method.' In this section of the report you want to explain to the reader what you did during the investigation, what data you collected and why, and how you went about analyzing your data. All of this information can be found in the 'Plan Your Investigation' section of your handout [investigation log]. Remember that you all planned and carried out different investigations, so do not assume that the reader will know what you did." Point to the image and say, "There are some sentence starters here to help you begin writing this part of the report." Ask the students, "Do you have any questions about what you need to do?"
9. Answer any questions that come up.
10. Tell the students, "Okay. Let's write."
11. Give the students 10 minutes to write the "Method" section of the report. As they work, move from student to student to check in, ask probing questions, and offer a suggestion if a student gets stuck.
12. If possible, use a document camera to project the "Argument" section of the draft report from the Investigation Handout (or the investigation log in their workbook) on a screen or board (or take a picture of it and project the picture on a screen or board).
13. Tell the students, "The last part of the report is called the 'Argument.' In this section of the report you want to share your claim, evidence, and justification of the evidence with the reader. All of this information can be found on your whiteboard." Point to the image and say, "There are some sentence starters here to help you begin writing this part of the report." Ask the students, "Do you have any questions about what you need to do?"
14. Answer any questions that come up.
15. Tell the students, "Okay. Let's write."
16. Give the students 10 minutes to write the "Argument" section of the report. As they work, move from student to student to check in, ask probing questions, and offer a suggestion if a student gets stuck.

Stage 7: Peer Review (30 minutes)

Your students will use either the Investigation Handout or the investigation log in the student workbook when doing the peer review. When you give the directions shown in quotes in the following steps, substitute "workbook" (as shown in brackets) for "Investigation Handout" if they are using the workbook.

1. Tell the students, "We are now going to review our reports to find ways to make them better. I'm going to come around and collect your Investigation Handout [workbook]. While I do that, please take out a pencil."
2. Collect the Investigation Handouts or workbooks from the students.
3. If possible, use a document camera to project the peer-review guide (PRG; see Appendix 4) on a screen or board (or take a picture of it and project the picture on a screen or board).
4. Tell the students, "We are going to use this peer-review guide to give each other feedback." Point to the image.
5. Give the following instructions:
   - *If using the Investigation Handout,* tell the students, "I'm going to ask you to work with a partner to do this. I'm going to give you and your partner a draft report to read and a peer-review guide to fill out. You two will then read the report together. Once you are done reading the report, I want you to answer each of the questions on the peer-review guide." Point to the review questions on the image of the PRG.
   - *If using the workbook,* tell the students, "I'm going to ask you to work with a partner to do this. I'm going to give you and your partner a draft report to read. You two will then read the report together. Once you are done reading the report, I want you to answer each of the questions on the peer-review guide that is right after the report in the investigation log." Point to the review questions on the image of the PRG.
6. Tell the students, "You can check 'yes,' 'almost,' or 'no' after each question." Point to the checkboxes on the image of the PRG.
7. Tell the students, "This will be your rating for this part of the report. Make sure you agree on the rating you give the author. If you mark 'almost' or 'no,' then you need to tell the author what he or she needs to do to get a 'yes.'" Point to the space for the reviewer feedback on the image of the PRG.
8. Tell the students, "It is really important for you to give the authors feedback that is helpful. That means you need to tell them exactly what they need to do to make their reports better." Ask the students, "Do you have any questions about what you need to do?"
9. Answer any questions that come up.
10. Tell the students, "Please sit with a partner who is not in your current group." Allow the students time to sit with a partner.
11. Give the following instructions:
    - *If using the Investigation Handout,* tell the students, "Okay, I am now going to give you one report to read and one peer-review guide to fill out." Pass out one

report to each pair. Make sure that the report you give a pair was not written by one of the students in that pair. Give each pair one PRG to fill out as a team.

- *If using the workbook,* tell the students, "Okay, I am now going to give you one report to read." Pass out a workbook to each pair. Make sure that the workbook you give a pair is not from one of the students in that pair.

12. Tell the students, "Okay, I'm going to give you 15 minutes to read the report I gave you and to fill out the peer-review guide. Go ahead and get started."
13. Give the students 15 minutes to work. As they work, move around from pair to pair to check in and see how things are going, answer questions, and offer advice.
14. After 15 minutes pass, tell the students, "Okay, time is up." *If using the Investigation Handout,* say, "Please give me the report and the peer-review guide that you filled out." *If using the workbook,* say, "Please give me the workbook that you have."
15. Collect the Investigation Handouts and the PRGs, or collect the workbooks if they are being used. Be sure you keep the handout and the PRG together.
16. Give the following instructions:
    - *If using the Investigation Handout,* tell the students, "Okay, I am now going to give you a different report to read and a new peer-review guide to fill out." Pass out another report to each pair. Make sure that this report was not written by one of the students in that pair. Give each pair a new PRG to fill out as a team.
    - *If using the workbook,* tell the students, "Okay, I am now going to give you a different report to read." Pass out a different workbook to each pair. Make sure that the workbook you give a pair is not from one of the students in that pair.
17. Tell the students, "Okay, I'm going to give you 15 minutes to read this new report and to fill out the peer-review guide. Go ahead and get started."
18. Give the students 15 minutes to work. As they work, move around from pair to pair to check in and see how things are going, answer questions, and offer advice.
19. After 15 minutes pass, tell the students, "Okay, time is up." *If using the Investigation Handout,* say, "Please give me the report and the peer-review guide that you filled out." *If using the workbook,* say, "Please give me the workbook that you have."
20. Collect the Investigation Handouts and the PRGs, or collect the workbooks if they are being used. Be sure you keep the handout and the PRG together.

Stage 8: Revise the Report (30 minutes)

Your students will use either the Investigation Handout or the investigation log in the student workbook when revising the report. Except where noted below, the directions are the same whether using the handout or the log.

1. Give the following instructions:
   - *If using the Investigation Handout,* tell the students, "You are now going to revise your investigation report based on the feedback you get from your classmates. Please take out a pencil while I hand back your draft report and the peer-review guide."
   - *If using the investigation log in the student workbook,* tell the students, "You are now going to revise your investigation report based on the feedback you get from your classmates. Please take out a pencil while I hand back your investigation logs."
2. *If using the Investigation Handout,* pass back the handout and the PRG to each student. *If using the investigation log,* pass back the log to each student.
3. Tell the students, "Please take a few minutes to read over the peer-review guide. You should use it to figure out what you need to change in your report and how you will change the report."
4. Allow the students time to read the PRG.
5. *If using the investigation log,* if possible use a document camera to project the "Write Your Final Report" section from the investigation log on a screen or board (or take a picture of it and project the picture on a screen or board).
6. Give the following instructions:
   - *If using the Investigation Handout,* tell the students, "Okay. Let's revise our reports. Please take out a piece of paper. I would like you to rewrite your report. You can use your draft report as a starting point, but use the feedback on the peer-review guide to help make it better."
   - *If using the investigation log,* tell the students, "Okay. Let's revise our reports. I would like you to rewrite your report in the section of the investigation log that says 'Write Your Final Report.'" Point to the image on the screen and tell the students, "You can use your draft report as a starting point, but use the feedback on the peer-review guide to help make your report better."

   Ask the students, "Do you have any questions about what you need to do?"
7. Answer any questions that come up.
8. Tell the students, "Okay. Let's write."
9. Give the students 30 minutes to rewrite their report. As they work, move from student to student to check in, ask probing questions, and offer a suggestion if a student gets stuck.

10. Give the following instructions:
    - *If using the Investigation Handout,* tell the students, "Okay. Time's up. I will now come around and collect your Investigation Handout, the peer-review guide, and your final report."
    - *If using the investigation log,* tell the students, "Okay. Time's up. I will now come around and collect your workbooks."
11. *If using the Investigation Handout,* collect all the Investigation Handouts, PRGs, and final reports. *If using the investigation log,* collect all the workbooks.
12. *If using the Investigation Handout,* use the "Teacher Score" columns in the PRG to grade the final report. *If using the investigation log,* use the "ADI Investigation Report Grading Rubric" in the investigation log to grade the final report. Whether you are using the handout or the log, you can give the students feedback about their writing in the "Teacher Comments" section.

How to Use the Checkout Questions

The Checkout Questions are an optional assessment. We recommend giving them to students one day after they finish stage 8 of the ADI investigation. The Checkout Questions can be used as a formative or summative assessment of student thinking. If you plan to use them as a formative assessment, we recommend that you look over the student answers to determine if you need to reteach the core idea and/or crosscutting concept from the investigation, but do not grade them. If you plan to use them as a summative assessment, we have included a "Teacher Scoring Rubric" at the end of the Checkout Questions that you can use to score a student's ability to apply the core idea in a new scenario and explain their use of a crosscutting concept. The rubric includes a 4-point scale that ranges from 0 (the student cannot apply the core idea correctly in all cases and cannot explain the [crosscutting concept] to 3 (the student can apply the core idea correctly in all cases and can fully explain the [crosscutting concept]). The Checkout Questions, regardless of how you decide to use them, are a great way to make student thinking visible so you can determine if the students have learned the core idea and the crosscutting concept.

A student who can apply the core idea correctly in all cases and can predict the motion of the ball in the half-pipe would give the following answers, respectively, to questions 1–3: J, C, and H. He or she should then be able to explain the pattern as follows: the ball moves a little bit less up the side of the ramp each time it changes direction until it comes to a stop at the bottom.

Connections to Standards

Table 3.2 (p. 130) highlights how the investigation can be used to address specific performance expectations from the *NGSS, Common Core State Standards (CCSS)* in English language arts (ELA) and in mathematics, and *English Language Proficiency (ELP) Standards.*

TABLE 3.2

Investigation 3 alignment with standards

***NGSS* performance expectations**	Strong alignment • 3-PS2-2: Make observations and/or measurements of an object's motion to provide evidence that a pattern can be used to predict future motion. Moderate alignment (this investigation can be used to build toward these performance expectations) • 3-PS2-1: Plan and conduct an investigation to provide evidence of the effects of balanced and unbalanced forces on the motion of an object.
***CCSS ELA*—Reading: Informational Text**	Key ideas and details • CCSS.ELA-LITERACY.RI.3.1: Ask and answer questions to demonstrate understanding of a text, referring explicitly to the text as the basis for the answers. • CCSS.ELA-LITERACY.RI.3.2: Determine the main idea of a text; recount the key details and explain how they support the main idea. • CCSS.ELA-LITERACY.RI.3.3: Describe the relationship between a series of historical events, scientific ideas or concepts, or steps in technical procedures in a text, using language that pertains to time, sequence, and cause/effect. Craft and structure • CCSS.ELA-LITERACY.RI.3.4: Determine the meaning of general academic and domain-specific words and phrases in a text relevant to a *grade 3 topic or subject area*. • CCSS.ELA-LITERACY.RI.3.5: Use text features and search tools (e.g., key words, sidebars, hyperlinks) to locate information relevant to a given topic efficiently. • CCSS.ELA-LITERACY.RI.3.6: Distinguish their own point of view from that of the author of a text. Integration of knowledge and ideas • CCSS.ELA-LITERACY.RI.3.7: Use information gained from illustrations (e.g., maps, photographs) and the words in a text to demonstrate understanding of the text (e.g., where, when, why, and how key events occur). • CCSS.ELA-LITERACY.RI.3.8: Describe the logical connection between particular sentences and paragraphs in a text (e.g., comparison, cause/effect, first/second/third in a sequence). • CCSS.ELA-LITERACY.RI.3.9: Compare and contrast the most important points and key details presented in two texts on the same topic. Range of reading and level of text complexity • CCSS.ELA-LITERACY.RI.3.10: By the end of the year, read and comprehend informational texts, including history/social studies, science, and technical texts, at the high end of the grades 2–3 text complexity band independently and proficiently.

Continued

Table 3.2 (*continued*)

***CCSS ELA*—Writing**	Text types and purposes
	• CCSS.ELA-LITERACY.W.3.1: Write opinion pieces on topics or texts, supporting a point of view with reasons.
	  o CCSS.ELA-LITERACY.W.3.1.A: Introduce the topic or text they are writing about, state an opinion, and create an organizational structure that lists reasons.
	  o CCSS.ELA-LITERACY.W.3.1.B: Provide reasons that support the opinion.
	  o CCSS.ELA-LITERACY.W.3.1.C: Use linking words and phrases (e.g., *because*, *therefore*, *since, for example*) to connect opinion and reasons.
	  o CCSS.ELA-LITERACY.W.3.1.D: Provide a concluding statement or section.
	• CCSS.ELA-LITERACY.W.3.2: Write informative or explanatory texts to examine a topic and convey ideas and information clearly.
	  o CCSS.ELA-LITERACY.W.3.2.A: Introduce a topic and group related information together; include illustrations when useful to aiding comprehension.
	  o CCSS.ELA-LITERACY.W.3.2.B: Develop the topic with facts, definitions, and details.
	  o CCSS.ELA-LITERACY.W.3.2.C: Use linking words and phrases (e.g., *also*, *another*, *and*, *more*, *but*) to connect ideas within categories of information.
	  o CCSS.ELA-LITERACY.W.3.2.D: Provide a concluding statement or section.
	Production and distribution of writing
	• CCSS.ELA-LITERACY.W.3.4: With guidance and support from adults, produce writing in which the development and organization are appropriate to task and purpose.
	• CCSS.ELA-LITERACY.W.3.5: With guidance and support from peers and adults, develop and strengthen writing as needed by planning, revising, and editing.
	• CCSS.ELA-LITERACY.W.3.6: With guidance and support from adults, use technology to produce and publish writing (using keyboarding skills) as well as to interact and collaborate with others.
	Research to build and present knowledge
	• CCSS.ELA-LITERACY.W.3.8: Recall information from experiences or gather information from print and digital sources; take brief notes on sources and sort evidence into provided categories.
	Range of writing
	• CCSS.ELA-LITERACY.W.3.10: Write routinely over extended time frames (time for research, reflection, and revision) and shorter time frames (a single sitting or a day or two) for a range of discipline-specific tasks, purposes, and audiences.

Continued

Table 3.2 (*continued*)

<table>
<tr><td>CCSS ELA—
Speaking and Listening</td><td>Comprehension and collaboration
• CCSS.ELA-LITERACY.SL.3.1: Engage effectively in a range of collaborative discussions (one-on-one, in groups, and teacher-led) with diverse partners on grade 3 topics and texts, building on others' ideas and expressing their own clearly.
o CCSS.ELA-LITERACY.SL.3.1.A: Come to discussions prepared, having read or studied required material; explicitly draw on that preparation and other information known about the topic to explore ideas under discussion.
o CCSS.ELA-LITERACY.SL.3.1.B: Follow agreed-upon rules for discussions (e.g., gaining the floor in respectful ways, listening to others with care, speaking one at a time about the topics and texts under discussion).
o CCSS.ELA-LITERACY.SL.3.1.C: Ask questions to check understanding of information presented, stay on topic, and link their comments to the remarks of others.
o CCSS.ELA-LITERACY.SL.3.1.D: Explain their own ideas and understanding in light of the discussion.
• CCSS.ELA-LITERACY.SL.3.2: Determine the main ideas and supporting details of a text read aloud or information presented in diverse media and formats, including visually, quantitatively, and orally.
• CCSS.ELA-LITERACY.SL.3.3: Ask and answer questions about information from a speaker, offering appropriate elaboration and detail.
Presentation of knowledge and ideas
• CCSS.ELA-LITERACY.SL.3.4: Report on a topic or text, tell a story, or recount an experience with appropriate facts and relevant, descriptive details, speaking clearly at an understandable pace.
• CCSS.ELA-LITERACY.SL.3.6: Speak in complete sentences when appropriate to task and situation in order to provide requested detail or clarification.</td></tr>
</table>

Continued

Table 3.2 *(continued)*

***CCSS Mathematics—* Operations and Algebraic Thinking**	Represent and solve problems involving multiplication and division. • CCSS.MATH.CONTENT.3.OA.A.2: Interpret whole-number quotients of whole numbers. • CCSS.MATH.CONTENT.3.OA.A.3: Use multiplication and division within 100 to solve word problems in situations involving equal groups, arrays, and measurement quantities Understand properties of multiplication and the relationship between multiplication and division. • CCSS.MATH.CONTENT.3.OA.B.5: Apply properties of operations as strategies to multiply and divide. • CCSS.MATH.CONTENT.3.OA.B.6: Understand division as an unknown-factor problem. Multiply and divide within 100. • CCSS.MATH.CONTENT.3.OA.C.7: Fluently multiply and divide within 100, using strategies such as the relationship between multiplication and division or properties of operations. Solve problems involving the four operations, and identify and explain patterns in arithmetic. • CCSS.MATH.CONTENT.3.OA.D.8: Solve two-step word problems using the four operations. Represent these problems using equations with a letter standing for the unknown quantity. Assess the reasonableness of answers using mental computation and estimation strategies including rounding. • CCSS.MATH.CONTENT.3.OA.D.9: Identify arithmetic patterns (including patterns in the addition table or multiplication table), and explain them using properties of operations.
***CCSS Mathematics—* Number and Operations in Base Ten**	Use place value understanding and properties of operations to perform multi-digit arithmetic. • CCSS.MATH.CONTENT.3.NBT.A.1: Use place value understanding to round whole numbers to the nearest 10 or 100. • CCSS.MATH.CONTENT.3.NBT.A.2: Fluently add and subtract within 1,000 using strategies and algorithms based on place value, properties of operations, and/or the relationship between addition and subtraction.

Continued

Table 3.2 *(continued)*

***CCSS Mathematics—* Measurement and Data**	Solve problems involving measurement and estimation. • CCSS.MATH.CONTENT.3.MD.A.1: Tell and write time to the nearest minute and measure time intervals in minutes. Solve word problems involving addition and subtraction of time intervals in minutes. Represent and interpret data. • CCSS.MATH.CONTENT.3.MD.B.3: Draw a scaled picture graph and a scaled bar graph to represent a data set with several categories. Solve one- and two-step "how many more" and "how many less" problems using information presented in scaled bar graphs. • CCSS.MATH.CONTENT.3.MD.B.4: Generate measurement data by measuring lengths using rulers marked with halves and fourths of an inch. Show the data by making a line plot, where the horizontal scale is marked off in appropriate units—whole numbers, halves, or quarters.
ELP Standards	Receptive modalities • ELP 1: Construct meaning from oral presentations and literary and informational text through grade-appropriate listening, reading, and viewing. • ELP 8: Determine the meaning of words and phrases in oral presentations and literary and informational text. Productive modalities • ELP 3: Speak and write about grade-appropriate complex literary and informational texts and topics. • ELP 4: Construct grade-appropriate oral and written claims and support them with reasoning and evidence. • ELP 7: Adapt language choices to purpose, task, and audience when speaking and writing. Interactive modalities • ELP 2: Participate in grade-appropriate oral and written exchanges of information, ideas, and analyses, responding to peer, audience, or reader comments and questions. • ELP 5: Conduct research and evaluate and communicate findings to answer questions or solve problems. • ELP 6: Analyze and critique the arguments of others orally and in writing. Linguistic structures of English • ELP 9: Create clear and coherent grade-appropriate speech and text. • ELP 10: Make accurate use of standard English to communicate in grade-appropriate speech and writing.

Investigation Handout

Investigation 3

Changes in Motion: Where Will the Marble Be Located Each Time It Changes Direction in a Half-Pipe?

Introduction

Objects often move in a predictable pattern. For example, when we bounce a ball, it moves up and down in a predictable pattern. Children playing on swings or seesaws also follow a predictable pattern as they move back and forth or up and down. Take a moment to explore what happens when you place a marble in a half-pipe. Be sure to place the marble at different spots in the half-pipe and pay attention to how the marble moves after each time you let go of it. As you use these materials, keep track of what you observe and what you are wondering about in the boxes below.

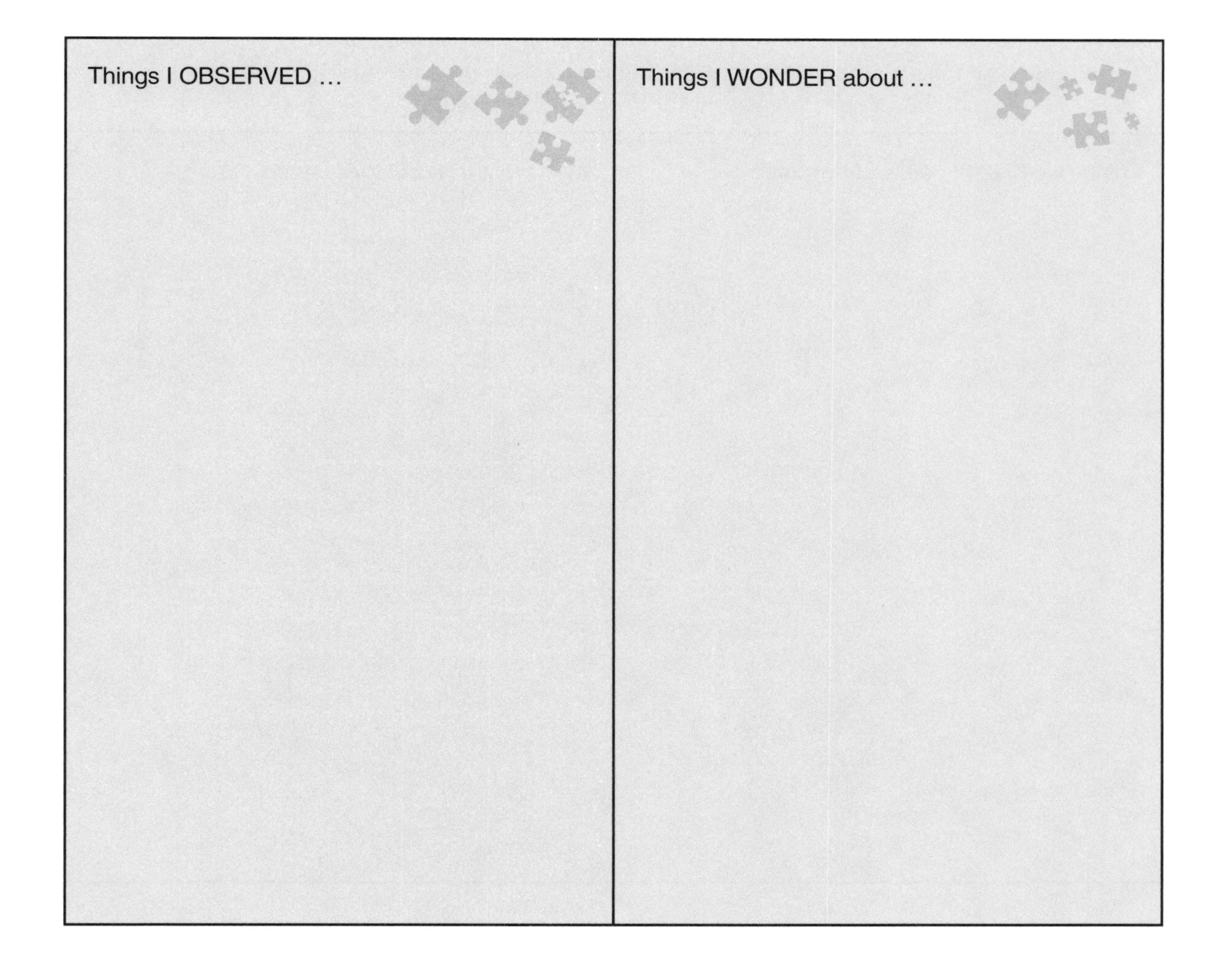

The motion of the marble in a half-pipe is another example of an object moving in a predictable pattern. Once a scientist discovers a pattern in the way an object moves, he or she can describe the motion of that object using numbers. This is important because scientists can predict the future motion of an object when they can use numbers to describe the pattern an object follows as it moves.

To be able to describe the motion of an object using numbers, a scientist needs to be able to measure how far an object moves from its initial position. Scientists call this measurement the *displacement* of an object. To measure the displacement of an object, a scientist must pick a reference point. He or she can then measure how far an object moves from this reference point in a given direction. A scientist can also record how long it takes for an object to move from one position to a different position. Scientists call this measurement the *speed* of an object. Speed is how fast an object travels. With this information, a scientist can use numbers to describe the motion of an object by reporting the direction an object moved (left, right, up, or down), how far it moved (its displacement), how long it took for the object to reach the new position (time), and its speed (how fast it moved).

In this investigation, you will examine how a marble moves back and forth in a half-pipe. Your goal is to find a way to describe the marble's motion using numbers. The half-pipe track has ruler tape attached to it to make it easier for you to measure how far the marble moves from the reference point on the track. The reference point is located at the middle of the half-pipe. It is marked with a circle. You will also have a stopwatch so you can keep track of time. With this information, you should be able to not only describe how the marble moves using numbers but also use this pattern to predict the motion of the marble when it is placed in different starting positions in the half-pipe.

Things we KNOW from what we read …	What we will NEED to figure out …

Your Task

Use what you know about the motion of objects and patterns to design and carry out an investigation to describe the pattern a marble follows as it rolls back and forth in a half-pipe. To create the pattern, you will need to use numbers to describe how the marble changes position over time. You will then need to show that you can use the pattern you developed to make accurate predictions about how the marble will move when it is placed in different starting positions in the half-pipe.

The *guiding question* of this investigation is, ***Where will the marble be located each time it changes direction in a half-pipe?***

Materials

You may use any of the following materials during your investigation:

- Safety glasses or goggles (required)
- 3 Plastic tracks
- 1 Marble
- Stopwatch
- Ruler tape

Safety Rules

Follow all normal safety rules. In addition, be sure to follow these rules:

- Wear sanitized safety glasses or goggles during setup, investigation activity, and cleanup.
- Do not throw objects or put any objects your mouth.
- Immediately pick up any slip or fall hazards (e.g., marbles) from the floor.
- Wash your hands with soap and water when you are done collecting data.

Plan Your Investigation

Prepare a plan for your investigation by filling out the chart that follows; this plan is called an *investigation proposal*. Before you start developing your plan, be sure to discuss the following questions with the other members of your group:

- What information should we collect so we can **describe** the motion of the ball?
- What types of **patterns** might we look for to help answer the guiding question?

Our guiding question:

This is a picture of how we will set up the equipment:

We will collect the following data:

These are the steps we will follow to collect data:

I approve of this investigation proposal.

Teacher's signature

Date

Collect Your Data

Keep a record of what you measure or observe during your investigation in the space below.

Analyze Your Data

You will need to analyze the data you collected before you can develop an answer to the guiding question. To do this, create a graph that shows how the position of the marble (distance from the reference point) changed as it went back and forth in the half-pipe.

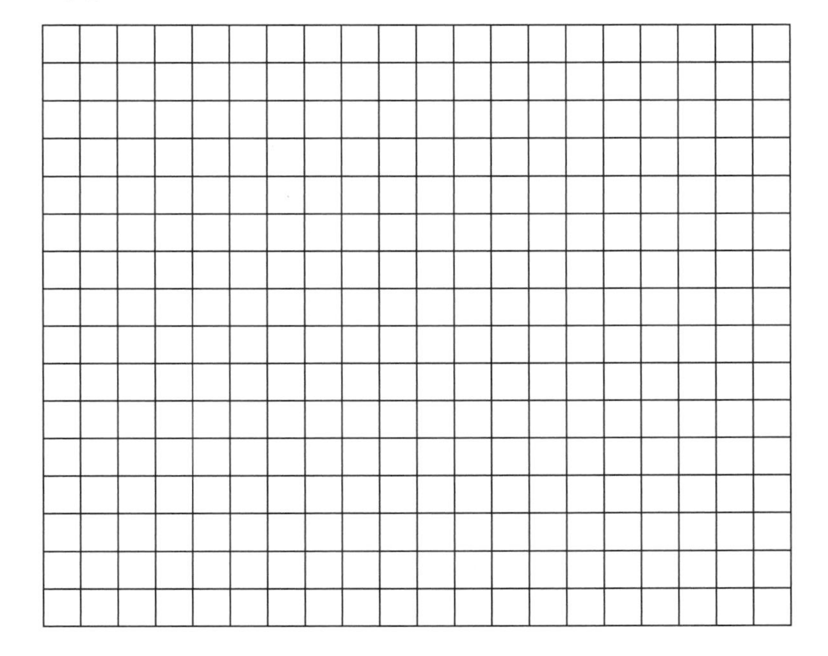

Draft Argument

Develop an argument on a whiteboard. It should include the following parts:

1. A *claim:* Your answer to the guiding question.
2. *Evidence:* An analysis of the data and an explanation of what the analysis means.
3. A *justification of the evidence:* Why your group thinks the evidence is important.

The Guiding Question:	
Our Claim:	
Our Evidence:	Our Justification of the Evidence:

Argumentation Session

Share your argument with your classmates. Be sure to ask them how to make your draft argument better. Keep track of their suggestions in the space below.

Draft Report

Prepare an *investigation report* to share what you have learned. Use the information in this handout and your group's final argument to write a *draft* of your investigation report.

Introduction

We have been studying ______________________________ in class.

Before we started this investigation, we explored ______________________________

____

We noticed ______________________________

____

____

My goal for this investigation was to figure out ______________________________

____

____

The guiding question was ______________________________

____

____

Method

To gather the data I needed to answer this question, I ______________________________

____

____

____

____

____

____

I then analyzed the data I collected by ____

____

____

Argument

My claim is ____

____

____

The figure below shows ____

____

____

____

____

____

____

____

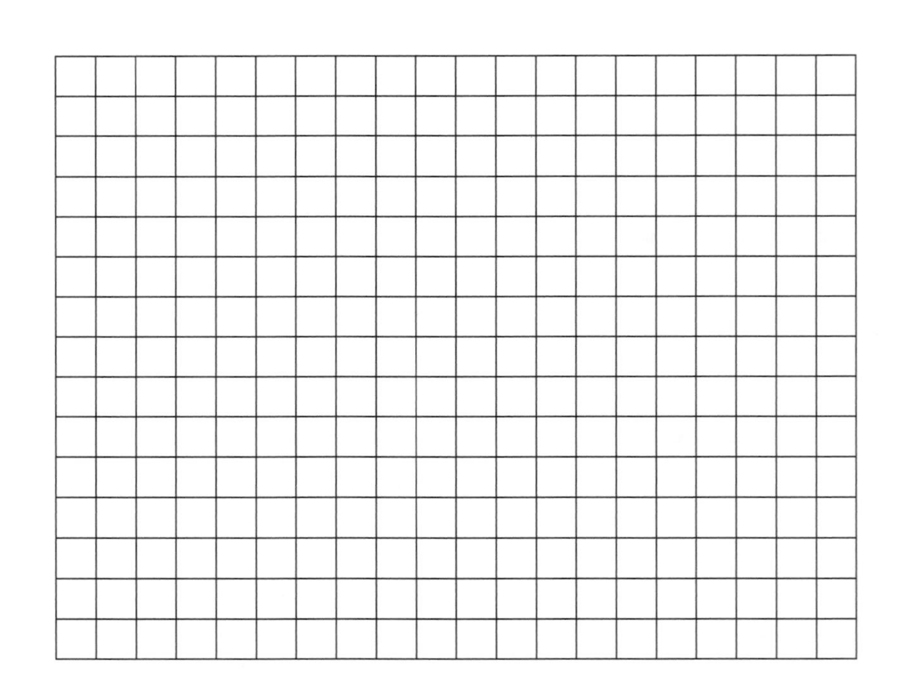

This evidence is important because ____

____

____

____

____

____

____

____

____

Review

Your friends need your help! Review the draft of their investigation reports and give them ideas about how to improve. Use the *peer-review guide* when doing your review.

Submit Your Final Report

Once you have received feedback from your friends about your draft report, create your final investigation report and hand it in to your teacher.

Checkout Questions

Investigation 3. Changes in Motion: Where Will the Marble Be Located Each Time It Changes Direction in a Half-Pipe?

The image below shows a half-pipe. Imagine someone places a ball at point A and then lets go of it. Your job is to predict where the ball will be located each time it changes direction as it moves.

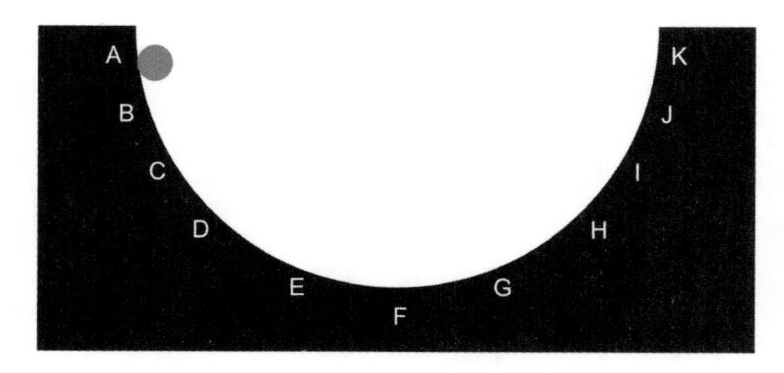

1. Where will the ball be when it changes direction for the first time? H I J K
2. Where will the ball be when it changes direction for the second time? A B C D
3. Where will the ball be when it changes direction for the third time? G H I J
4. Explain your thinking. What pattern from your investigation did you use to decide where the ball would be each time it changed direction as it moved back and forth in the half-pipe?

____

____

____

____

____

____

____

____

Teacher Scoring Rubric for the Checkout Questions

Level	Description
3	The student can apply the core idea correctly in all cases and can fully explain the pattern.
2	The student can apply the core idea correctly in all cases but cannot fully explain the pattern.
1	The student cannot apply the core idea correctly in all cases but can fully explain the pattern.
0	The student cannot apply the core idea correctly in all cases and cannot explain the pattern.

Investigation 4

Balanced and Unbalanced Forces: How Do Balanced and Unbalanced Forces Acting on an Object Affect the Motion of That Object?

Purpose

The purpose of this investigation is to give students an opportunity to use the disciplinary core idea (DCI) of PS2.A: Forces and Motion and the crosscutting concept (CC) of Cause and Effect from *A Framework for K–12 Science Education* (NRC 2012) to figure out how balanced and unbalanced forces acting on a cart affect the motion of that cart. Students will also learn about how scientists assume that there is order and consistency in nature during the reflective discussion.

The Disciplinary Core Idea

Students in third grade should understand the following about Forces and Motion and be able to use this DCI to figure out how changing the strength and direction of the forces acting on a cart affects the motion of that cart:

> *Objects pull or push each other when they collide or are connected. Pushes and pulls can have different strengths and directions. Pushing or pulling on an object can change the speed or direction of its motion and can start or stop it. An object sliding on a surface or sitting on a slope experiences a pull due to friction on the object due to the surface that opposes the object's motion. (NRC 2012, p. 115)*

The Crosscutting Concept

Students in third grade should understand the following about the CC Cause and Effect:

> *Repeating patterns in nature, or events that occur together with regularity, are clues that scientists can use to start exploring causal, or cause-and-effect, relationships. … Any application of science, or any engineered solution to a problem, is dependent on understanding the cause-and-effect relationships between events; the quality of the application or solution often can be improved as knowledge of the relevant relationships is improved. (NRC 2012, p. 87)*

Students in third grade should be given opportunities to begin to "look for and analyze patterns—whether in their observations of the world or in the relationships between different quantities in data (e.g., the sizes of plants over time)"; "they can also begin to consider what might be causing these patterns and relationships and design tests that gather more evidence to support or refute their ideas" (NRC 2012, pp. 88–89).

Students should be encouraged to use their developing understanding of cause-and-effect relationships as a tool or a way of thinking about a phenomenon during this investigation to help them figure out how changing the strength and direction of the forces acting on a cart affects the motion of that cart.

What Students Figure Out

Any force that acts on an object has both a strength and a direction. An object (such as a cart) can have multiple forces acting on it at the same time. When all the forces acting on an object are *balanced* (equal in terms of strength but acting in opposite directions), the object will remain at rest if the object is at rest or continue moving in the same direction at the same speed if it is in motion. When the forces acting on an object are *unbalanced* (unequal in strength or not in opposite directions), the motion of the object will change in terms of speed or direction.

Timeline

The time needed to complete this investigation is 260 minutes (4 hours and 20 minutes). The amount of instructional time needed for each stage of the investigation is as follows:

- *Stage 1.* Introduce the task and the guiding question: 35 minutes
- *Stage 2.* Design a method and collect data: 50 minutes
- *Stage 3.* Create a draft argument: 35 minutes
- *Stage 4.* Argumentation session: 30 minutes
- *Stage 5.* Reflective discussion: 15 minutes
- *Stage 6.* Write a draft report: 35 minutes
- *Stage 7.* Peer review: 30 minutes
- *Stage 8.* Revise the report: 30 minutes

This investigation can be completed in one day or over eight days (one day for each stage) during your designated science time in the daily schedule.

Materials and Preparation

The materials needed for this investigation are listed in Table 4.1. The carts and pulleys can be purchased from a science education supply company such as Ward's Science, Flinn Scientific, or Fisher Scientific. The other items can be purchased from a big-box retail store such as Wal-Mart or Target or through an online retailer such as Amazon. The materials for this investigation can also be purchased as a complete kit (which includes enough materials for 24 students, or six groups) at *www.argumentdriveninquiry.com*.

TABLE 4.1

Materials for Investigation 4

Item	Quantity
Safety glasses or goggles	1 per student
Cart (such as Hall's Carriage)	1 per group
Bench clamp pulleys	2 per group
Paper clips	2 per group
Washers	10 per group
Kite string, 6'	1 per group
Whiteboard, 2' × 3'*	1 per group
Investigation Handout	1 per student
Peer-review guide and teacher scoring rubric	1 per student
Checkout Questions (optional)	1 per student

*As an alternative, students can use computer and presentation software such as Microsoft PowerPoint or Apple Keynote to create their arguments.

Be sure to use a set routine for distributing and collecting the materials. One option is to set up the materials for each group in a kit that you can deliver to each group. A second option is to have all the materials on a table or cart at a central location. You can then assign a member of each group to be the "materials manager." This individual is responsible for collecting all the materials his or her group needs from the table or cart during class and for returning all the materials at the end of the class.

Figure 4.1 provides a diagram of how the students can set up the equipment for this investigation. Students can attach a 3-foot piece of kite string to each side of the cart. They can then bend the paper clips into a J-shaped hook and tie one to each end of the string. The students can then add washers to the paper clips. The bench clamp pulleys can be attached to the side of the table.

FIGURE 4.1

How to set up the equipment for Investigation 4

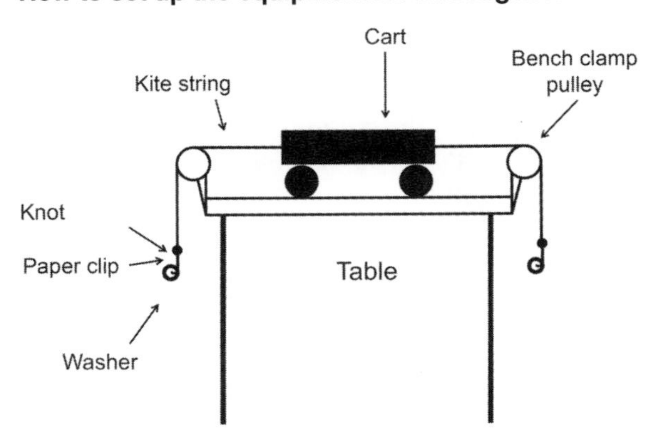

Safety Precautions

Remind students to follow all normal safety rules. In addition, tell the students to take the following safety precautions:

- Wear sanitized safety glasses or goggles during setup, investigation activity, and cleanup.

- Do not throw objects or put any objects in their mouth.
- Keep fingers and toes away from moving objects.
- Wash their hands with soap and water when done collecting the data.

Lesson Plan by Stage

Stage 1: Introduce the Task and the Guiding Question (35 minutes)

1. Ask the students to sit in six groups, with three or four students in each group.
2. Ask students to clear off their desks except for a pencil (and their *Student Workbook for Argument-Driven Inquiry in Third-Grade Science* if they have one).
3. Pass out an Investigation Handout to each student (or ask students to turn to Investigation Log 4 in their workbook).
4. Read the first paragraph of the "Introduction" aloud to the class. Ask the students to follow along as you read.
5. Pass out a cart to each group. Each cart should have a piece of string attached to it so students can pull on the string from different directions.
6. Remind students of the safety rules and explain the safety precautions for this investigation.
7. Tell students to pull on the strings and record their observations and questions in the "OBSERVED/WONDER" chart in the "Introduction" section of their Investigation Handout (or the investigation log in their workbook).
8. Ask students to share *what they observed* as they pulled on the cart from different directions.
9. Ask students to share *what questions they have* about the motion of the cart.
10. Tell the students, "Some of your questions might be answered by reading the rest of the 'Introduction.'"
11. Ask the students to read the rest of the "Introduction" on their own *or* ask them to follow along as you read aloud.
12. Once the students have read the rest of the "Introduction," ask them to fill out the "KNOW/NEED" chart on their Investigation Handout (or in their investigation log) as a group.
13. Ask students to share what they learned from the reading. Add these ideas to a class "know / need to figure out" chart.
14. Ask students to share what they think they will need to figure out based on what they read. Add these ideas to the class "know / need to figure out" chart.
15. Tell the students, "It looks like we have something to figure out. Let's see what we will need to do during our investigation."
16. Read the task and the guiding question aloud.

17. Tell the students, "I have lots of materials here that you can use."
18. Introduce the students to the materials available for them to use during the investigation by showing them how to use the pulleys and washers to change the pulling force on each side of the cart. Students can change the amount of pulling force acting on each side of the cart by hanging different numbers of washers from each string (see Figure 4.1, p. 147).

Stage 2: Design a Method and Collect Data (50 minutes)

1. Tell the students, "I am now going to give you and the other members of your group about 15 minutes to plan your investigation. Before you begin, I want you all to take a couple of minutes to discuss the following questions with the rest of your group."
2. Show the following questions on the screen or board:
   - What types of *patterns* might we look for to help answer the guiding question?
   - What information do we need to find *a cause-and-effect relationship?*
3. Tell the students, "Please take a few minutes to come up with an answer to these questions." Give the students two or three minutes to discuss these two questions.
4. Ask two or three different groups to share their answers. Be sure to highlight or write down any important ideas on the board so students can refer to them later.
5. If possible, use a document camera to project an image of the graphic organizer for this investigation on a screen or board (or take a picture of it and project the picture on a screen or board). Tell the students, "I now want you all to plan out your investigation. To do that, you will need to create an investigation proposal by filling out this graphic organizer."
6. Point to the box labeled "Our guiding question:" and tell the students, "You can put the question we are trying to answer in this box." Then ask, "Where can we find the guiding question?"
7. Wait for a student to answer where to find the guiding question (the answer is "in the handout").
8. Point to the box labeled "We will collect the following data:" and tell the students, "You can list the measurements or observations that you will need to collect during the investigation in this box."
9. Point to the box labeled "This is a picture of how we will set up the equipment:" and tell the students, "You can draw a picture of how you plan to set up the equipment during the investigation so you can collect the data you need in this box."
10. Point to the box labeled "These are the steps we will follow to collect data:" and tell the students, "You can list what you are going to do to collect the data

you need and what you will do with your data once you have it. Be sure to give enough detail that I could do your investigation for you."

11. Ask the students, "Do you have any questions about what you need to do?"
12. Answer any questions that come up.
13. Tell the students, "Once you are done, raise your hand and let me know. I'll then come by and look over your proposal and give you some feedback. You may not begin collecting data until I have approved your proposal by signing it. You need to have your proposal done in the next 15 minutes."
14. Give the students 15 minutes to work in their groups on their investigation proposal. As they work, move from group to group to check in, ask probing questions, and offer a suggestion if a group gets stuck.

What should a student-designed investigation look like?

The students' investigation proposal should include the following information:

- The guiding question is "How do balanced and unbalanced forces acting on an object affect the motion of that object?"
- The data that the student should collect are (1) number of washers hanging from each side of the cart and (2) direction the cart moves.
- To collect the data, the students might conduct a series of tests in which they hold the cart in the middle of the table, add a varying number of washers to the strings attached to the right and left sides of the cart, let go of the cart, and observe the direction in which the cart moves. The students could start with one washer on each string, then in the second experiment add one washer to the string on the right side and two washers to the string on the left side, and so on. But this is just an example of how the students could collect data; there should be a lot of variation in the student-designed investigations.

15. As each group finishes its investigation proposal, be sure to read it over and determine if it will be productive or not. If you feel the investigation will be productive (not necessarily what you would do or what the other groups are doing), sign your name on the proposal and let the group start collecting data. If the plan needs to be changed, offer some suggestions or ask some probing questions, and have the group make the changes before you approve it.
16. Pass out the materials or have one student from each group collect the materials they need from a central supply table or cart for the groups that have an approved proposal.
17. Remind students of the safety rules and precautions for this investigation.

18. Tell the students to collect their data and record their observations or measurements in the "Collect Your Data" box in their Investigation Handout (or the investigation log in their workbook).
19. Give the students 10 minutes to collect their data.
20. Be sure to collect the materials from each group before asking them to analyze their data.

Stage 3: Create a Draft Argument (35 minutes)

1. Tell the students, "Now that we have all this data, we need to analyze the data so we can figure out an answer to the guiding question."
2. If possible, project an image of the "Analyze Your Data" text and box for this investigation on a screen or board using a document camera (or take a picture of it and project the picture on a screen or board). Point to the box and tell the students, "You can create a table or graph as a way to analyze your data. You can make your table or graph in this box."
3. Ask the students, "What information do we need to include in a table or graph?"
4. Tell the students, "Please take a few minutes to discuss this question with your group, and be ready to share."
5. Give the students five minutes to discuss.
6. Ask two or three different groups to share their answers. Be sure to highlight or write down any important ideas on the board so students can refer to them later.
7. Tell the students, "I am now going to give you and the other members of your group about 10 minutes to create your table or graph." If the students are having trouble making a table or graph, you can take a few minutes to provide a mini-lesson about how to create a table or graph from a bunch of observations or measurements (this strategy is called just-in-time instruction because it is offered only when students get stuck).

What should a table or graph look like for this investigation?

There are a number of different ways that students can analyze the observations or measurements they collect during this investigation. One of the most straightforward ways is to create a table with three columns. The first column can be labeled "cart moves to the left" or "L," the second column can be labeled "cart moves to the right" or "R," and the third column can be labeled "cart does not move" or "S." The students can then include the appropriate configurations of cart and hanging washers (number of washers or the amount of mass they added to each side of the cart) that they tested in each column. This allows them to show a clear pattern as part of their analysis of the data. An example of this type of table can be seen in Figure 4.2. There are other options for analyzing the data they collected. Students often come up with some unique ways of analyzing their data, so be sure to give them some voice and choice during this stage.

8. Give the students 10 minutes to analyze their data. As they work, move from group to group to check in, ask probing questions, and offer suggestions.
9. Tell the students, "I am now going to give you and the other members of your group 15 minutes to create an argument to share what you have learned and convince others that they should believe you. Before you do that, we need to take a few minutes to discuss what you need to include in your argument."
10. If possible, use a document camera to project the "Argument Presentation on a Whiteboard" image from the Investigation Handout (or the investigation log in their workbook) on a screen or board (or take a picture of it and project the picture on a screen or board).
11. Point to the box labeled "The Guiding Question:" and tell the students, "You can put the question we are trying to answer here on your whiteboard."
12. Point to the box labeled "Our Claim:" and tell the students, "You can put your claim here on your whiteboard. The claim is your answer to the guiding question."
13. Point to the box labeled "Our Evidence:" and tell the students, "You can put the evidence that you are using to support your claim here on your whiteboard. Your evidence will need to include the analysis you just did and an explanation of what your analysis means or shows. Scientists always need to support their claims with evidence."
14. Point to the box labeled "Our Justification of the Evidence:" and tell the students, "You can put your justification of your evidence here on your whiteboard. Your justification needs to explain why your evidence is important. Scientists often use core ideas to explain why the evidence they are using matters. Core ideas are important concepts that scientists use to help them make sense of what happens during an investigation."
15. Ask the students, "What are some core ideas that we read about earlier that might help us explain why the evidence we are using is important?"
16. Ask students to share some of the core ideas from the "Introduction" section of the Investigation Handout (or the investigation log in the workbook). List these core ideas on the board.

FIGURE 4.2

Example of an argument

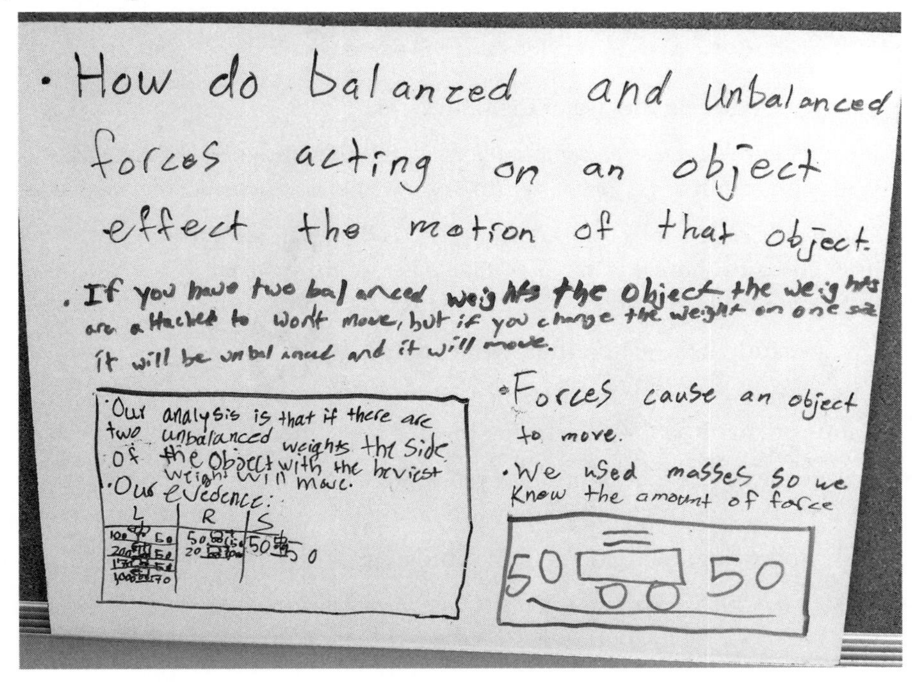

17. Tell the students, "That is great. I would like to see everyone try to include these core ideas in your justification of the evidence. Your goal is to use these core ideas to help explain why your evidence matters and why the rest of us should pay attention to it."
18. Ask the students, "Do you have any questions about what you need to do?"
19. Answer any questions that come up.
20. Tell the students, "Okay, go ahead and start working on your arguments. You need to have your argument done in the next 15 minutes. It doesn't need to be perfect. We just need something down on the whiteboards so we can share our ideas."
21. Give the students 15 minutes to work in their groups on their arguments. As they work, move from group to group to check in, ask probing questions, and offer a suggestion if a group gets stuck. Figure 4.2 shows an example of an argument created by students for this investigation.

Stage 4: Argumentation Session (30 minutes)

The argumentation session can be conducted in a whole-class presentation format, a gallery walk format, or a modified gallery walk format. We recommend using a whole-class presentation format for the first investigation, but try to transition to either the gallery walk or modified gallery walk format as soon as possible because that will maximize student voice and choice inside the classroom. The following list shows the steps for the three formats; unless otherwise noted, the steps are the same for all three formats.

1. Begin by introducing the use of the whiteboard.
   - *If using the whole-class presentation format,* tell the students, "We are now going to share our arguments. Please set up your whiteboard so everyone can see them."
   - *If using the gallery walk or modified gallery walk format,* tell the students, "We are now going to share our arguments. Please set up your whiteboard so they are facing the walls."
2. Allow the students to set up their whiteboards.
   - *If using the whole-class presentation format,* the whiteboards should be set up on stands or chairs so they are facing toward the center of the room.
   - *If using the gallery walk or modified gallery walk format,* the whiteboards should be set up on stands or chairs so they are facing toward the outside of the room.
3. Give the following instructions to the students:
   - *If using the whole-class presentation format or the modified gallery walk format,* tell the students, "Okay, before we get started I want to explain what we are going to do next. I'm going to ask some of you to present your arguments to your classmates. If you are presenting your argument, your job is to share your group's claim, evidence, and justification of the evidence. The rest of you will be reviewers. If you are a reviewer, your job is to listen to the presenters, ask the presenters questions if you do not understand something, and then offer them some suggestions about ways to make their argument better. After we have a chance to learn from each other, I'm going to give you some time to revise your arguments and make them better."
   - *If using the gallery walk format,* tell the students, "Okay, before we get started I want to explain what we are going to do next. You are going to have an opportunity to read the arguments that were created by other groups. Your group will go to a different group's argument. I'll give you a few minutes to read it and review it. Your job is to offer them some suggestions about ways to make their argument better. You can use sticky notes to give them suggestions. Please be specific about what you want to change and be specific about how you think they should change it. After we have a chance to learn from each other, I'm going to give you some time to revise your arguments and make them better."
4. Use a document camera to project the "Ways to IMPROVE our argument …" box from the Investigation Handout (or the investigation log in their workbook)

on a screen or board (or take a picture of it and project the picture on a screen or board).

- *If using the whole-class presentation format or the modified gallery walk format,* point to the box and tell the students, "If you are a presenter, you can write down the suggestions you get from the reviewers here. If you are a reviewer, and you see a good idea from another group, you can write down that idea here. Once we are done with the presentations, I will give you a chance to use these suggestions or ideas to improve your arguments.
- *If using the gallery walk format,* point to the box and tell the students, "If you see good ideas from another group, you can write them down here. Once we are done reviewing the different arguments, I will give you a chance to use these ideas to improve your own arguments. It is important to share ideas like this."

Ask the students, "Do you have any questions about what you need to do?"

5. Answer any questions that come up.
6. Give the following instructions:
   - *If using the whole-class presentation format,* tell the students, "Okay. Let's get started."
   - *If using the gallery walk format,* tell the students, "Okay, I'm now going to tell you which argument to go to and review.
   - *If using the modified gallery walk format,* tell the students, "Okay, I'm now going to assign you to be a presenter or a reviewer." Assign one or two students from each group to be presenters and one or two students from each group to be reviewers.
7. Begin the review of the arguments.
   - *If using the whole-class presentation format,* have four or five groups present their argument one at a time. Give each group only two to three minutes to present their argument. Then give the class two to three minutes to ask them questions and offer suggestions. Be sure to encourage as much participation from the students as possible.
   - *If using the gallery walk format,* tell the students, "Okay. Let's get started. Each group, move one argument to the left. Don't move to the next argument until I tell you to move. Once you get there, read the argument and then offer suggestions about how to make it better. I will put some sticky notes next to each argument. You can use the sticky notes to leave your suggestions." Give each group about three to four minutes to read the arguments, talk, and offer suggestions.

     a. Tell the students, "Okay. Let's rotate. Move one group to the left."
     b. Again, give each group three or four minutes to read, talk, and offer suggestions.

c. Repeat this process for two more rotations.

- *If using the modified gallery walk format,* tell the students, "Okay. Let's get started. Reviewers, move one group to the left. Don't move to the next group until I tell you to move. Presenters, go ahead and share your argument with the reviewers when they get there." Give each group of presenters and reviewers about three to four minutes to talk.

  a. Tell the students, "Okay. Let's rotate. Reviewers, move one group to the left."
  b. Again, give each group of presenters and reviewers about three or four minutes to talk.
  c. Repeat this process for two more rotations.

8. Tell the students to return to their workstations.
9. Give the following instructions about revising the argument:
   - *If using the whole-class presentation format,* tell the students, "I'm now going to give you about 10 minutes to revise your argument. Take a few minutes to talk in your groups and determine what you want to change to make your argument better. Once you have decided what to change, go ahead and make the changes to your whiteboard."
   - *If using the gallery walk format,* tell the students, "I'm now going to give you about 10 minutes to revise your argument. Take a few minutes to read the suggestions that were left at your argument. Then talk in your groups and determine what you want to change to make your argument better. Once you have decided what to change, go ahead and make the changes to your whiteboard."
   - *If using the modified gallery walk format,* "I'm now going to give you about 10 minutes to revise your argument. Please return to your original groups." Wait for the students to move back into their original groups and then tell the students, "Okay, take a few minutes to talk in your groups and determine what you want to change to make your argument better. Once you have decided what to change, go ahead and make the changes to your whiteboard."
10. Ask the students, "Do you have any questions about what you need to do?"
11. Answer any questions that come up.
12. Tell the students, "Okay. Let's get started."
13. Give the students 10 minutes to work in their groups on their arguments. As they work, move from group to group to check in, ask probing questions, and offer a suggestion if a group gets stuck.

Stage 5: Reflective Discussion (15 minutes)

1. Tell the students, "We are now going to take a minute to talk about what we did and what we have learned."
2. Place a cart on a table. Ask the students, "What do you all see going on here?"
3. Allow students to share their ideas.
4. Ask the students, "What forces are acting on this cart?"
5. Allow students to share their ideas. Keep probing until someone mentions gravity.
6. Ask the students, "If there is gravity acting on this cart, why doesn't it move?"
7. Allow students to share their ideas. Keep probing until someone mentions the table.
8. Tell the students, "The cart doesn't move because the force of gravity pushing down on the cart is the same as the force from the table pushing up on the cart. These two forces are equal in strength but opposite in direction."
9. Push on the cart so it starts to roll. Ask the students, "Why did the cart start to move?"
10. Allow students to share their ideas. Keep probing until someone mentions that you applied a force to it.
11. Ask the students, "How could someone keep this cart from moving while I'm pushing on it?"
12. Allow students to share their ideas. Keep probing until someone mentions that they could push on it as well.
13. Ask the students, "Could someone show me where you would have to push on the cart to keep it from moving if I push on it right here?"
14. Allow students to share their ideas. Keep probing until someone mentions pushing in the opposite direction. Invite that student to come up and demonstrate.
15. Tell the students, "Okay, let's make sure we are on the same page. Forces are pushes or pulls. Any force that acts on an object has both a strength and a direction. A cart has multiple forces acting on it at the same time. When all the forces acting on the cart are equal in terms of strength but act in opposite directions, the cart will not move. When the forces acting on a cart are not balanced, which means they are unequal in strength or not in opposite directions, the cart will move. The difference between balanced and unbalanced forces is a really important core idea in science."
16. Ask the students, "Does anyone have any questions about this core idea?"
17. Answer any questions that come up.

18. Tell the students, "We also looked for a cause-and-effect relationship during our investigation." Then ask, "Can anyone tell me why it is useful to look for a cause-and-effect relationship?"
19. Allow students to share their ideas.
20. Tell the students, "Cause-and-effect relationships are important because they allow us to predict what will happen in the future."
21. Ask the students, "What was the cause and what was the effect that we uncovered today?"
22. Allow students to share their ideas. Keep probing until someone mentions that the cause was changing the amount of pulling force acting on the cart and the effect was the direction the cart moved.
23. Ask the students, "Can anyone tell me what will happen to the cart if we add more pulling force to one side of it?"
24. Allow students to share their ideas. Keep probing until someone mentions that the cart will move in the direction of the greatest pulling force.
25. Tell the students, "That is great, and if we know that we can predict what will happen the next time we apply a force to an object."
26. Tell the students, "We are now going take a minute to talk about what went well and what didn't go so well during our investigation. We need to talk about this because you are going to be planning and carrying out your own investigations like this a lot this year, and I want to help you all get better at it."
27. Show an image of the question "What made your investigation scientific?" on the screen. Tell the students, "Take a few minutes to talk about how you would answer this question with the other people in your group. Be ready to share with the rest of the class." Give the students two to three minutes to talk in their group.
28. Ask the students, "What do you all think? Who would like to share an idea?"
29. Allow students to share their ideas. Be sure to expand on their ideas about what makes an investigation scientific.
30. Show an image of the question "What made your investigation not so scientific?" on the screen. Tell the students, "Take a few minutes to talk about how you would answer this question with the other people in your group. Be ready to share with the rest of the class." Give the students two to three minutes to talk in their group.
31. Ask the students, "What do you all think? Who would like to share an idea?"
32. Allow students to share their ideas. Be sure to expand on their ideas about what makes an investigation less scientific.

33. Show an image of the question "What rules can we put into place to help us make sure our next investigation is more scientific?" on the screen. Tell the students, "Take a few minutes to talk about how you would answer this question with the other people in your group. Be ready to share with the rest of the class." Give the students two to three minutes to talk in their group.
34. Ask the students, "What do you all think? Who would like to share an idea?"
35. Allow students to share their ideas. Once they have shared their ideas, offer a suggestion for a possible class rule.
36. Ask the students, "What do you all think? Should we make this a rule?"
37. If the students agree, write the rule on the board or make a class "Rules for Scientific Investigation" chart so you can refer to it during the next investigation.
38. Tell the students, "We are now going take a minute to talk about how scientists think about the world."
39. Show an image of the question "Why can scientists explain or predict how an object will move no matter where it is in the world?" on the screen. Tell the students, "Take a few minutes to talk about how you would answer this question with the other people in your group. Be ready to share with the rest of the class." Give the students two to three minutes to talk in their group.
40. Ask the students, "What do you all think? Who would like to share an idea?"
41. Allow students to share their ideas.
42. Tell the students, "Okay, let's make sure we are all on the same page. Scientists can explain or make predictions about how an object will or will not move no matter where it is in the world because the basic laws of nature are the same everywhere."
43. Ask the students, "For example, what will happen to a ball if there are balanced forces acting on it?"
44. Allow students to share their ideas.
45. Ask the students, "Does it matter if that ball is here in our classroom or if it is in Africa, Australia, or Asia?"
46. Allow students to share their ideas. Tell the students, "The ball will not move when balanced forces are acting on it in all these different places because the laws of nature are the same everywhere. This is an important idea in science. This is one of the things scientists try to keep in mind when they are trying to figure something out."
47. Ask the students, "Does anyone have any questions about how scientists think about the world?"
48. Answer any questions that come up.

Stage 6: Write a Draft Report (35 minutes)

Your students will use either the Investigation Handout or the investigation log in the student workbook when writing the draft report. When you give the directions shown in quotes in the following steps, substitute "investigation log" (as shown in brackets) for "handout" if they are using the workbook.

1. Tell the students, "You are now going to write an investigation report to share what you have learned. Please take out a pencil and turn to the 'Draft Report' section of your handout [investigation log]."
2. If possible, use a document camera to project the "Introduction" section of the draft report from the Investigation Handout (or the investigation log in their workbook) on a screen or board (or take a picture of it and project the picture on a screen or board).
3. Tell the students, "The first part of the report is called the 'Introduction.' In this section of the report you want to explain to the reader what you were investigating, why you were investigating it, and what question you were trying to answer. All of this information can be found in the text at the beginning of your handout [investigation log]." Point to the image and say, "There are some sentence starters here to help you begin writing the report." Ask the students, "Do you have any questions about what you need to do?"
4. Answer any questions that come up.
5. Tell the students, "Okay. Let's write."
6. Give the students 10 minutes to write the "Introduction" section of the report. As they work, move from student to student to check in, ask probing questions, and offer a suggestion if a student gets stuck.
7. If possible, use a document camera to project the "Method" section of the draft report from the Investigation Handout (or the investigation log in their workbook) on a screen or board (or take a picture of it and project the picture on a screen or board).
8. Tell the students, "The second part of the report is called the 'Method.' In this section of the report you want to explain to the reader what you did during the investigation, what data you collected and why, and how you went about analyzing your data. All of this information can be found in the 'Plan Your Investigation' section of your handout [investigation log]. Remember that you all planned and carried out different investigations, so do not assume that the reader will know what you did." Point to the image and say, "There are some sentence starters here to help you begin writing this part of the report." Ask the students, "Do you have any questions about what you need to do?"
9. Answer any questions that come up.
10. Tell the students, "Okay. Let's write."

11. Give the students 10 minutes to write the "Method" section of the report. As they work, move from student to student to check in, ask probing questions, and offer a suggestion if a student gets stuck.
12. If possible, use a document camera to project the "Argument" section of the draft report from the Investigation Handout (or the investigation log in their workbook) on a screen or board (or take a picture of it and project the picture on a screen or board).
13. Tell the students, "The last part of the report is called the 'Argument.' In this section of the report you want to share your claim, evidence, and justification of the evidence with the reader. All of this information can be found on your whiteboard." Point to the image and say, "There are some sentence starters here to help you begin writing this part of the report." Ask the students, "Do you have any questions about what you need to do?"
14. Answer any questions that come up.
15. Tell the students, "Okay. Let's write."
16. Give the students 10 minutes to write the "Argument" section of the report. As they work, move from student to student to check in, ask probing questions, and offer a suggestion if a student gets stuck.

Stage 7: Peer Review (30 minutes)

Your students will use either the Investigation Handout or the investigation log in the student workbook when doing the peer review. When you give the directions shown in quotes in the following steps, substitute "workbook" (as shown in brackets) for "Investigation Handout" if they are using the workbook.

1. Tell the students, "We are now going to review our reports to find ways to make them better. I'm going to come around and collect your Investigation Handout [workbook]. While I do that, please take out a pencil."
2. Collect the Investigation Handouts or workbooks from the students.
3. If possible, use a document camera to project the peer-review guide (PRG; see Appendix 4) on a screen or board (or take a picture of it and project the picture on a screen or board).
4. Tell the students, "We are going to use this peer-review guide to give each other feedback." Point to the image.
5. Give the following instructions:
    - *If using the Investigation Handout,* tell the students, "I'm going to ask you to work with a partner to do this. I'm going to give you and your partner a draft report to read and a peer-review guide to fill out. You two will then read the report together. Once you are done reading the report, I want you to answer each of

the questions on the peer-review guide." Point to the review questions on the image of the PRG.

- *If using the workbook,* tell the students, "I'm going to ask you to work with a partner to do this. I'm going to give you and your partner a draft report to read. You two will then read the report together. Once you are done reading the report, I want you to answer each of the questions on the peer-review guide that is right after the report in the investigation log." Point to the review questions on the image of the PRG.

6. Tell the students, "You can check 'yes,' 'almost,' or 'no' after each question." Point to the checkboxes on the image of the PRG.
7. Tell the students, "This will be your rating for this part of the report. Make sure you agree on the rating you give the author. If you mark 'almost' or 'no,' then you need to tell the author what he or she needs to do to get a 'yes.'" Point to the space for the reviewer feedback on the image of the PRG.
8. Tell the students, "It is really important for you to give the authors feedback that is helpful. That means you need to tell them exactly what they need to do to make their reports better." Ask the students, "Do you have any questions about what you need to do?"
9. Answer any questions that come up.
10. Tell the students, "Please sit with a partner who is not in your current group." Allow the students time to sit with a partner.
11. Give the following instructions:
    - *If using the Investigation Handout,* tell the students, "Okay, I am now going to give you one report to read and one peer-review guide to fill out." Pass out one report to each pair. Make sure that the report you give a pair was not written by one of the students in that pair. Give each pair one PRG to fill out as a team.
    - *If using the workbook,* tell the students, "Okay, I am now going to give you one report to read." Pass out a workbook to each pair. Make sure that the workbook you give a pair is not from one of the students in that pair.
12. Tell the students, "Okay, I'm going to give you 15 minutes to read the report I gave you and to fill out the peer-review guide. Go ahead and get started."
13. Give the students 15 minutes to work. As they work, move around from pair to pair to check in and see how things are going, answer questions, and offer advice.
14. After 15 minutes pass, tell the students, "Okay, time is up." *If using the Investigation Handout,* say, "Please give me the report and the peer-review guide that you filled out." *If using the workbook,* say, "Please give me the workbook that you have."

15. Collect the Investigation Handouts and the PRGs, or collect the workbooks if they are being used. Be sure you keep the handout and the PRG together.
16. Give the following instructions:
    - *If using the Investigation Handout,* tell the students, "Okay, I am now going to give you a different report to read and a new peer-review guide to fill out." Pass out another report to each pair. Make sure that this report was not written by one of the students in that pair. Give each pair a new PRG to fill out as a team.
    - *If using the workbook,* tell the students, "Okay, I am now going to give you a different report to read." Pass out a different workbook to each pair. Make sure that the workbook you give a pair is not from one of the students in that pair.
17. Tell the students, "Okay, I'm going to give you 15 minutes to read this new report and to fill out the peer-review guide. Go ahead and get started."
18. Give the students 15 minutes to work. As they work, move around from pair to pair to check in and see how things are going, answer questions, and offer advice.
19. After 15 minutes pass, tell the students, "Okay, time is up." *If using the Investigation Handout,* say, "Please give me the report and the peer-review guide that you filled out." *If using the workbook,* say, "Please give me the workbook that you have."
20. Collect the Investigation Handouts and the PRGs, or collect the workbooks if they are being used. Be sure you keep the handout and the PRG together.

Stage 8: Revise the Report (30 minutes)

Your students will use either the Investigation Handout or the investigation log in the student workbook when revising the report. Except where noted below, the directions are the same whether using the handout or the log.

1. Give the following instructions:
    - *If using the Investigation Handout,* tell the students, "You are now going to revise your investigation report based on the feedback you get from your classmates. Please take out a pencil while I hand back your draft report and the peer-review guide."
    - *If using the investigation log in the student workbook,* tell the students, "You are now going to revise your investigation report based on the feedback you get from your classmates. Please take out a pencil while I hand back your investigation logs."
2. *If using the Investigation Handout,* pass back the handout and the PRG to each student. *If using the investigation log,* pass back the log to each student.

3. Tell the students, "Please take a few minutes to read over the peer-review guide. You should use it to figure out what you need to change in your report and how you will change the report."
4. Allow the students time to read the PRG.
5. *If using the investigation log,* if possible use a document camera to project the "Write Your Final Report" section from the investigation log on a screen or board (or take a picture of it and project the picture on a screen or board).
6. Give the following instructions:
   - *If using the Investigation Handout,* tell the students, "Okay. Let's revise our reports. Please take out a piece of paper. I would like you to rewrite your report. You can use your draft report as a starting point, but use the feedback on the peer-review guide to help make it better."
   - *If using the investigation log,* tell the students, "Okay. Let's revise our reports. I would like you to rewrite your report in the section of the investigation log that says 'Write Your Final Report.'" Point to the image on the screen and tell the students, "You can use your draft report as a starting point, but use the feedback on the peer-review guide to help make your report better."

   Ask the students, "Do you have any questions about what you need to do?"
7. Answer any questions that come up.
8. Tell the students, "Okay. Let's write."
9. Give the students 30 minutes to rewrite their report. As they work, move from student to student to check in, ask probing questions, and offer a suggestion if a student gets stuck.
10. Give the following instructions:
    - *If using the Investigation Handout,* tell the students, "Okay. Time's up. I will now come around and collect your Investigation Handout, the peer-review guide, and your final report."
    - *If using the investigation log,* tell the students, "Okay. Time's up. I will now come around and collect your workbooks."
11. *If using the Investigation Handout,* collect all the Investigation Handouts, PRGs, and final reports. *If using the investigation log,* collect all the workbooks.
12. *If using the Investigation Handout,* use the "Teacher Score" columns in the PRG to grade the final report. *If using the investigation log,* use the "ADI Investigation Report Grading Rubric" in the investigation log to grade the final report. Whether you are using the handout or the log, you can give the students feedback about their writing in the "Teacher Comments" section.

How to Use the Checkout Questions

The Checkout Questions are an optional assessment. We recommend giving them to students one day after they finish stage 8 of the ADI investigation. The Checkout Questions can be used as a formative or summative assessment of student thinking. If you plan to use them as a formative assessment, we recommend that you look over the student answers to determine if you need to reteach the core idea and/or crosscutting concept from the investigation, but do not grade them. If you plan to use them as a summative assessment, we have included a "Teacher Scoring Rubric" at the end of the Checkout Questions that you can use to score a student's ability to apply the core idea in a new scenario and explain their use of a crosscutting concept. The rubric includes a 4-point scale that ranges from 0 (the student cannot apply the core idea correctly in all cases and cannot explain the [crosscutting concept] to 3 (the student can apply the core idea correctly in all cases and can fully explain the [crosscutting concept]). The Checkout Questions, regardless of how you decide to use them, are a great way to make student thinking visible so you can determine if the students have learned the core idea and the crosscutting concept.

A student who can apply the core idea correctly in all cases and can explain the cause-and-effect relationships would give the following answers for the images in question 1: A—It won't move, B—Left, C—Right, D—It won't move, and E—Left. He or she should then be able to explain that the cart will move toward the greatest pulling force but will not move if the forces acting on it are balanced.

Connections to Standards

Table 4.2 (p. 166) highlights how the investigation can be used to address specific performance expectations from the *NGSS, Common Core State Standards (CCSS)* in English language arts (ELA) and in mathematics, and *English Language Proficiency (ELP) Standards.*

TABLE 4.2

Investigation 4 alignment with standards

***NGSS* performance expectation**	Strong alignment • 3-PS2-1: Plan and conduct an investigation to provide evidence of the effects of balanced and unbalanced forces on the motion of an object.
***CCSS ELA*—Reading: Informational Text**	Key ideas and details • CCSS.ELA-LITERACY.RI.3.1: Ask and answer questions to demonstrate understanding of a text, referring explicitly to the text as the basis for the answers. • CCSS.ELA-LITERACY.RI.3.2: Determine the main idea of a text; recount the key details and explain how they support the main idea. • CCSS.ELA-LITERACY.RI.3.3: Describe the relationship between a series of historical events, scientific ideas or concepts, or steps in technical procedures in a text, using language that pertains to time, sequence, and cause/effect. Craft and structure • CCSS.ELA-LITERACY.RI.3.4: Determine the meaning of general academic and domain-specific words and phrases in a text relevant to a *grade 3 topic or subject area*. • CCSS.ELA-LITERACY.RI.3.5: Use text features and search tools (e.g., key words, sidebars, hyperlinks) to locate information relevant to a given topic efficiently. • CCSS.ELA-LITERACY.RI.3.6: Distinguish their own point of view from that of the author of a text. Integration of knowledge and ideas • CCSS.ELA-LITERACY.RI.3.7: Use information gained from illustrations (e.g., maps, photographs) and the words in a text to demonstrate understanding of the text (e.g., where, when, why, and how key events occur). • CCSS.ELA-LITERACY.RI.3.8: Describe the logical connection between particular sentences and paragraphs in a text (e.g., comparison, cause/effect, first/second/third in a sequence). • CCSS.ELA-LITERACY.RI.3.9: Compare and contrast the most important points and key details presented in two texts on the same topic. Range of reading and level of text complexity • CCSS.ELA-LITERACY.RI.3.10: By the end of the year, read and comprehend informational texts, including history/social studies, science, and technical texts, at the high end of the grades 2–3 text complexity band independently and proficiently.

Continued

Table 4.2 (*continued*)

***CCSS ELA*—Writing**	Text types and purposes • CCSS.ELA-LITERACY.W.3.1: Write opinion pieces on topics or texts, supporting a point of view with reasons. o CCSS.ELA-LITERACY.W.3.1.A: Introduce the topic or text they are writing about, state an opinion, and create an organizational structure that lists reasons. o CCSS.ELA-LITERACY.W.3.1.B: Provide reasons that support the opinion. o CCSS.ELA-LITERACY.W.3.1.C: Use linking words and phrases (e.g., *because*, *therefore*, *since, for example*) to connect opinion and reasons. o CCSS.ELA-LITERACY.W.3.1.D: Provide a concluding statement or section. • CCSS.ELA-LITERACY.W.3.2: Write informative or explanatory texts to examine a topic and convey ideas and information clearly. o CCSS.ELA-LITERACY.W.3.2.A: Introduce a topic and group related information together; include illustrations when useful to aiding comprehension. o CCSS.ELA-LITERACY.W.3.2.B: Develop the topic with facts, definitions, and details. o CCSS.ELA-LITERACY.W.3.2.C: Use linking words and phrases (e.g., *also*, *another*, *and*, *more*, *but*) to connect ideas within categories of information. o CCSS.ELA-LITERACY.W.3.2.D: Provide a concluding statement or section. Production and distribution of writing • CCSS.ELA-LITERACY.W.3.4: With guidance and support from adults, produce writing in which the development and organization are appropriate to task and purpose. • CCSS.ELA-LITERACY.W.3.5: With guidance and support from peers and adults, develop and strengthen writing as needed by planning, revising, and editing. • CCSS.ELA-LITERACY.W.3.6: With guidance and support from adults, use technology to produce and publish writing (using keyboarding skills) as well as to interact and collaborate with others. Research to build and present knowledge • CCSS.ELA-LITERACY.W.3.8: Recall information from experiences or gather information from print and digital sources; take brief notes on sources and sort evidence into provided categories. Range of writing • CCSS.ELA-LITERACY.W.3.10: Write routinely over extended time frames (time for research, reflection, and revision) and shorter time frames (a single sitting or a day or two) for a range of discipline-specific tasks, purposes, and audiences.

Continued

Table 4.2 *(continued)*

CCSS ELA— **Speaking and Listening**	Comprehension and collaboration • CCSS.ELA-LITERACY.SL.3.1: Engage effectively in a range of collaborative discussions (one-on-one, in groups, and teacher-led) with diverse partners on *grade 3 topics and texts*, building on others' ideas and expressing their own clearly. o CCSS.ELA-LITERACY.SL.3.1.A: Come to discussions prepared, having read or studied required material; explicitly draw on that preparation and other information known about the topic to explore ideas under discussion. o CCSS.ELA-LITERACY.SL.3.1.B: Follow agreed-upon rules for discussions (e.g., gaining the floor in respectful ways, listening to others with care, speaking one at a time about the topics and texts under discussion). o CCSS.ELA-LITERACY.SL.3.1.C: Ask questions to check understanding of information presented, stay on topic, and link their comments to the remarks of others. o CCSS.ELA-LITERACY.SL.3.1.D: Explain their own ideas and understanding in light of the discussion. • CCSS.ELA-LITERACY.SL.3.2: Determine the main ideas and supporting details of a text read aloud or information presented in diverse media and formats, including visually, quantitatively, and orally. • CCSS.ELA-LITERACY.SL.3.3: Ask and answer questions about information from a speaker, offering appropriate elaboration and detail. Presentation of knowledge and ideas • CCSS.ELA-LITERACY.SL.3.4: Report on a topic or text, tell a story, or recount an experience with appropriate facts and relevant, descriptive details, speaking clearly at an understandable pace. • CCSS.ELA-LITERACY.SL.3.6: Speak in complete sentences when appropriate to task and situation in order to provide requested detail or clarification.
CCSS Mathematics— **Operations and Algebraic Thinking**	Solve problems involving the four operations, and identify and explain patterns in arithmetic. • CCSS.MATH.CONTENT.3.OA.D.8: Solve two-step word problems using the four operations. Represent these problems using equations with a letter standing for the unknown quantity. Assess the reasonableness of answers using mental computation and estimation strategies including rounding. • CCSS.MATH.CONTENT.3.OA.D.9: Identify arithmetic patterns (including patterns in the addition table or multiplication table), and explain them using properties of operations.
CCSS Mathematics— **Number and Operations in Base Ten**	Use place value understanding and properties of operations to perform multi-digit arithmetic. • CCSS.MATH.CONTENT.3.NBT.A.2: Fluently add and subtract within 1,000 using strategies and algorithms based on place value, properties of operations, and/or the relationship between addition and subtraction.

Continued

Table 4.2 (*continued*)

CCSS Mathematics— **Measurement and Data**	Solve problems involving measurement and estimation. • CCSS.MATH.CONTENT.3.MD.A.2: Measure and estimate liquid volumes and masses of objects using standard units of grams (g), kilograms (kg), and liters (l). Add, subtract, multiply, or divide to solve one-step word problems involving masses or volumes that are given in the same units, e.g., by using drawings (such as a beaker with a measurement scale) to represent the problem. Represent and interpret data. • CCSS.MATH.CONTENT.3.MD.B.3: Draw a scaled picture graph and a scaled bar graph to represent a data set with several categories. Solve one- and two-step "how many more" and "how many less" problems using information presented in scaled bar graphs. • CCSS.MATH.CONTENT.3.MD.B.4: Generate measurement data by measuring lengths using rulers marked with halves and fourths of an inch. Show the data by making a line plot, where the horizontal scale is marked off in appropriate units—whole numbers, halves, or quarters.
ELP Standards	Receptive modalities • ELP 1: Construct meaning from oral presentations and literary and informational text through grade-appropriate listening, reading, and viewing. • ELP 8: Determine the meaning of words and phrases in oral presentations and literary and informational text. Productive modalities • ELP 3: Speak and write about grade-appropriate complex literary and informational texts and topics. • ELP 4: Construct grade-appropriate oral and written claims and support them with reasoning and evidence. • ELP 7: Adapt language choices to purpose, task, and audience when speaking and writing. Interactive modalities • ELP 2: Participate in grade-appropriate oral and written exchanges of information, ideas, and analyses, responding to peer, audience, or reader comments and questions. • ELP 5: Conduct research and evaluate and communicate findings to answer questions or solve problems. • ELP 6: Analyze and critique the arguments of others orally and in writing. Linguistic structures of English • ELP 9: Create clear and coherent grade-appropriate speech and text. • ELP 10: Make accurate use of standard English to communicate in grade-appropriate speech and writing.

Investigation 4

Balanced and Unbalanced Forces: How Do Balanced and Unbalanced Forces Acting on an Object Affect the Motion of That Object?

Introduction

A *force* is a push or a pull. A force can cause an object to move, stop moving, or change how it is moving. Take a moment to explore what happens when you apply more than one pulling force to a cart. You can apply more than one pulling force to a cart by pulling on two strings that are attached to it at the same time. Keep track of what you observe and what you are wondering about in the boxes below.

Things I OBSERVED …	Things I WONDER about …

An object will often have more than one force acting on it at the same time. The forces acting on an object can be either balanced or unbalanced. *Balanced forces* are two forces that are the same size but are acting on the object in opposite directions. Scientists describe balanced forces as being equal and opposite. *Unbalanced forces* are not equal and opposite. One example of an unbalanced force is two different-size forces acting on an object in opposite directions. A second example of an unbalanced force is two same-size forces acting on an object in the same direction. When you try to determine if all the forces acting on an object are balanced or unbalanced, it is important to remember that forces that act in the same direction combine by addition and forces that act in opposite directions combine by subtraction.

In this investigation, your goal is to determine how the motion of an object will change when balanced and unbalanced forces act on it. You will be able to observe the relationship between the forces acting an object (a cause) and how the motion of the object changes (the effect) by keeping track of the direction a cart moves when different masses are hung from each side of it. The direction the cart moves will represent the change in motion. The amount of mass hung from each side of the cart will represent the size of the force pulling in opposite directions. You can use this method to examine the relationship between the forces acting on an object and how the motion of the object changes because the two pulling forces will be either balanced or unbalanced, depending on how much mass you decide to hang from each side of the cart.

Things we KNOW from what we read …	What we will NEED to figure out …

Your Task

Use what you know about forces, motion, patterns, and cause and effect to design and carry out an investigation to determine how a cart moves when balanced and unbalanced forces act on it.

The *guiding question* of this investigation is, ***How do balanced and unbalanced forces acting on an object affect the motion of that object?***

Materials

You may use any of the following materials during your investigation:

- Safety glasses or goggles (required)
- Cart
- 2 bench clamp pulleys
- Kite string (6 feet)
- 2 paper clips
- 10 washers

Safety Rules

Follow all normal safety rules. In addition, be sure to follow these rules:

- Wear safety glasses or goggles during setup, investigation activity, and cleanup.
- Do not throw objects or put any objects in your mouth.
- Keep your fingers and toes away from moving objects.
- Wash your hands with soap and water when you are done collecting the data.

Plan Your Investigation

Prepare a plan for your investigation by filling out the chart that follows; this plan is called an *investigation proposal*. Before you start developing your plan, be sure to discuss the following questions with the other members of your group:

- What types of **patterns** might we look for to help answer the guiding question?
- What information do we need to find a **cause-and-effect relationship?**

Our guiding question:

This is a picture of how we will set up the equipment:

We will collect the following data:

These are the steps we will follow to collect data:

I approve of this investigation proposal.

Teacher's signature

Date

Collect Your Data

Keep a record of what you measure or observe during your investigation in the space below.

Analyze Your Data

You will need to analyze the data you collected before you can develop an answer to the guiding question. In the space below, create a table or a graph.

Draft Argument

Develop an argument on a whiteboard. It should include the following parts:

1. A *claim:* Your answer to the guiding question.
2. *Evidence:* An analysis of the data and an explanation of what the analysis means.
3. A *justification of the evidence:* Why your group thinks the evidence is important.

The Guiding Question:	
Our Claim:	
Our Evidence:	Our Justification of the Evidence:

Argumentation Session

Share your argument with your classmates. Be sure to ask them how to make your draft argument better. Keep track of their suggestions in the space below.

Ways to IMPROVE our argument …

Draft Report

Prepare an *investigation report* to share what you have learned. Use the information in this handout and your group's final argument to write a *draft* of your investigation report.

Introduction

We have been studying ____ in class.

Before we started this investigation, we explored ____

____

____

We noticed ____

____

____

My goal for this investigation was to figure out ____

____

____

The guiding question was ____

____

____

Method

To gather the data I needed to answer this question, I ____

____

____

____

____

____

I then analyzed the data I collected by ____

____

____

____

Argument

My claim is ____

____

____

____

The ______________________________ below shows __________________________

____

____

____

____

____

____

This evidence is important because ____

____

____

____

____

____

____

____

____

Review

Your friends need your help! Review the draft of their investigation reports and give them ideas about how to improve. Use the *peer-review guide* when doing your review.

Submit Your Final Report

Once you have received feedback from your friends about your draft report, create your final investigation report and hand it in to your teacher.

Checkout Questions

Investigation 4. Balanced and Unbalanced Forces: How Do Balanced and Unbalanced Forces Acting on an Object Affect the Motion of That Object?

1. Shown below are five different carts. Each cart has a different amount of mass hanging from each side of it. Imagine that you are holding on to each of these carts so they cannot move. Which way will each cart move when you let go of it?

A.

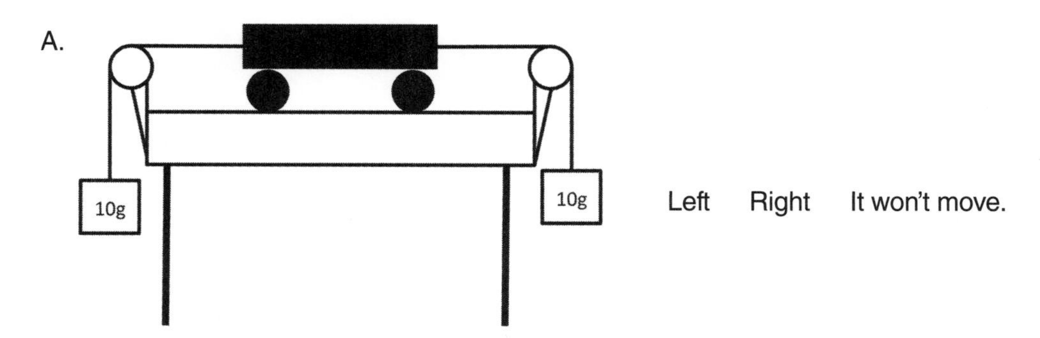

Left Right It won't move.

B.

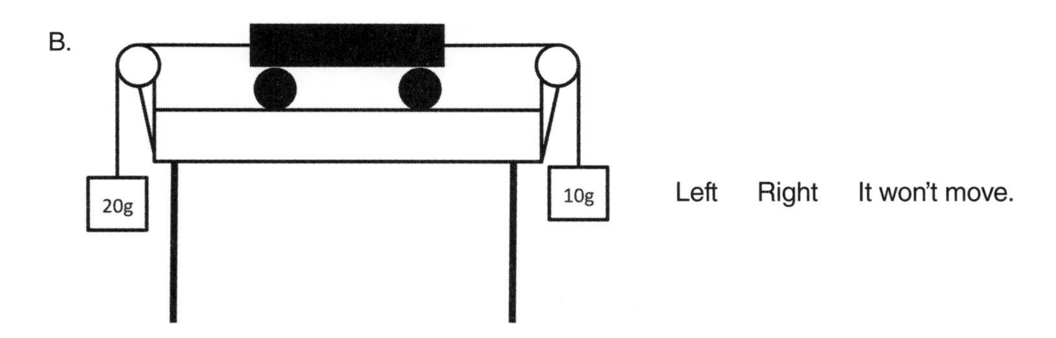

Left Right It won't move.

C.

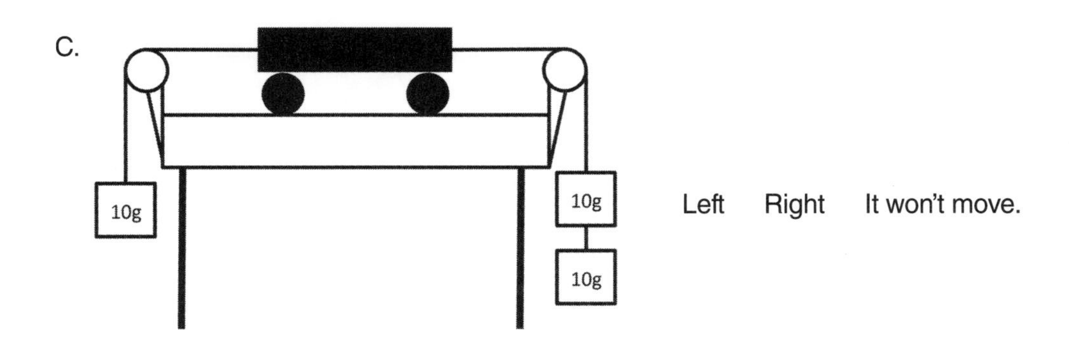

Left Right It won't move.

D.

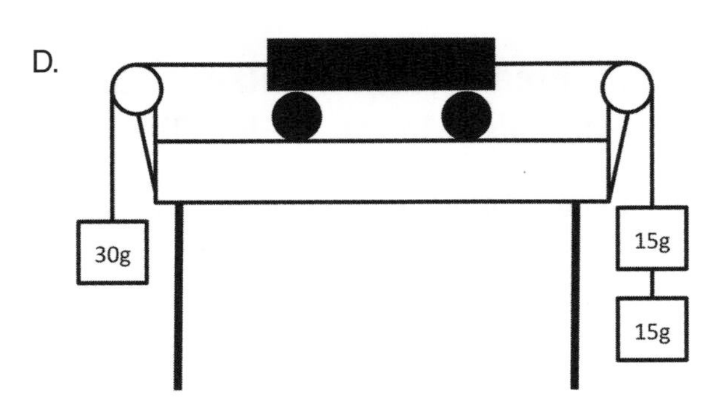

Left Right It won't move.

E.

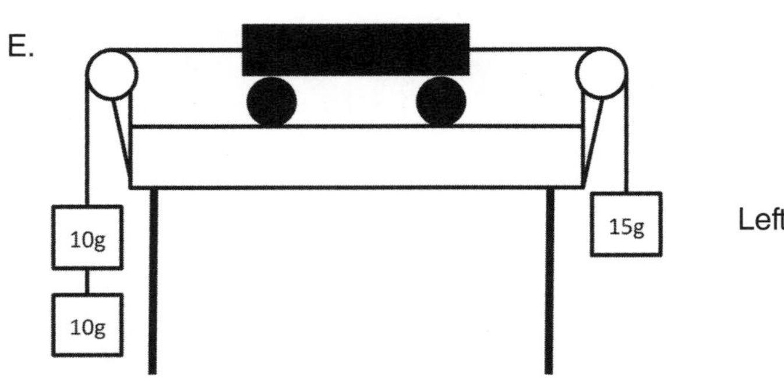

Left Right It won't move.

2. Explain your thinking. What *cause-and-effect relationship* did you use to determine which way the cart would move after you let go of it?

____

____

____

____

____

____

____

Teacher Scoring Rubric for the Checkout Questions

Level	Description
3	The student can apply the core idea correctly in all cases and can fully explain the cause-and-effect relationship.
2	The student can apply the core idea correctly in all cases but cannot fully explain the cause-and-effect relationship.
1	The student cannot apply the core idea correctly in all cases but can fully explain the cause-and-effect relationship.
0	The student cannot apply the core idea correctly in all cases and cannot explain the cause-and-effect relationship.

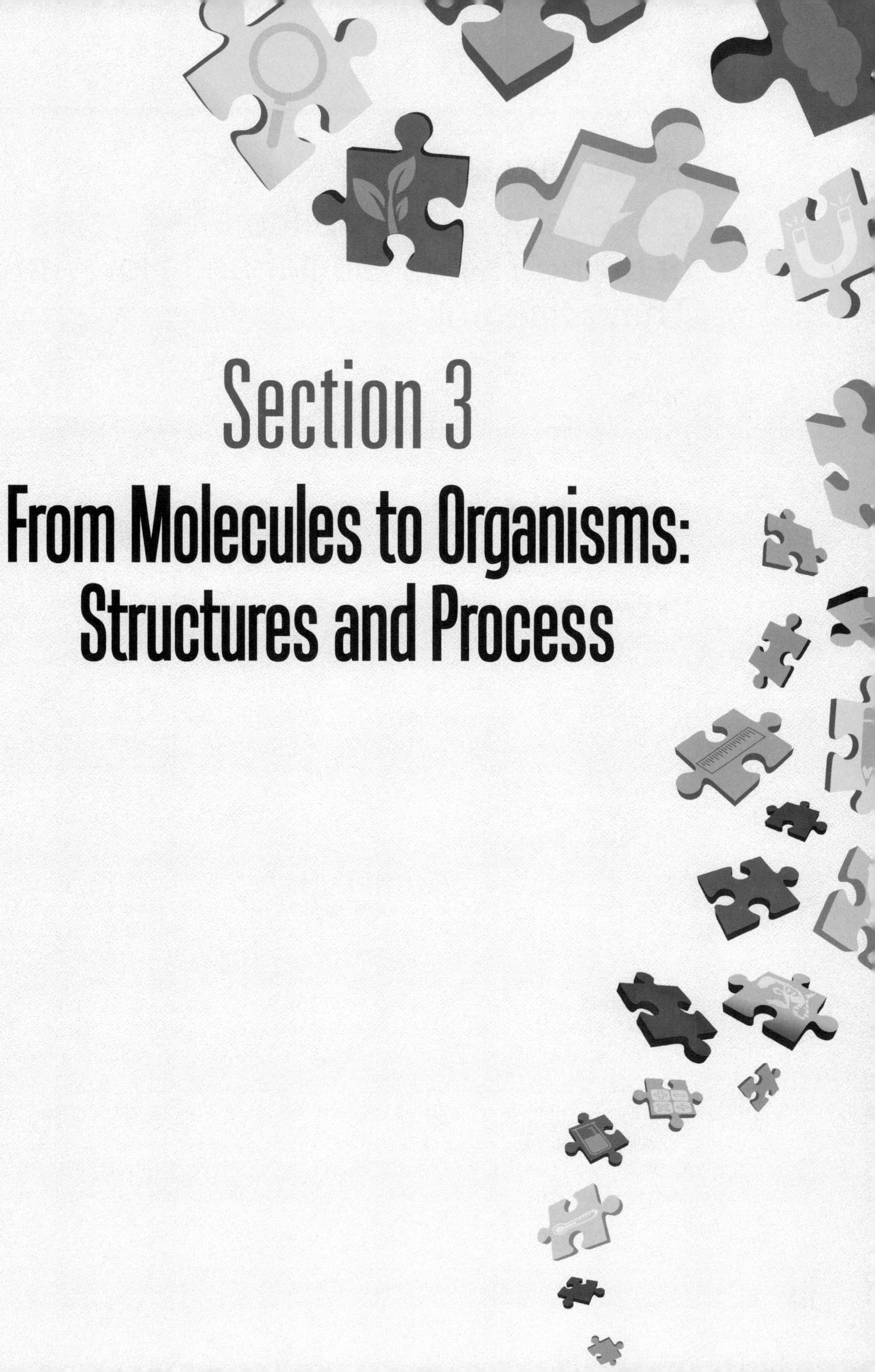

Section 3

From Molecules to Organisms: Structures and Process

Investigation 5

Life Cycles: How Are the Life Cycles of Living Things Similar and How Are They Different?

Purpose

The purpose of this investigation is to give students an opportunity to use the disciplinary core idea (DCI) of LS1.B: Growth and Development of Organisms and the crosscutting concept (CC) of Patterns from *A Framework for K–12 Science Education* (NRC 2012) to figure out the similarities and differences in the life cycles of several different organisms. Students will also learn about the types of questions scientists investigate during the reflective discussion.

The Disciplinary Core Idea

Students in third grade should understand the following about Growth and Development of Organisms and be able to use this DCI to figure out the similarities and differences in the life cycles of several different organisms:

> *Plants and animals have predictable characteristics at different stages of development. Plants and animals grow and change. Adult plants and animals can have young. In many kinds of animals, parents and the offspring themselves engage in behaviors that help the offspring to survive. (NRC 2012, p. 146)*

The Crosscutting Concept

Students in third grade should understand the following about the CC Patterns:

> *Noticing patterns is often a first step to organizing and asking scientific questions about why and how the patterns occur. One major use of pattern recognition is in classification, which depends on careful observation of similarities and differences; objects can be classified into groups on the basis of similarities of visible or microscopic features or on the basis of similarities of function. Such classification is useful in codifying relationships and organizing a multitude of objects or processes into a limited number of groups. (NRC 2012, p. 85)*

Students in third grade should also be given opportunities to "investigate the characteristics that allow classification of animal types (e.g., mammals, fish, insects), of plants (e.g., trees, shrubs, grasses), or of materials (e.g., wood, rock, metal, plastic)" (NRC 2012, p. 86).

Students should be encouraged to use their developing understanding of patterns as a tool or a way of thinking about a phenomenon during this investigation to help them figure out how the life cycles of plants and animals are similar and how they are different.

What Students Figure Out

Plants and animals have life cycles that include being born (called sprouting in plants), growing, developing into adults, reproducing, and eventually dying. How an organism grows and develops into an adult is unique to each type of living thing. Some animals, for example, go through an incomplete metamorphosis as they grow and develop into an adult (such as grasshoppers and dragonflies); others go through a complete metamorphosis (such as frogs, silkworms, and butterflies). Reproduction is also an important part of the life cycle of every kind of organism. Without reproduction, there would be no beginning to a life cycle and there would be no living things on Earth because all living things eventually die.

Timeline

The time needed to complete this investigation is 260 minutes (4 hours and 20 minutes). The amount of instructional time needed for each stage of the investigation is as follows:

- *Stage 1.* Introduce the task and the guiding question: 35 minutes
- *Stage 2.* Design a method and collect data: 50 minutes
- *Stage 3.* Create a draft argument: 35 minutes
- *Stage 4.* Argumentation session: 30 minutes
- *Stage 5.* Reflective discussion: 15 minutes
- *Stage 6.* Write a draft report: 35 minutes
- *Stage 7.* Peer review: 30 minutes
- *Stage 8.* Revise the report: 30 minutes

This investigation can be completed in one day or over eight days (one day for each stage) during your designated science time in the daily schedule.

Materials and Preparation

The materials needed for this investigation are listed in Table 5.1 (p. 184). The items can be purchased through an online retailer such as Amazon, RealBug (*www.realcoolbug.com*), or Everything Insects (*www.everythinginsects.com*). The materials for this investigation can also be purchased as a complete kit (which includes enough materials for 24 students, or six groups) at *www.argumentdriveninquiry.com*.

TABLE 5.1

Materials for Investigation 5

Item	Quantity
Bean germination (in acrylic)	1 per class
Peanut germination (in acrylic)	1 per class
Life cycle of cabbage worm / cabbage butterfly (in acrylic)	1 per class
Life cycle of dragonfly (in acrylic)	1 per class
Life cycle of fern (in acrylic)	1 per class
Life cycle of frog (in acrylic)	1 per class
Life cycle of grasshopper (in acrylic)	1 per class
Life cycle of silkworm (in acrylic)	1 per class
Whiteboard, 2' × 3'*	1 per group
Investigation Handout	1 per student
Peer-review guide and teacher scoring rubric	1 per student
Checkout Questions (optional)	1 per student

* As an alternative, students can use computer and presentation software such as Microsoft PowerPoint or Apple Keynote to create their arguments.

Be sure to use a set routine for distributing and collecting the materials. One option is to set up all the materials on a table or cart at a central location. You can then assign a member of each group to be the "materials manager." This individual is responsible for collecting all the materials his or her group needs from the table or cart during class and for returning all the materials at the end of the class. We also recommend that you have each group collect data about one organism at a time so that you need to purchase only one specimen of each plant or animal listed in Table 5.1.

Safety Precautions

Remind students to follow all normal safety rules. In addition, tell the students to take the following safety precautions:

- Do not throw objects or put any objects in their mouth.
- Wash their hands with soap and water when done collecting the data.

Lesson Plan by Stage

Stage 1: Introduce the Task and the Guiding Question (35 minutes)

1. Ask the students to sit in six groups, with three or four students in each group.

2. Ask students to clear off their desks except for a pencil (and their *Student Workbook for Argument-Driven Inquiry in Third-Grade Science* if they have one).
3. Pass out an Investigation Handout to each student (or ask students to turn to Investigation Log 5 in their workbook).
4. Read the first paragraph of the "Introduction" aloud to the class. Ask the students to follow along as you read.
5. Pass out the acrylic life cycle of one organism to each group.
6. Remind students of the safety rules and precautions for this investigation.
7. Tell students to record their observations and questions about the organism in the "OBSERVED/WONDER" chart in the "Introduction" section of their Investigation Handout (or the investigation log in their workbook).
8. Ask students to share *what they observed* about the organism.
9. Ask students to share *what questions they have* about the organism.
10. Tell the students, "Some of your questions might be answered by reading the rest of the 'Introduction.'"
11. Ask the students to read the rest of the "Introduction" on their own *or* ask them to follow along as you read aloud.
12. Once the students have read the rest of the "Introduction," ask them to fill out the "KNOW/NEED" chart on their Investigation Handout (or in their investigation log) as a group.
13. Ask students to share what they learned from the reading. Add these ideas to a class "know / need to figure out" chart.
14. Ask students to share what they think they will need to figure out based on what they read. Add these ideas to the class "know / need to figure out" chart.
15. Tell the students, "It looks like we have something to figure out. Let's see what we will need to do during our investigation."
16. Read the task and the guiding question aloud.
17. Tell the students, "I have lots of different plants or animals here that you will be able to study during your investigation."
18. Introduce the students to the organisms available for them to use during the investigation by holding each one up and then telling them the name of each one.

Stage 2: Design a Method and Collect Data (50 minutes)

1. Tell the students, "I am now going to give you and the other members of your group about 15 minutes to plan your investigation. Before you begin, I want

you all to take a couple of minutes to discuss the following questions with the rest of your group."

2. Show the following questions on the screen or board:
   - What information should we collect so we can *compare and contrast* the life cycles?
   - What types of *patterns* might we look for to help answer the guiding question?
3. Tell the students, "Please take a few minutes to come up with an answer to these questions." Give the students two or three minutes to discuss these two questions.
4. Ask two or three different groups to share their answers. Be sure to highlight or write down any important ideas on the board so students can refer to them later.
5. If possible, use a document camera to project an image of the graphic organizer for this investigation on a screen or board (or take a picture of it and project the picture on a screen or board). Tell the students, "I now want you all to plan out your investigation. To do that, you will need to create an investigation proposal by filling out this graphic organizer."
6. Point to the box labeled "Our guiding question:" and tell the students, "You can put the question we are trying to answer in this box." Then ask, "Where can we find the guiding question?"
7. Wait for a student to answer where to find the guiding question (the answer is "in the handout").
8. Point to the box labeled "We will collect the following data:" and tell the students, "You can list the measurements or observations that you will need to collect during the investigation in this box."
9. Point to the box labeled "These are the steps we will follow to collect data:" and tell the students, "You can list what you are going to do to collect the data you need and what you will do with your data once you have it. Be sure to give enough detail that I could do your investigation for you."
10. Ask the students, "Do you have any questions about what you need to do?"
11. Answer any questions that come up.
12. Tell the students, "Once you are done, raise your hand and let me know. I'll then come by and look over your proposal and give you some feedback. You may not begin collecting data until I have approved your proposal by signing it. You need to have your proposal done in the next 15 minutes."
13. Give the students 15 minutes to work in their groups on their investigation proposal. As they work, move from group to group to check in, ask probing questions, and offer a suggestion if a group gets stuck.

What should a student-designed investigation look like?

The students' investigation proposal should include the following information:

- The guiding question is "How are the life cycles of living things similar and how are they different?"
- The data that the student should collect are (1) name of organism, (2) stages of the life cycle for that organism, and (3) characteristics of the organism during each stage of its life cycle.
- To collect the data, the students could follow these steps (but this is just an example, and there should be a lot of variation in the student-designed investigation):
    1. Lay out the acrylic block(s) that shows the life cycle of an organism on the table.
    2. Count the number of stages in that organism's life cycle.
    3. Identify any changes in the organism as it moves through each stage.
    4. Repeat steps 1–3 for each organism.
    5. Identify things that make the life cycle of each organism similar and things that make them different.

14. As each group finishes its investigation proposal, be sure to read it over and determine if it will be productive or not. If you feel the investigation will be productive (not necessarily what you would do or what the other groups are doing), sign your name on the proposal and let the group start collecting data. If the plan needs to be changed, offer some suggestions or ask some probing questions, and have the group make the changes before you approve it.
15. Pass out the materials or have one student from each group collect the materials they need from a central supply table or cart for the groups that have an approved proposal.
16. Tell the students to collect their data and record their observations or measurements in the "Collect Your Data" box in their Investigation Handout (or the investigation log in their workbook).
17. Give the students 10 minutes to collect their data.
18. Be sure to collect the materials from each group before asking them to analyze their data.

Stage 3: Create a Draft Argument (35 Minutes)

1. Tell the students, "Now that we have all this data, we need to analyze the data so we can figure out an answer to the guiding question."
2. If possible, project an image of the "Analyze Your Data" section for this investigation on a screen or board using a document camera (or take a picture of it and project the picture on a screen or board). Point to the section and tell the students, "You can create a graph as a way to analyze your data. You can make your graph in this section."
3. Ask the students, "What information do we need to include in a graph?"
4. Tell the students, "Please take a few minutes to discuss this question with your group, and be ready to share."
5. Give the students five minutes to discuss.
6. Ask two or three different groups to share their answers. Be sure to highlight or write down any important ideas on the board so students can refer to them later.
7. Tell the groups of students, "I am now going to give you and the other members of your group about 10 minutes to create your graph." If the students are having trouble making a graph, you can take a few minutes to provide a minilesson about how to create a graph from a bunch of observations or measurements (this strategy is called just-in-time instruction because it is offered only when students get stuck).

What should a table or graph look like for this investigation?

There are a number of different ways that students can analyze the observations or measurements they collect during this investigation. One of the most straightforward ways is to create a bar graph. This bar graph should have the names of the different stages of a life cycle that the students identified on the horizontal or *x*-axis and the number of different types of organisms that go through this stage on the vertical or *y*-axis. An example of this type of graph can be seen in Figure 5.1 (p. 190). There are other options for analyzing the data they collected. For example, students can also provide qualitative descriptions of what they observed and then organize these observations into a figure that includes pictures. Students often come up with some unique ways of analyzing their data, so be sure to give them some voice and choice during this stage.

8. Give the students 10 minutes to analyze their data. As they work, move from group to group to check in, ask probing questions, and offer suggestions.
9. Tell the students, "I am now going to give you and the other members of your group 15 minutes to create an argument to share what you have learned and convince others that they should believe you. Before you do that, we need to take a few minutes to discuss what you need to include in your argument."
10. If possible, use a document camera to project the "Argument Presentation on a Whiteboard" image from the Investigation Handout (or the investigation log in their workbook) on a screen or board (or take a picture of it and project the picture on a screen or board).
11. Point to the box labeled "The Guiding Question:" and tell the students, "You can put the question we are trying to answer here on your whiteboard."
12. Point to the box labeled "Our Claim:" and tell the students, "You can put your claim here on your whiteboard. The claim is your answer to the guiding question."
13. Point to the box labeled "Our Evidence:" and tell the students, "You can put the evidence that you are using to support your claim here on your whiteboard. Your evidence will need to include the analysis you just did and an explanation of what your analysis means or shows. Scientists always need to support their claims with evidence."
14. Point to the box labeled "Our Justification of the Evidence:" and tell the students, "You can put your justification of your evidence here on your whiteboard. Your justification needs to explain why your evidence is important. Scientists often use core ideas to explain why the evidence they are using matters. Core ideas are important concepts that scientists use to help them make sense of what happens during an investigation."
15. Ask the students, "What are some core ideas that we read about earlier that might help us explain why the evidence we are using is important?"
16. Ask students to share some of the core ideas from the "Introduction" section of the Investigation Handout (or the investigation log in the workbook). List these core ideas on the board.
17. Tell the students, "That is great. I would like to see everyone try to include these core ideas in your justification of the evidence. Your goal is to use these core ideas to help explain why your evidence matters and why the rest of us should pay attention to it."
18. Ask the students, "Do you have any questions about what you need to do?"
19. Answer any questions that come up.
20. Tell the students, "Okay, go ahead and start working on your arguments. You need to have your argument done in the next 15 minutes. It doesn't need to

be perfect. We just need something down on the whiteboards so we can share our ideas."

21. Give the students 15 minutes to work in their groups on their arguments. As they work, move from group to group to check in, ask probing questions, and offer a suggestion if a group gets stuck. Figure 5.1 shows an example of an argument created by students for this investigation.

FIGURE 5.1

Example of an argument

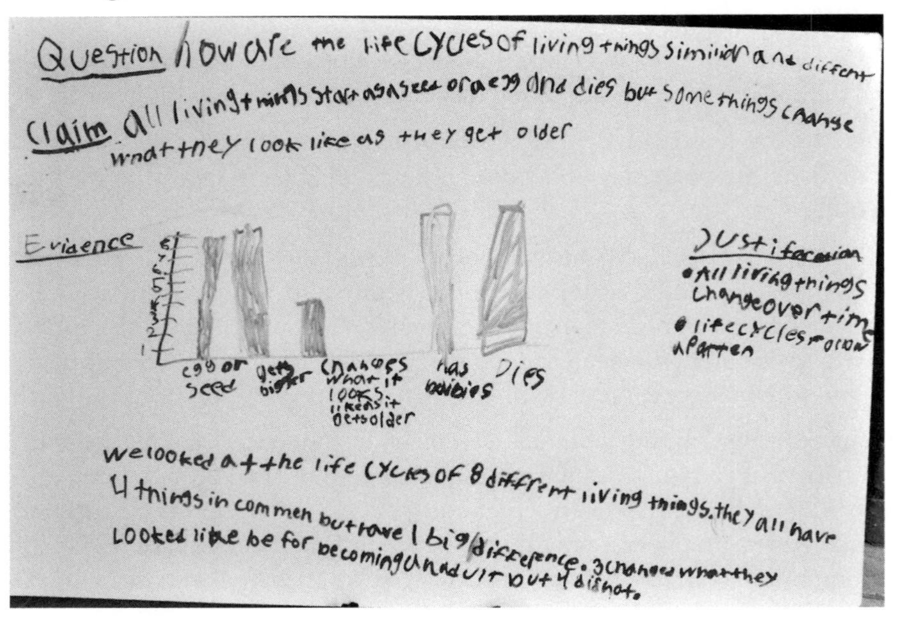

Stage 4: Argumentation Session (30 minutes)

The argumentation session can be conducted in a whole-class presentation format, a gallery walk format, or a modified gallery walk format. We recommend using a whole-class presentation format for the first investigation, but try to transition to either the gallery walk or modified gallery walk format as soon as possible because that will maximize student voice and choice inside the classroom. The following list shows the steps for the three formats; unless otherwise noted, the steps are the same for all three formats.

1. Begin by introducing the use of the whiteboard.
    - *If using the whole-class presentation format,* tell the students, "We are now going to share our arguments. Please set up your whiteboard so everyone can see them."
    - *If using the gallery walk or modified gallery walk format,* tell the students, "We are now going to share our arguments. Please set up your whiteboard so they are facing the walls."

2. Allow the students to set up their whiteboards.
   - *If using the whole-class presentation format,* the whiteboards should be set up on stands or chairs so they are facing toward the center of the room.
   - *If using the gallery walk or modified gallery walk format,* the whiteboards should be set up on stands or chairs so they are facing toward the outside of the room.
3. Give the following instructions to the students:
   - *If using the whole-class presentation format or the modified gallery walk format,* tell the students, "Okay, before we get started I want to explain what we are going to do next. I'm going to ask some of you to present your arguments to your classmates. If you are presenting your argument, your job is to share your group's claim, evidence, and justification of the evidence. The rest of you will be reviewers. If you are a reviewer, your job is to listen to the presenters, ask the presenters questions if you do not understand something, and then offer them some suggestions about ways to make their argument better. After we have a chance to learn from each other, I'm going to give you some time to revise your arguments and make them better."
   - *If using the gallery walk format,* tell the students, "Okay, before we get started I want to explain what we are going to do next. You are going to have an opportunity to read the arguments that were created by other groups. Your group will go to a different group's argument. I'll give you a few minutes to read it and review it. Your job is to offer them some suggestions about ways to make their argument better. You can use sticky notes to give them suggestions. Please be specific about what you want to change and be specific about how you think they should change it. After we have a chance to learn from each other, I'm going to give you some time to revise your arguments and make them better."
4. Use a document camera to project the "Ways to IMPROVE our argument …" box from the Investigation Handout (or the investigation log in their workbook) on a screen or board (or take a picture of it and project the picture on a screen or board).
   - *If using the whole-class presentation format or the modified gallery walk format,* point to the box and tell the students, "If you are a presenter, you can write down the suggestions you get from the reviewers here. If you are a reviewer, and you see a good idea from another group, you can write down that idea here. Once we are done with the presentations, I will give you a chance to use these suggestions or ideas to improve your arguments.
   - *If using the gallery walk format,* point to the box and tell the students, "If you see good ideas from another group, you can write them down here. Once we are done reviewing the different arguments, I will give you a chance to use these ideas to improve your own arguments. It is important to share ideas like this."

5. Ask the students, "Do you have any questions about what you need to do?"
6. Answer any questions that come up.
7. Give the following instructions:
   - *If using the whole-class presentation format,* tell the students, "Okay. Let's get started."
   - *If using the gallery walk format,* tell the students, "Okay, I'm now going to tell you which argument to go to and review.
   - *If using the modified gallery walk format,* tell the students, "Okay, I'm now going to assign you to be a presenter or a reviewer." Assign one or two students from each group to be presenters and one or two students from each group to be reviewers.
8. Begin the review of the arguments.
   - *If using the whole-class presentation format,* have four or five groups present their argument one at a time. Give each group only two to three minutes to present their argument. Then give the class two to three minutes to ask them questions and offer suggestions. Be sure to encourage as much participation from the students as possible.
   - *If using the gallery walk format,* tell the students, "Okay. Let's get started. Each group, move one argument to the left. Don't move to the next argument until I tell you to move. Once you get there, read the argument and then offer suggestions about how to make it better. I will put some sticky notes next to each argument. You can use the sticky notes to leave your suggestions." Give each group about three to four minutes to read the arguments, talk, and offer suggestions.

     a. Tell the students, "Okay. Let's rotate. Move one group to the left."
     b. Again, give each group three or four minutes to read, talk, and offer suggestions.
     c. Repeat this process for two more rotations.
   - *If using the modified gallery walk format,* tell the students, "Okay. Let's get started. Reviewers, move one group to the left. Don't move to the next group until I tell you to move. Presenters, go ahead and share your argument with the reviewers when they get there." Give each group of presenters and reviewers about three to four minutes to talk.

     a. Tell the students, "Okay. Let's rotate. Reviewers, move one group to the left."
     b. Again, give each group of presenters and reviewers about three or four minutes to talk.
     c. Repeat this process for two more rotations.
9. Tell the students to return to their workstations.

10. Give the following instructions about revising the argument:
    - *If using the whole-class presentation format,* tell the students, "I'm now going to give you about 10 minutes to revise your argument. Take a few minutes to talk in your groups and determine what you want to change to make your argument better. Once you have decided what to change, go ahead and make the changes to your whiteboard."
    - *If using the gallery walk format,* tell the students, "I'm now going to give you about 10 minutes to revise your argument. Take a few minutes to read the suggestions that were left at your argument. Then talk in your groups and determine what you want to change to make your argument better. Once you have decided what to change, go ahead and make the changes to your whiteboard."
    - *If using the modified gallery walk format,* "I'm now going to give you about 10 minutes to revise your argument. Please return to your original groups." Wait for the students to move back into their original groups and then tell the students, "Okay, take a few minutes to talk in your groups and determine what you want to change to make your argument better. Once you have decided what to change, go ahead and make the changes to your whiteboard."
11. Ask the students, "Do you have any questions about what you need to do?"
12. Answer any questions that come up.
13. Tell the students, "Okay. Let's get started."
14. Give the students 10 minutes to work in their groups on their arguments. As they work, move from group to group to check in, ask probing questions, and offer a suggestion if a group gets stuck.

Stage 5: Reflective Discussion (15 minutes)

1. Tell the students, "We are now going to take a minute to talk about what we did and what we have learned."
2. Show Figure 5.2 (p. 194) on the screen. This image shows the life cycles of a butterfly and a beetle. In both life cycles, the insect begins as an egg. In a life cycle that includes a complete metamorphosis, the insect passes through four distinct phases, which produce an adult that does not resemble the larvae. In a life cycle that includes an incomplete metamorphosis, an insect does not go through a full transformation but instead transitions from a nymph to an adult by molting its exoskeleton whenever it becomes too tight.
3. Ask the students, "What do you all see going on here?"
4. Allow students to share their ideas.
5. Ask the students, "What is similar about these two life cycles?"

FIGURE 5.2

Life cycles of a butterfly and a beetle

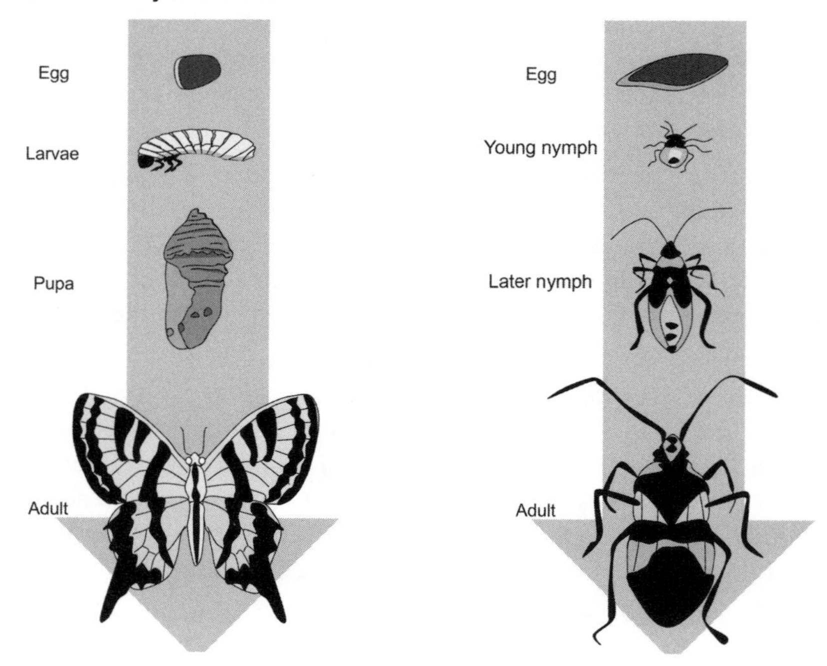

A full-color version of this figure is available on the book's Extras page at *www.nsta.org/adi-3rd*.

6. Allow students to share their ideas.
7. Ask the students, "What is different about these two life cycles?"
8. Allow students to share their ideas.
9. Ask the students, "What is missing from the images of life cycles for these two animals?"
10. Allow students to share their ideas. Ask probing questions until someone mentions reproduction.
11. Tell the students, "Okay, let's make sure we are on the same page. Plants and animals have life cycles that include being born (called sprouting in plants), growing, developing into adults, reproducing, and eventually dying. How an organism grows and develops into an adult is unique to each type of living thing. Some animals, for example, go through an incomplete metamorphosis as they grow and develop into an adult like beetles do [point to the image of the beetle], and some go through a complete metamorphosis like butterflies do [point to the image of the butterfly]. Reproduction is also an important part of the life cycle of every kind of organism. Without reproduction, there would be no beginning to a life cycle and there would be no living things on Earth

because all living things eventually die. The fact that living things have life cycles is a really important core idea in science."

12. Ask the students, "Does anyone have any questions about this core idea?"
13. Answer any questions that come up.
14. Tell the students, "We also looked for patterns during our investigation." Then ask, "Can anyone tell me why we needed to look for patterns?"
15. Allow students to share their ideas.
16. Tell the students, "Patterns are really important in science. Scientists look for patterns all the time. In fact, they even use patterns to help classify different things just like we did."
17. Tell the students, "We are now going take a minute to talk about what went well and what didn't go so well during our investigation. We need to talk about this because you are going to be planning and carrying out your own investigations like this a lot this year, and I want to help you all get better at it."
18. Show an image of the question "What made your investigation scientific?" on the screen. Tell the students, "Take a few minutes to talk about how you would answer this question with the other people in your group. Be ready to share with the rest of the class." Give the students two to three minutes to talk in their group.
19. Ask the students, "What do you all think? Who would like to share an idea?"
20. Allow students to share their ideas. Be sure to expand on their ideas about what makes an investigation scientific.
21. Show an image of the question "What made your investigation not so scientific?" on the screen. Tell the students, "Take a few minutes to talk about how you would answer this question with the other people in your group. Be ready to share with the rest of the class." Give the students two to three minutes to talk in their group.
22. Ask the students, "What do you all think? Who would like to share an idea?"
23. Allow students to share their ideas. Be sure to expand on their ideas about what makes an investigation less scientific.
24. Show an image of the question "What rules can we put into place to help us make sure our next investigation is more scientific?" on the screen. Tell the students, "Take a few minutes to talk about how you would answer this question with the other people in your group. Be ready to share with the rest of the class." Give the students two to three minutes to talk in their group.
25. Ask the students, "What do you all think? Who would like to share an idea?"
26. Allow students to share their ideas. Once they have shared their ideas, offer a suggestion for a possible class rule.

27. Ask the students, "What do you all think? Should we make this a rule?"
28. If the students agree, write the rule on the board or make a class "Rules for Scientific Investigation" chart so you can refer to it during the next investigation.
29. Tell the students, "We are now going take a minute to talk about what makes science different from other subjects."
30. Show an image of the question "What types of questions do scientists investigate?" on the screen. Tell the students, "Take a few minutes to talk about how you would answer this question with the other people in your group. Be ready to share with the rest of the class." Give the students two to three minutes to talk in their group.
31. Ask the students, "What do you all think? Who would like to share an idea?"
32. Allow students to share their ideas.
33. Tell the students, "Okay, let's make sure we are all on the same page. Scientists ask questions about how the natural world works. They don't ask questions about what people like or don't like or what is good or bad."
34. Show Figure 5.3 and the question "Is this a beautiful picture?" on the screen.
35. Ask the students, "Is this a scientific question? Why or why not?"
36. Allow students to share their ideas.

FIGURE 5.3

A sunset at Ocean Beach in San Francisco, California

A full-color version of this figure is available on the book's Extras page at *www.nsta.org/adi-3rd*.

37. Tell the students, "This is not a scientific question because it is asking for a judgment."
38. Show Figure 5.3 again on the screen along with the question "Why does the sky turn red at sunset?"
39. Ask the students, "Is this a scientific question? Why or why not?"
40. Allow students to share their ideas.
41. Tell the students, "This is a scientific question because it is asking for an explanation about how or why something in the natural world happens."
42. Ask the students, "Does anyone have any questions about what types of questions scientists do and do not ask?"
43. Answer any questions that come up.

Stage 6: Write a Draft Report (35 minutes)

Your students will use either the Investigation Handout or the investigation log in the student workbook when writing the draft report. When you give the directions shown in quotes in the following steps, substitute "investigation log" (as shown in brackets) for "handout" if they are using the workbook.

1. Tell the students, "You are now going to write an investigation report to share what you have learned. Please take out a pencil and turn to the 'Draft Report' section of your handout [investigation log]."
2. If possible, use a document camera to project the "Introduction" section of the draft report from the Investigation Handout (or the investigation log in their workbook) on a screen or board (or take a picture of it and project the picture on a screen or board).
3. Tell the students, "The first part of the report is called the 'Introduction.' In this section of the report you want to explain to the reader what you were investigating, why you were investigating it, and what question you were trying to answer. All of this information can be found in the text at the beginning of your handout [investigation log]." Point to the image and say, "There are some sentence starters here to help you begin writing the report." Ask the students, "Do you have any questions about what you need to do?"
4. Answer any questions that come up.
5. Tell the students, "Okay. Let's write."
6. Give the students 10 minutes to write the "Introduction" section of the report. As they work, move from student to student to check in, ask probing questions, and offer a suggestion if a student gets stuck.
7. If possible, use a document camera to project the "Method" section of the draft report from the Investigation Handout (or the investigation log in their

workbook) on a screen or board (or take a picture of it and project the picture on a screen or board).

8. Tell the students, "The second part of the report is called the 'Method.' In this section of the report you want to explain to the reader what you did during the investigation, what data you collected and why, and how you went about analyzing your data. All of this information can be found in the 'Plan Your Investigation' section of your handout [investigation log]. Remember that you all planned and carried out different investigations, so do not assume that the reader will know what you did." Point to the image and say, "There are some sentence starters here to help you begin writing this part of the report." Ask the students, "Do you have any questions about what you need to do?"
9. Answer any questions that come up.
10. Tell the students, "Okay. Let's write."
11. Give the students 10 minutes to write the "Method" section of the report. As they work, move from student to student to check in, ask probing questions, and offer a suggestion if a student gets stuck.
12. If possible, use a document camera to project the "Argument" section of the draft report from the Investigation Handout (or the investigation log in their workbook) on a screen or board (or take a picture of it and project the picture on a screen or board).
13. Tell the students, "The last part of the report is called the 'Argument.' In this section of the report you want to share your claim, evidence, and justification of the evidence with the reader. All of this information can be found on your whiteboard." Point to the image and say, "There are some sentence starters here to help you begin writing this part of the report." Ask the students, "Do you have any questions about what you need to do?"
14. Answer any questions that come up.
15. Tell the students, "Okay. Let's write."
16. Give the students 10 minutes to write the "Argument" section of the report. As they work, move from student to student to check in, ask probing questions, and offer a suggestion if a student gets stuck.

Stage 7: Peer Review (30 minutes)

Your students will use either the Investigation Handout or the investigation log in the student workbook when doing the peer review. When you give the directions shown in quotes in the following steps, substitute "workbook" (as shown in brackets) for "Investigation Handout" if they are using the workbook.

1. Tell the students, "We are now going to review our reports to find ways to make them better. I'm going to come around and collect your Investigation Handout [workbook]. While I do that, please take out a pencil."
2. Collect the Investigation Handouts or workbooks from the students.
3. If possible, use a document camera to project the peer-review guide (PRG; see Appendix 4) on a screen or board (or take a picture of it and project the picture on a screen or board).
4. Tell the students, "We are going to use this peer-review guide to give each other feedback." Point to the image.
5. Give the following instructions:
   - *If using the Investigation Handout,* tell the students, "I'm going to ask you to work with a partner to do this. I'm going to give you and your partner a draft report to read and a peer-review guide to fill out. You two will then read the report together. Once you are done reading the report, I want you to answer each of the questions on the peer-review guide." Point to the review questions on the image of the PRG.
   - *If using the workbook,* tell the students, "I'm going to ask you to work with a partner to do this. I'm going to give you and your partner a draft report to read. You two will then read the report together. Once you are done reading the report, I want you to answer each of the questions on the peer-review guide that is right after the report in the investigation log." Point to the review questions on the image of the PRG.
6. Tell the students, "You can check 'yes,' 'almost,' or 'no' after each question." Point to the checkboxes on the image of the PRG.
7. Tell the students, "This will be your rating for this part of the report. Make sure you agree on the rating you give the author. If you mark 'almost' or 'no,' then you need to tell the author what he or she needs to do to get a 'yes.'" Point to the space for the reviewer feedback on the image of the PRG.
8. Tell the students, "It is really important for you to give the authors feedback that is helpful. That means you need to tell them exactly what they need to do to make their reports better." Ask the students, "Do you have any questions about what you need to do?"
9. Answer any questions that come up.
10. Tell the students, "Please sit with a partner who is not in your current group." Allow the students time to sit with a partner.
11. Give the following instructions:
    - *If using the Investigation Handout,* tell the students, "Okay, I am now going to give you one report to read and one peer-review guide to fill out." Pass out one

report to each pair. Make sure that the report you give a pair was not written by one of the students in that pair. Give each pair one PRG to fill out as a team.

- *If using the workbook,* tell the students, "Okay, I am now going to give you one report to read." Pass out a workbook to each pair. Make sure that the workbook you give a pair is not from one of the students in that pair.

12. Tell the students, "Okay, I'm going to give you 15 minutes to read the report I gave you and to fill out the peer-review guide. Go ahead and get started."
13. Give the students 15 minutes to work. Be sure to move around from pair to pair to check in and see how things are going, answer questions, and offer advice.
14. After 15 minutes pass, tell the students, "Okay, time is up." *If using the Investigation Handout,* say, "Please give me the report and the peer-review guide that you filled out." *If using the workbook,* say, "Please give me the workbook that you have."
15. Collect the Investigation Handouts and the PRGs, or collect the workbooks if they are being used. Be sure you keep the handout and the PRG together.
16. Give the following instructions:
    - *If using the Investigation Handout,* tell the students, "Okay, I am now going to give you a different report to read and a new peer-review guide to fill out." Pass out another report to each pair. Make sure that this report was not written by one of the students in that pair. Give each pair a new PRG to fill out as a team.
    - *If using the workbook,* tell the students, "Okay, I am now going to give you a different report to read." Pass out a different workbook to each pair. Make sure that the workbook you give a pair is not from one of the students in that pair.
17. Tell the students, "Okay, I'm going to give you 15 minutes to read this new report and to fill out the peer-review guide. Go ahead and get started."
18. Give the students 15 minutes to work. As they work, move around from pair to pair to check in and see how things are going, answer questions, and offer advice.
19. After 15 minutes pass, tell the students, "Okay, time is up." *If using the Investigation Handout,* say, "Please give me the report and the peer-review guide that you filled out." *If using the workbook,* say, "Please give me the workbook that you have."
20. Collect the Investigation Handouts and the PRGs, or collect the workbooks if they are being used. Be sure you keep the handout and the PRG together.

Stage 8: Revise the Report (30 minutes)

Your students will use either the Investigation Handout or the investigation log in the student workbook when revising the report. Except where noted below, the directions are the same whether using the handout or the log.

1. Give the following instructions:
   - *If using the Investigation Handout,* tell the students, "You are now going to revise your investigation report based on the feedback you get from your classmates. Please take out a pencil while I hand back your draft report and the peer-review guide."
   - *If using the investigation log in the student workbook,* tell the students, "You are now going to revise your investigation report based on the feedback you get from your classmates. Please take out a pencil while I hand back your investigation logs."
2. *If using the Investigation Handout,* pass back the handout and the PRG to each student. *If using the investigation log,* pass back the log to each student.
3. Tell the students, "Please take a few minutes to read over the peer-review guide. You should use it to figure out what you need to change in your report and how you will change the report."
4. Allow the students time to read the PRG.
5. *If using the investigation log,* if possible use a document camera to project the "Write Your Final Report" section from the investigation log on a screen or board (or take a picture of it and project the picture on a screen or board).
6. Give the following instructions:
   - *If using the Investigation Handout,* tell the students, "Okay. Let's revise our reports. Please take out a piece of paper. I would like you to rewrite your report. You can use your draft report as a starting point, but use the feedback on the peer-review guide to help make it better."
   - *If using the investigation log,* tell the students, "Okay. Let's revise our reports. I would like you to rewrite your report in the section of the investigation log that says 'Write Your Final Report.'" Point to the image on the screen and tell the students, "You can use your draft report as a starting point, but use the feedback on the peer-review guide to help make your report better."

   Ask the students, "Do you have any questions about what you need to do?"
7. Answer any questions that come up.
8. Tell the students, "Okay. Let's write."
9. Give the students 30 minutes to rewrite their report. As they work, move from student to student to check in, ask probing questions, and offer a suggestion if a student gets stuck.
10. Give the following instructions:
    - *If using the Investigation Handout,* tell the students, "Okay. Time's up. I will now come around and collect your Investigation Handout, peer-review guide, and your final report."

- *If using the investigation log,* tell the students, "Okay. Time's up. I will now come around and collect your workbooks."

11. *If using the Investigation Handout,* collect all the Investigation Handouts, PRGs, and final reports. *If using the investigation log,* collect all the workbooks.
12. *If using the Investigation Handout,* use the "Teacher Score" columns in the PRG to grade the final report. *If using the investigation log,* use the "ADI Investigation Report Grading Rubric" in the investigation log to grade the final report. Whether you are using the handout or the log, you can give the students feedback about their writing in the "Teacher Comments" section.

How to Use the Checkout Questions

The Checkout Questions are an optional assessment. We recommend giving them to students one day after they finish stage 8 of the ADI investigation. The Checkout Questions can be used as a formative or summative assessment of student thinking. If you plan to use them as a formative assessment, we recommend that you look over the student answers to determine if you need to reteach the core idea and/or crosscutting concept from the investigation, but do not grade them. If you plan to use them as a summative assessment, we have included a "Teacher Scoring Rubric" at the end of the Checkout Questions that you can use to score a student's ability to apply the core idea in a new scenario and explain their use of a crosscutting concept. The rubric includes a 4-point scale that ranges from 0 (the student cannot apply the core idea correctly in all cases and cannot explain the [crosscutting concept]) to 3 (the student can apply the core idea correctly in all cases and can fully explain the [crosscutting concept]). The Checkout Questions, regardless of how you decide to use them, are a great way to make student thinking visible so you can determine if the students have learned the core idea and the crosscutting concept.

A student who can apply the core idea correctly in all cases and can explain the pattern would check all the boxes for questions 1–4 and check the boxes next to silkworm, frog, and butterfly for question 5. He or she should then be able to explain in question 6 that all living things are born, grow, reproduce, and die, but only some completely change what they look like at some point in their life cycle.

Connections to Standards

Table 5.2 highlights how the investigation can be used to address specific performance expectations from the *NGSS, Common Core State Standards (CCSS)* in English language arts (ELA) and in mathematics, and *English Language Proficiency (ELP) Standards.*

TABLE 5.2

Investigation 5 alignment with standards

***NGSS* performance expectation**	Strong alignment • 3-LS1-1: Develop models to describe that organisms have unique and diverse life cycles but all have in common birth, growth, reproduction, and death.
***CCSS ELA*—Reading: Informational Text**	Key ideas and details • CCSS.ELA-LITERACY.RI.3.1: Ask and answer questions to demonstrate understanding of a text, referring explicitly to the text as the basis for the answers. • CCSS.ELA-LITERACY.RI.3.2: Determine the main idea of a text; recount the key details and explain how they support the main idea. • CCSS.ELA-LITERACY.RI.3.3: Describe the relationship between a series of historical events, scientific ideas or concepts, or steps in technical procedures in a text, using language that pertains to time, sequence, and cause/effect. Craft and structure • CCSS.ELA-LITERACY.RI.3.4: Determine the meaning of general academic and domain-specific words and phrases in a text relevant to a *grade 3 topic or subject area.* • CCSS.ELA-LITERACY.RI.3.5: Use text features and search tools (e.g., key words, sidebars, hyperlinks) to locate information relevant to a given topic efficiently. • CCSS.ELA-LITERACY.RI.3.6: Distinguish their own point of view from that of the author of a text. Integration of knowledge and ideas • CCSS.ELA-LITERACY.RI.3.7: Use information gained from illustrations (e.g., maps, photographs) and the words in a text to demonstrate understanding of the text (e.g., where, when, why, and how key events occur). • CCSS.ELA-LITERACY.RI.3.8: Describe the logical connection between particular sentences and paragraphs in a text (e.g., comparison, cause/effect, first/second/third in a sequence). • CCSS.ELA-LITERACY.RI.3.9: Compare and contrast the most important points and key details presented in two texts on the same topic. Range of reading and level of text complexity • CCSS.ELA-LITERACY.RI.3.10: By the end of the year, read and comprehend informational texts, including history/social studies, science, and technical texts, at the high end of the grades 2–3 text complexity band independently and proficiently.

Continued

Table 5.2 *(continued)*

***CCSS ELA*—Writing**	Text types and purposes
	• CCSS.ELA-LITERACY.W.3.1: Write opinion pieces on topics or texts, supporting a point of view with reasons.
	  ○ CCSS.ELA-LITERACY.W.3.1.A: Introduce the topic or text they are writing about, state an opinion, and create an organizational structure that lists reasons.
	  ○ CCSS.ELA-LITERACY.W.3.1.B: Provide reasons that support the opinion.
	  ○ CCSS.ELA-LITERACY.W.3.1.C: Use linking words and phrases (e.g., *because*, *therefore*, *since, for example*) to connect opinion and reasons.
	  ○ CCSS.ELA-LITERACY.W.3.1.D: Provide a concluding statement or section.
	• CCSS.ELA-LITERACY.W.3.2: Write informative or explanatory texts to examine a topic and convey ideas and information clearly.
	  ○ CCSS.ELA-LITERACY.W.3.2.A: Introduce a topic and group related information together; include illustrations when useful to aiding comprehension.
	  ○ CCSS.ELA-LITERACY.W.3.2.B: Develop the topic with facts, definitions, and details.
	  ○ CCSS.ELA-LITERACY.W.3.2.C: Use linking words and phrases (e.g., *also*, *another*, *and*, *more*, *but*) to connect ideas within categories of information.
	  ○ CCSS.ELA-LITERACY.W.3.2.D: Provide a concluding statement or section.
	Production and distribution of writing
	• CCSS.ELA-LITERACY.W.3.4: With guidance and support from adults, produce writing in which the development and organization are appropriate to task and purpose.
	• CCSS.ELA-LITERACY.W.3.5: With guidance and support from peers and adults, develop and strengthen writing as needed by planning, revising, and editing.
	• CCSS.ELA-LITERACY.W.3.6: With guidance and support from adults, use technology to produce and publish writing (using keyboarding skills) as well as to interact and collaborate with others.
	Research to build and present knowledge
	• CCSS.ELA-LITERACY.W.3.8: Recall information from experiences or gather information from print and digital sources; take brief notes on sources and sort evidence into provided categories.
	Range of writing
	• CCSS.ELA-LITERACY.W.3.10: Write routinely over extended time frames (time for research, reflection, and revision) and shorter time frames (a single sitting or a day or two) for a range of discipline-specific tasks, purposes, and audiences.

Continued

Table 5.2 *(continued)*

CCSS ELA— **Speaking and Listening**	Comprehension and collaboration • CCSS.ELA-LITERACY.SL.3.1: Engage effectively in a range of collaborative discussions (one-on-one, in groups, and teacher-led) with diverse partners on *grade 3 topics and texts*, building on others' ideas and expressing their own clearly. o CCSS.ELA-LITERACY.SL.3.1.A: Come to discussions prepared, having read or studied required material; explicitly draw on that preparation and other information known about the topic to explore ideas under discussion. o CCSS.ELA-LITERACY.SL.3.1.B: Follow agreed-upon rules for discussions (e.g., gaining the floor in respectful ways, listening to others with care, speaking one at a time about the topics and texts under discussion). o CCSS.ELA-LITERACY.SL.3.1.C: Ask questions to check understanding of information presented, stay on topic, and link their comments to the remarks of others. o CCSS.ELA-LITERACY.SL.3.1.D: Explain their own ideas and understanding in light of the discussion. • CCSS.ELA-LITERACY.SL.3.2: Determine the main ideas and supporting details of a text read aloud or information presented in diverse media and formats, including visually, quantitatively, and orally. • CCSS.ELA-LITERACY.SL.3.3: Ask and answer questions about information from a speaker, offering appropriate elaboration and detail. Presentation of knowledge and ideas • CCSS.ELA-LITERACY.SL.3.4: Report on a topic or text, tell a story, or recount an experience with appropriate facts and relevant, descriptive details, speaking clearly at an understandable pace. • CCSS.ELA-LITERACY.SL.3.6: Speak in complete sentences when appropriate to task and situation in order to provide requested detail or clarification.
CCSS Mathematics— **Measurement and Data**	Represent and interpret data • CCSS.MATH.CONTENT.3.MD.B.3: Draw a scaled picture graph and a scaled bar graph to represent a data set with several categories. Solve one- and two-step "how many more" and "how many less" problems using information presented in scaled bar graphs.

Continued

Table 5.2 *(continued)*

ELP Standards	Receptive modalities • ELP 1: Construct meaning from oral presentations and literary and informational text through grade-appropriate listening, reading, and viewing. • ELP 8: Determine the meaning of words and phrases in oral presentations and literary and informational text. Productive modalities • ELP 3: Speak and write about grade-appropriate complex literary and informational texts and topics. • ELP 4: Construct grade-appropriate oral and written claims and support them with reasoning and evidence. • ELP 7: Adapt language choices to purpose, task, and audience when speaking and writing. Interactive modalities • ELP 2: Participate in grade-appropriate oral and written exchanges of information, ideas, and analyses, responding to peer, audience, or reader comments and questions. • ELP 5: Conduct research and evaluate and communicate findings to answer questions or solve problems. • ELP 6: Analyze and critique the arguments of others orally and in writing. Linguistic structures of English • ELP 9: Create clear and coherent grade-appropriate speech and text. • ELP 10: Make accurate use of standard English to communicate in grade-appropriate speech and writing.

Investigation 5

Life Cycles: How Are the Life Cycles of Living Things Similar and How Are They Different?

Introduction

All living things change over time. A tree starts life as a seed, tadpoles become frogs, and puppies turn into dogs. Take a few minutes to examine how a living thing changes over time as it ages. Keep track of what you observe and what you are wondering about in the boxes below.

A living thing will follow a predictable pattern of change as it ages. Scientists call this pattern of change a *life cycle.* A life cycle is a series of stages or events that a living thing will go through during its life. Every type of living thing has a specific life cycle.

Think about the life of bird as an example. A bird starts its life within an egg. When a bird hatches from its egg, it is called a hatchling. While a bird lives in a nest and is fed by its parents, it is called a nestling. When a young bird develops its first flight feathers, it is called a fledgling. A young bird

that can fly but is not ready to have offspring (another word for "babies") yet is called a juvenile. A full-grown bird that is ready to have offspring is called an adult. The stage of a bird's life that involves mating and then caring for eggs, hatchlings, and fledglings is called parenthood. After parenthood, a bird will grow old and die. All birds go through these same stages of life. This series of stages of growth and development is the unique life cycle of birds. Other living things, such as butterflies, frogs, and corn, will also follow a predictable pattern of growth and development as they age.

In this investigation you will observe the life cycles of several different kinds of animals and several different kinds of plants. Your goal is to determine what makes the life cycles of these living things similar and what makes them different. To accomplish this goal, you will need to look for patterns in the way different types of living things grow and develop over time. Scientists often look for patterns in nature like this and then use these patterns to classify living things into groups. You can therefore use patterns about how living things grow and develop over time to help determine what is similar about life cycles and what makes them unique.

Things we KNOW from what we read …	What we will NEED to figure out …

Your Task

Use what you know about plants, animals, and patterns to design and carry out an investigation to learn more about the life cycles of different plants and animals.

The *guiding question* of this investigation is, **How are the life cycles of living things similar and how are they different?**

Materials

You may use any of the following materials during your investigation:

- Bean germination in acrylic
- Peanut germination in acrylic
- Life cycle of cabbage worm / cabbage butterfly in acrylic
- Life cycle of dragonfly in acrylic
- Life cycle of fern in acrylic
- Life cycle of frog in acrylic
- Life cycle of grasshopper in acrylic
- Life cycle of silkworm in acrylic

Safety Rules

Follow all normal safety rules. In addition, be sure to follow these rules:

- Do not throw objects or put any objects in your mouth.
- Wash your hands with soap and water when done collecting the data.

Plan Your Investigation

Prepare a plan for your investigation by filling out the chart that follows; this plan is called an *investigation proposal.* Before you start developing your plan, be sure to discuss the following questions with the other members of your group:

- What information should we collect so we can **compare and contrast** the life cycles?
- What types of **patterns** might we look for to help answer the guiding question?

Our guiding question:

We will collect the following data:

These are the steps we will follow to collect data:

I approve of this investigation proposal.

Teacher's signature

Date

Collect Your Data

Keep a record of what you measure or observe during your investigation in the space below.

Analyze Your Data

You will need to analyze the data you collected before you can develop an answer to the guiding question. In the space below, you can create a graph that shows how many organisms have the same stage of a life cycle.

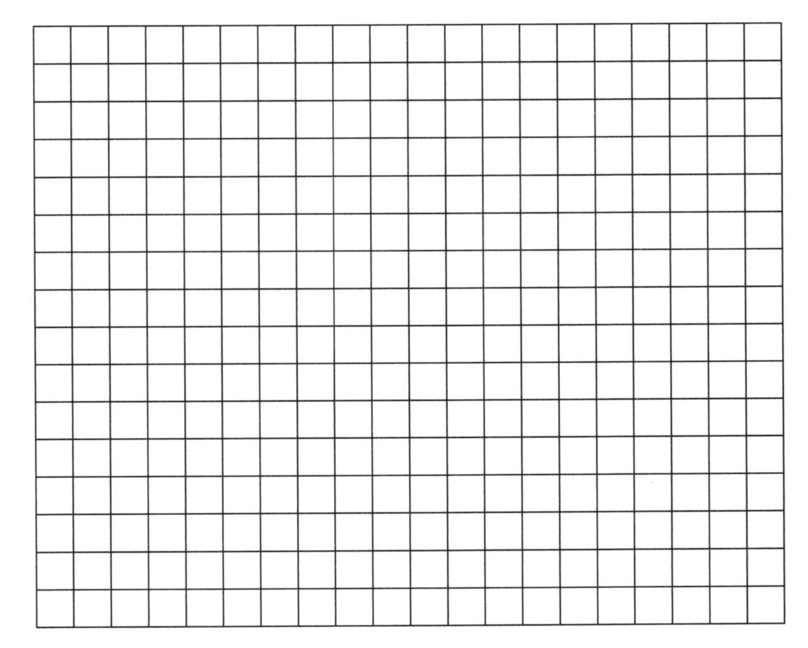

Draft Argument

Develop an argument on a whiteboard. It should include the following parts:

1. A *claim:* Your answer to the guiding question.
2. *Evidence:* An analysis of the data and an explanation of what the analysis means.
3. A *justification of the evidence:* Why your group thinks the evidence is important.

The Guiding Question:	
Our Claim:	
Our Evidence:	Our Justification of the Evidence:

Argumentation Session

Share your argument with your classmates. Be sure to ask them how to make your draft argument better. Keep track of their suggestions in the space below.

Ways to IMPROVE our argument …

Draft Report

Prepare an *investigation report* to share what you have learned. Use the information in this handout and your group's final argument to write a *draft* of your investigation report.

Introduction

We have been studying ______________________________ in class.

Before we started this investigation, we explored ______________________________

____

____

We noticed ______________________________

____

____

My goal for this investigation was to figure out ______________________________

____

____

____

The guiding question was ______________________________

____

____

____

Method

To gather the data I needed to answer this question, I ______________________________

____

____

____

____

I then analyzed the data I collected by ______________________________

____

____

____

____

Argument

My claim is ______________________________

____

____

The graph below shows ______________________________

____

____

____

____

____

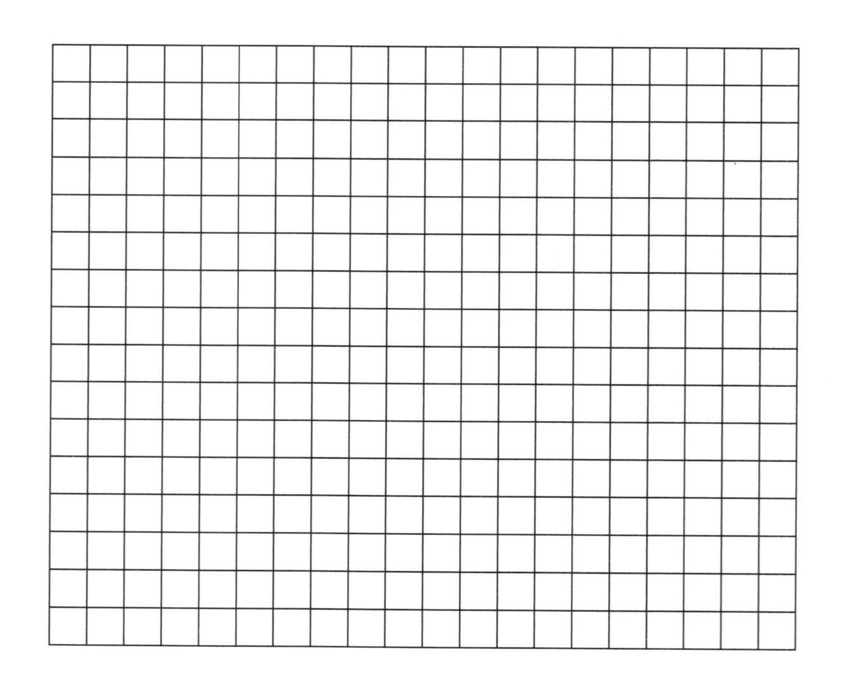

This evidence is important because ____

____

____

____

____

____

____

____

____

____

Review

Your friends need your help! Review the draft of their investigation reports and give them ideas about how to improve. Use the *peer-review guide* when doing your review.

Submit Your Final Report

Once you have received feedback from your friends about your draft report, create your final investigation report and hand it in to your teacher.

Checkout Questions

Investigation 5. Life Cycles: How Are the Life Cycles of Living Things Similar and How Are They Different?

1. Listed below are some living things. Place an *X* next to the living things that are born at some point in their life cycle.

☐ Corn	☐ Whale
☐ Mouse	☐ Cactus
☐ Tree	☐ Lizard

2. Listed below are some living things. Place an *X* next to the living things that die at some point in their life cycle.

☐ Panda	☐ Ladybug
☐ Snakes	☐ Ants
☐ Daisies	☐ Cheetah

3. Listed below are some living things. Place an *X* next to the living things that grow at some point during their life cycle.

☐ Spider	☐ Salmon
☐ Bald eagle	☐ Clown fish
☐ Grasshopper	☐ Grass

4. Listed below are some living things. Place an *X* next to the living things that reproduce at some point during their life cycle.

☐ Bean	☐ Tree
☐ Silkworm	☐ Dragonfly
☐ Frog	☐ Alligator

5. Listed below are some living things. Place an *X* next to the living things that go through a complete metamorphosis at some point during their life cycle.

☐ Bean ☐ Grasshopper

☐ Silkworm ☐ Dragonfly

☐ Frog ☐ Butterfly

6. Explain your thinking. What *pattern* from your investigation did you use to decide if these living things are born, grow, reproduce, and die during their life cycle?

____

____

____

____

____

____

____

____

____

Teacher Scoring Rubric for the Checkout Questions

Level	Description
3	The student can apply the core idea correctly in all cases and can fully explain the pattern.
2	The student can apply the core idea correctly in all cases but cannot fully explain the pattern.
1	The student cannot apply the core idea correctly in all cases but can fully explain the pattern.
0	The student cannot apply the core idea correctly in all cases and cannot explain the pattern.

Investigation 6

Life in Groups: Why Do Wolves Live in Groups?

Purpose

The purpose of this investigation is to give students an opportunity to use the disciplinary core idea (DCI) of LS2.D: Social Interactions and Group Behavior and the crosscutting concept (CC) of Cause and Effect from *A Framework for K–12 Science Education* (NRC 2012) to figure out why wolves live in groups. Students will also learn about how scientists use different methods to answer different types of questions during the reflective discussion.

The Disciplinary Core Idea

Students in third grade should understand the following about Social Interactions and Group Behavior and be able to use this DCI to figure out if living in a group makes it easier for wolves to get the food they need to survive:

> *Being part of a group helps animals obtain food, defend themselves, and cope with changes. Groups may serve different functions and vary dramatically in size. (NRC 2012, p. 156)*

The Crosscutting Concept

Students in third grade should understand the following about the CC Cause and Effect:

> *Repeating patterns in nature, or events that occur together with regularity, are clues that scientists can use to start exploring causal, or cause-and-effect, relationships. … Any application of science, or any engineered solution to a problem, is dependent on understanding the cause-and-effect relationships between events; the quality of the application or solution often can be improved as knowledge of the relevant relationships is improved. (NRC 2012, p. 87)*

Students in third grade should be given opportunities to begin to "look for and analyze patterns—whether in their observations of the world or in the relationships between different quantities in data (e.g., the sizes of plants over time)"; "they can also begin to consider what might be causing these patterns and relationships and design tests that gather more evidence to support or refute their ideas" (NRC 2012, pp. 88–89).

Students should be encouraged to use their developing understanding of cause-and-effect relationships as a tool or a way of thinking about a phenomenon during this investigation to help them figure out if living in a group (the cause) makes it easier for wolves to get the food they need to survive (the effect).

What Students Figure Out

Groups of wolves (called packs) are able to hunt larger prey than wolves hunting on their own. Wolves also tend to be more successful (in catching and killing an animal) when they hunt as part of a pack. Wolf packs are hierarchies with dominant members, consist of both male and females, and tend to be stable over time.

Timeline

The time needed to complete this investigation is 280 minutes (4 hours and 40 minutes). The amount of instructional time needed for each stage of the investigation is as follows:

- *Stage 1.* Introduce the task and the guiding question: 35 minutes
- *Stage 2.* Design a method and collect data: 70 minutes
- *Stage 3.* Create a draft argument: 35 minutes
- *Stage 4.* Argumentation session: 30 minutes
- *Stage 5.* Reflective discussion: 15 minutes
- *Stage 6.* Write a draft report: 35 minutes
- *Stage 7.* Peer review: 30 minutes
- *Stage 8.* Revise the report: 30 minutes

This investigation can be completed in one day or over eight days (one day for each stage) during your designated science time in the daily schedule.

Materials and Preparation

The materials needed for this investigation are listed in Table 6.1 (p. 220). In addition to these materials, students will need to watch several videos of wolves hunting different prey. These videos, which are available online, are listed below together with the URLs for accessing them; this list is also included in the Investigation 6 supplementary materials on the book's Extras page at *www.nsta.org/adi-3rd,* and any changes that occur in the video titles and/or URLs will be made on the Extras page.

Video watched in stage 1:

- "Musk Ox vs. Wolves / National Geographic": *www.youtube.com/watch?v=pb6Rke7jiTc*

Videos watched in stage 2:

- "Wolves Hunting Caribou—Planet Earth—BBC Wildlife" (1 wolf): *https://youtu.be/A0E6geAq1k8*
- "Wolf Hunting Elk in Yellowstone / BBC" (4 wolves): *https://youtu.be/bfGP4Xbme3o*
- "Wolf Hunts Caribou—Nature's Epic Journeys: Episode 2 Preview" (1 wolf): *www.youtube.com/watch?v=NdVIxS8tgYM*

- "Pack of Wolves Hunt a Bison / Frozen Planet / BBC Earth" (25 wolves): *www.youtube.com/watch?v=8wl8ZxAaB2E*
- "Baby Bison Takes on Wolf and Wins / American's National Parks" (1 wolf): *https://youtu.be/K6TnWW1s4hE*
- "Nature / Wolves Hunting Buffalo / Cold Warriors: Wolves and Buffalo / PBS" (3 wolves): *https://youtu.be/tCG1I-Ssgww*
- "Clash: Encounters of Bears and Wolves / Clip 2 / PBS" (several wolves): *www.youtube.com/watch?v=ZNrEOZ4xCGY*
- "Bison and Her Calf Battle Wolves / North America" (3 wolves): *www.youtube.com/watch?v=GtG-9ftqoHw*

Students will need a computer or tablet that can access the internet and YouTube to view these videos. We recommend at least one computer or tablet per group, but each student can use a computer or tablet on his or her own if there are enough available. We also recommend making photocopies of the list of videos (you can use the list in the supplementary materials), bookmarking the URLs on the class computers or tablets, or creating a Google document with hyperlinks to the videos that you can share with the students to make it easier for the students to find and access the videos during the investigation. Some schools restrict web browsing, so be sure to check to see if students can access each video from a school computer before you begin the investigation.

TABLE 6.1

Materials for Investigation 6

Item	Quantity
Computer or tablet that can access the internet and YouTube	1 per group
Whiteboard, 2' × 3'*	1 per group
Investigation Handout	1 per student
Peer-review guide and teacher scoring rubric	1 per student
Checkout Questions (optional)	1 per student

* As an alternative, students can use computer and presentation software such as Microsoft PowerPoint or Apple Keynote to create their arguments.

Safety Precautions

Remind students to follow all normal safety rules.

Lesson Plan by Stage

Stage 1: Introduce the Task and the Guiding Question (35 minutes)

1. Ask the students to sit in six groups, with three or four students in each group.
2. Ask students to clear off their desks except for a pencil (and their *Student Workbook for Argument-Driven Inquiry in Third-Grade Science* if they have one).
3. Pass out an Investigation Handout to each student (or ask students to turn to Investigation Log 6 in their workbook).
4. Read the first paragraph of the "Introduction" aloud to the class. Ask the students to follow along as you read.
5. Show the video "Musk Ox vs. Wolves / National Geographic" (available at *www.youtube.com/watch?v=pb6Rke7jiTc).*
6. Tell students to record their observations and questions about the video in the "OBSERVED/WONDER" chart in the "Introduction" section of their Investigation Handout (or the investigation log in their workbook).
7. Ask students to share *what they observed* as they watched the video.
8. Ask students to share *what questions they have* about what they observed as they watched the video.
9. Tell the students, "Some of your questions might be answered by reading the rest of the 'Introduction.'"
10. Ask the students to read the rest of the "Introduction" on their own *or* ask them to follow along as you read aloud.
11. Once the students have read the rest of the "Introduction," ask them to fill out the "KNOW/NEED" chart on their Investigation Handout (or in their investigation log) as a group.
12. Ask students to share what they learned from the reading. Add these ideas to a class "know / need to figure out" chart.
13. Ask students to share what they think they will need to figure out based on what they read. Add these ideas to the class "know / need to figure out" chart.
14. Tell the students, "It looks like we have something to figure out. Let's see what we will need to do during our investigation."
15. Read the task and the guiding question aloud.
16. Tell the students, "I have lots of videos that you can use."
17. Introduce the students to the videos available for them to use during the investigation by showing them how to access them on the computers or tablets.

Stage 2: Design a Method and Collect Data (70 minutes)

1. Tell the students, "I am now going to give you and the other members of your group about 15 minutes to plan your investigation. Before you begin, I want you all to take a couple of minutes to discuss the following questions with the rest of your group."
2. Show the following questions on the screen or board:
   - What types of *patterns* might we look for to help answer the guiding question?
   - What information do we need to find a *cause-and-effect relationship?*
3. Tell the students, "Please take a few minutes to come up with an answer to these questions." Give the students two or three minutes to discuss these two questions.
4. Ask two or three different groups to share their answers. Be sure to highlight or write down any important ideas on the board so students can refer to them later.
5. If possible, use a document camera to project an image of the graphic organizer for this investigation on a screen or board (or take a picture of it and project the picture on a screen or board). Tell the students, "I now want you all to plan out your investigation. To do that, you will need to create an investigation proposal by filling out this graphic organizer."
6. Point to the box labeled "Our guiding question:" and tell the students, "You can put the question we are trying to answer in this box." Then ask, "Where can we find the guiding question?"
7. Wait for a student to answer where to find the guiding question (the answer is "in the handout").
8. Point to the box labeled "We will collect the following data from the videos:" and tell the students, "You can list the observations that you will need to collect during the investigation in this box."
9. Point to the box labeled "These are the steps we will follow to collect data as we watch the videos:" and tell the students, "You can list what you are going to do to collect the data you need and what you will do with your data once you have it. Be sure to give enough detail that I could do your investigation for you."
10. Ask the students, "Do you have any questions about what you need to do?"
11. Answer any questions that come up.
12. Tell the students, "Once you are done, raise your hand and let me know. I'll then come by and look over your proposal and give you some feedback. You may not begin collecting data until I have approved your proposal by signing it. You need to have your proposal done in the next 15 minutes."

13. Give the students 15 minutes to work in their groups on their investigation proposal. As they work, move from group to group to check in, ask probing questions, and offer a suggestion if a group gets stuck.

What should a student-designed investigation look like?

The students' investigation proposal should include the following information:

- The guiding question is "Why do wolves live in groups?"
- The data that the student should collect are (1) number of wolves, (2) type of prey, and (3) if the hunt was successful or not.
- To collect the data, the students could follow these steps (but this is just an example, and there should be a lot of variation in the student-designed investigation):
  1. Watch a video of a wolf hunt.
  2. Make observations about the number of wolves hunting, what the wolves are hunting, and if they catch what they are hunting or not.
  3. Repeat steps 1 and 2 for each video.
  4. Determine if living in a group (the cause) makes it easier for wolves to get the food they need to survive (the effect).

14. As each group finishes its investigation proposal, be sure to read it over and determine if it will be productive or not. If you feel the investigation will be productive (not necessarily what you would do or what the other groups are doing), sign your name on the proposal and let the group start collecting data. If the plan needs to be changed, offer some suggestions or ask some probing questions, and have the group make the changes before you approve it.
15. Give the students about 30 minutes to watch the videos and collect their data.
16. Be sure to remind the students to collect their data and record their observations or measurements in the "Collect Your Data" box in their Investigation Handout (or the investigation log in their workbook).

Stage 3: Create a Draft Argument (35 minutes)

1. Tell the students, "Now that we have all this data, we need to analyze the data so we can figure out an answer to the guiding question."
2. If possible, project an image of the "Analyze Your Data" text and box for this investigation on a screen or board using a document camera (or take a picture

of it and project the picture on a screen or board). Point to the box and tell the students, "You can create a table or a graph as a way to analyze your data. You can make your table or graph in this box."

3. Ask the students, "What information do we need to include in a table or graph?"
4. Tell the students, "Please take a few minutes to discuss this question with your group, and be ready to share."
5. Give the students five minutes to discuss.
6. Ask two or three different groups to share their answers. Be sure to highlight or write down any important ideas on the board so students can refer to them later.
7. Tell the students, "I am now going to give you and the other members of your group about 10 minutes to create a table or graph." If the students are having trouble making a table or graph, you can take a few minutes to provide a mini-lesson about how to create a table or graph from a bunch of observations or measurements (this strategy is called just-in-time instruction because it is offered only when students get stuck).

What should a table or graph look like for this investigation?

There are a number of different ways that students can analyze the observations they collect during this investigation. One of the most straightforward ways is to create a table with three columns. The first column can be labeled "number of wolves," the second column can be labeled "type of prey," and the third column can be labeled "hunt successful or not." The students can then include the appropriate observations from each video in each row (e.g., the first row would include data from the first video they watched, the second row would include data from the second video they watched, and so on). This allows them to show a clear pattern as part of their analysis of the data. An example of this type of table can be seen in Figure 6.1 (p. 226). There are other many possible ways to analyze the data for this investigation. Students often come up with some unique ways of analyzing their data, so be sure to give them some voice and choice during this stage.

8. Give the students 10 minutes to analyze their data. As they work, move from group to group to check in, ask probing questions, and offer suggestions.
9. Tell the students, "I am now going to give you and the other members of your group 15 minutes to create an argument to share what you have learned and

convince others that they should believe you. Before you do that, we need to take a few minutes to discuss what you need to include in your argument."

10. If possible, use a document camera to project the "Argument Presentation on a Whiteboard" image from the Investigation Handout (or the investigation log in their workbook) on a screen or board (or take a picture of it and project the picture on a screen or board).
11. Point to the box labeled "The Guiding Question:" and tell the students, "You can put the question we are trying to answer here on your whiteboard."
12. Point to the box labeled "Our Claim:" and tell the students, "You can put your claim here on your whiteboard. The claim is your answer to the guiding question."
13. Point to the box labeled "Our Evidence:" and tell the students, "You can put the evidence that you are using to support your claim here on your whiteboard. Your evidence will need to include the analysis you just did and an explanation of what your analysis means or shows. Scientists always need to support their claims with evidence."
14. Point to the box labeled "Our Justification of the Evidence:" and tell the students, "You can put your justification of your evidence here on your whiteboard. Your justification needs to explain why your evidence is important. Scientists often use core ideas to explain why the evidence they are using matters. Core ideas are important concepts that scientists use to help them make sense of what happens during an investigation."
15. Ask the students, "What are some core ideas that we read about earlier that might help us explain why the evidence we are using is important?"
16. Ask students to share some of the core ideas from the "Introduction" section of the Investigation Handout (or the investigation log in the workbook). List these core ideas on the board.
17. Tell the students, "That is great. I would like to see everyone try to include these core ideas in your justification of the evidence. Your goal is to use these core ideas to help explain why your evidence matters and why the rest of us should pay attention to it."
18. Ask the students, "Do you have any questions about what you need to do?"
19. Answer any questions that come up.
20. Tell the students, "Okay, go ahead and start working on your arguments. You need to have your argument done in the next 15 minutes. It doesn't need to be perfect. We just need something down on the whiteboards so we can share our ideas."
21. Give the students 15 minutes to work in their groups on their arguments. As they work, move from group to group to check in, ask probing questions, and

offer a suggestion if a group gets stuck. Figure 6.1 shows an example of an argument created by students for this investigation.

FIGURE 6.1

Example of an argument

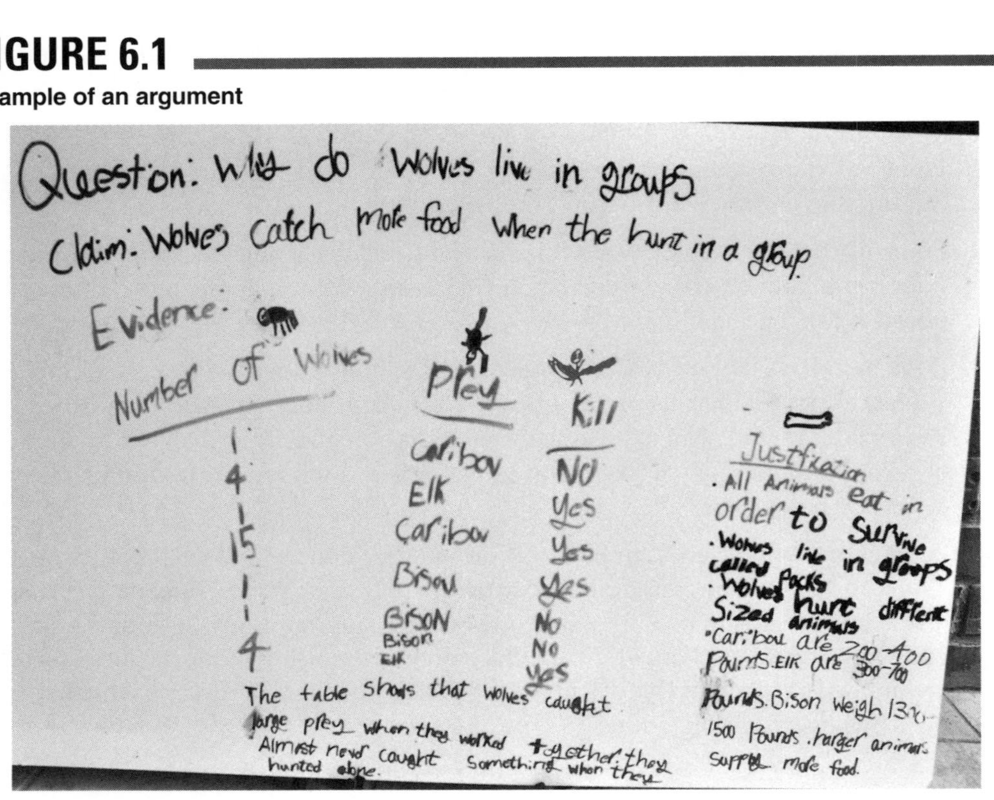

Stage 4: Argumentation Session (30 minutes)

The argumentation session can be conducted in a whole-class presentation format, a gallery walk format, or a modified gallery walk format. We recommend using a whole-class presentation format for the first investigation, but try to transition to either the gallery walk or modified gallery walk format as soon as possible because that will maximize student voice and choice inside the classroom. The following list shows the steps for the three formats; unless otherwise noted, the steps are the same for all three formats.

1. Begin by introducing the use of the whiteboard.
   - *If using the whole-class presentation format,* tell the students, "We are now going to share our arguments. Please set up your whiteboard so everyone can see them."
   - *If using the gallery walk or modified gallery walk format,* tell the students, "We are now going to share our arguments. Please set up your whiteboard so they are facing the walls."
2. Allow the students to set up their whiteboards.

- *If using the whole-class presentation format,* the whiteboards should be set up on stands or chairs so they are facing toward the center of the room.
- *If using the gallery walk or modified gallery walk format,* the whiteboards should be set up on stands or chairs so they are facing toward the outside of the room.

3. Give the following instructions to the students:
   - *If using the whole-class presentation format or the modified gallery walk format,* tell the students, "Okay, before we get started I want to explain what we are going to do next. I'm going to ask some of you to present your arguments to your classmates. If you are presenting your argument, your job is to share your group's claim, evidence, and justification of the evidence. The rest of you will be reviewers. If you are a reviewer, your job is to listen to the presenters, ask the presenters questions if you do not understand something, and then offer them some suggestions about ways to make their argument better. After we have a chance to learn from each other, I'm going to give you some time to revise your arguments and make them better."
   - *If using the gallery walk format,* tell the students, "Okay, before we get started I want to explain what we are going to do next. You are going to have an opportunity to read the arguments that were created by other groups. Your group will go to a different group's argument. I'll give you a few minutes to read it and review it. Your job is to offer them some suggestions about ways to make their argument better. You can use sticky notes to give them suggestions. Please be specific about what you want to change and be specific about how you think they should change it. After we have a chance to learn from each other, I'm going to give you some time to revise your arguments and make them better."
4. Use a document camera to project the "Ways to IMPROVE our argument …" box from the Investigation Handout (or the investigation log in their workbook) on a screen or board (or take a picture of it and project the picture on a screen or board).
   - *If using the whole-class presentation format or the modified gallery walk format,* point to the box and tell the students, "If you are a presenter, you can write down the suggestions you get from the reviewers here. If you are a reviewer, and you see a good idea from another group, you can write down that idea here. Once we are done with the presentations, I will give you a chance to use these suggestions or ideas to improve your arguments.
   - *If using the gallery walk format,* point to the box and tell the students, "If you see good ideas from another group, you can write them down here. Once we are done reviewing the different arguments, I will give you a chance to use these ideas to improve your own arguments. It is important to share ideas like this."

   Ask the students, "Do you have any questions about what you need to do?"

5. Answer any questions that come up.
6. Give the following instructions:
   - *If using the whole-class presentation format,* tell the students, "Okay. Let's get started."
   - *If using the gallery walk format,* tell the students, "Okay, I'm now going to tell you which argument to go to and review.
   - *If using the modified gallery walk format,* tell the students, "Okay, I'm now going to assign you to be a presenter or a reviewer." Assign one or two students from each group to be presenters and one or two students from each group to be reviewers.
7. Begin the review of the arguments.
   - *If using the whole-class presentation format,* have four or five groups present their argument one at a time. Give each group only two to three minutes to present their argument. Then give the class two to three minutes to ask them questions and offer suggestions. Be sure to encourage as much participation from the students as possible.
   - *If using the gallery walk format,* tell the students, "Okay. Let's get started. Each group, move one argument to the left. Don't move to the next argument until I tell you to move. Once you get there, read the argument and then offer suggestions about how to make it better. I will put some sticky notes next to each argument. You can use the sticky notes to leave your suggestions." Give each group about three to four minutes to read the arguments, talk, and offer suggestions.

     a. Tell the students, "Okay. Let's rotate. Move one group to the left."
     b. Again, give each group three or four minutes to read, talk, and offer suggestions.
     c. Repeat this process for two more rotations.
   - *If using the modified gallery walk format,* tell the students, "Okay. Let's get started. Reviewers, move one group to the left. Don't move to the next group until I tell you to move. Presenters, go ahead and share your argument with the reviewers when they get there." Give each group of presenters and reviewers about three to four minutes to talk.

     a. Tell the students, "Okay. Let's rotate. Reviewers, move one group to the left."
     b. Again, give each group of presenters and reviewers about three or four minutes to talk.
     c. Repeat this process for two more rotations.
8. Tell the students to return to their workstations.
9. Give the following instructions about revising the argument:

- *If using the whole-class presentation format,* tell the students, "I'm now going to give you about 10 minutes to revise your argument. Take a few minutes to talk in your groups and determine what you want to change to make your argument better. Once you have decided what to change, go ahead and make the changes to your whiteboard."
- *If using the gallery walk format,* tell the students, "I'm now going to give you about 10 minutes to revise your argument. Take a few minutes to read the suggestions that were left at your argument. Then talk in your groups and determine what you want to change to make your argument better. Once you have decided what to change, go ahead and make the changes to your whiteboard."
- *If using the modified gallery walk format,* "I'm now going to give you about 10 minutes to revise your argument. Please return to your original groups." Wait for the students to move back into their original groups and then tell the students, "Okay, take a few minutes to talk in your groups and determine what you want to change to make your argument better. Once you have decided what to change, go ahead and make the changes to your whiteboard."

Ask the students, "Do you have any questions about what you need to do?"

10. Answer any questions that come up.
11. Tell the students, "Okay. Let's get started."
12. Give the students 10 minutes to work in their groups on their arguments. As they work, move from group to group to check in, ask probing questions, and offer a suggestion if a group gets stuck.

Stage 5: Reflective Discussion (15 minutes)

1. Tell the students, "We are now going to take a minute to talk about what we did and what we have learned."
2. Show Figure 6.2 on a screen. Ask the students, "What do you all see going on here?"
3. Allow students to share their ideas.

FIGURE 6.2

Wolves hunting an elk

4. Ask the students, "Why would it be beneficial for the wolves to hunt together?"
5. Allow students to share their ideas. Keep probing until someone mentions that they are able to catch larger prey or catch food more often.
6. Ask the students, "What would be some drawbacks of hunting together?"
7. Allow students to share their ideas. Keep probing until someone mentions that they have to split the food.
8. Ask the students, "What are some other benefits of living in a group like this?"
9. Allow students to share their ideas.
10. Tell the students, "Okay, let's make sure we are on the same page. Being part of a group helps animals obtain food, defend themselves, and cope with changes. Wolves, for example, tend to be more successful at catching food when they hunt as part of a pack. Groups may serve different functions and vary dramatically in size. Wolf packs, for example, tend to include 5 to 10 members and include both males and females. These packs have a wolf that is the leader of the pack and wolves that follow the leader, and these relationships tend to be stable over time. The fact that living things often live in groups as a way to help them survive is a really important core idea in science."
11. Ask the students, "Does anyone have any questions about this core idea?"
12. Answer any questions that come up.
13. Tell the students, "We also looked for a cause-and-effect relationship during our investigation." Then ask, "Can anyone tell me why it is useful to look for cause-and-effect relationships?"
14. Allow students to share their ideas.
15. Tell the students, "Cause-and-effect relationships are important because they allow us to predict what will happen in the future."
16. Ask the students, "What was the cause and what was the effect that we uncovered today?"
17. Allow students to share their ideas. Keep probing until someone mentions that the cause was living in a group and the effect was ability to obtain food.
18. Tell the students, "That is great, and if we know that you can predict what will happen the next time we see a wolf pack hunting prey on a video."
19. Tell the students, "We are now going take a minute to talk about what went well and what didn't go so well during our investigation. We need to talk about this because you are going to be planning and carrying out your own investigations like this a lot this year, and I want to help you all get better at it."
20. Show an image of the question "What made your investigation scientific?" on the screen. Tell the students, "Take a few minutes to talk about how you would answer this question with the other people in your group. Be ready to share

with the rest of the class." Give the students two to three minutes to talk in their group.

21. Ask the students, "What do you all think? Who would like to share an idea?"
22. Allow three or four students to share their ideas. Be sure to expand on their ideas about what makes an investigation scientific.
23. Show an image of the question "What made your investigation not so scientific?" on the screen. Tell the students, "Take a few minutes to talk about how you would answer this question with the other people in your group. Be ready to share with the rest of the class." Give the students two to three minutes to talk in their group.
24. Ask the students, "What do you all think? Who would like to share an idea?"
25. Allow students to share their ideas. Be sure to expand on their ideas about what makes an investigation less scientific.
26. Show an image of the question "What rules can we put into place to help us make sure our next investigation is more scientific?" on the screen. Tell the students, "Take a few minutes to talk about how you would answer this question with the other people in your group. Be ready to share with the rest of the class." Give the students two to three minutes to talk in their group.
27. Ask the students, "What do you all think? Who would like to share an idea?"
28. Allow students to share their ideas. Once they have shared their ideas, offer a suggestion for a possible class rule.
29. Ask the students, "What do you all think? Should we make this a rule?"
30. If the students agree, write the rule on the board or make a class "Rules for Scientific Investigation" chart so you can refer to it during the next investigation.
31. Tell the students, "We are now going take a minute to talk about what scientists do to investigate the natural world."
32. Show an image of the question "Do all scientists follow the same method?" on the screen. Tell the students, "Take a few minutes to talk about how you would answer this question with the other people in your group. Be ready to share with the rest of the class." Give the students two to three minutes to talk in their group.
33. Ask the students, "What do you all think? Who would like to share an idea?"
34. Allow students to share their ideas.
35. Tell the students, "Okay, let's make sure we are all on the same page. Scientists use lots of different methods to answer different types of questions. Sometimes they need to go out into the field and watch what animals do. The videos you watched came from scientists working out in the field. Some scientists design experiments, and others analyze data collected by other scientists. There is no

one method that all scientists use, and the method used by scientists depends on what they are studying and what type of question they are asking. This is an important thing to understand about science."

36. Ask the students, "Does anyone have any questions about what scientists do to investigate the natural world?"
37. Answer any questions that come up.

Stage 6: Write a Draft Report (35 minutes)

Your students will use either the Investigation Handout or the investigation log in the student workbook when writing the draft report. When you give the directions shown in quotes in the following steps, substitute "investigation log" (as shown in brackets) for "handout" if they are using the workbook.

1. Tell the students, "You are now going to write an investigation report to share what you have learned. Please take out a pencil and turn to the 'Draft Report' section of your handout [investigation log]."
2. If possible, use a document camera to project the "Introduction" section of the draft report from the Investigation Handout (or the investigation log in their workbook) on a screen or board (or take a picture of it and project the picture on a screen or board).
3. Tell the students, "The first part of the report is called the 'Introduction.' In this section of the report you want to explain to the reader what you were investigating, why you were investigating it, and what question you were trying to answer. All of this information can be found in the text at the beginning of your handout [investigation log]." Point to the image and say, "There are some sentence starters here to help you begin writing the report." Ask the students, "Do you have any questions about what you need to do?"
4. Answer any questions that come up.
5. Tell the students, "Okay. Let's write."
6. Give the students 10 minutes to write the "Introduction" section of the report. As they work, move from student to student to check in, ask probing questions, and offer a suggestion if a student gets stuck.
7. If possible, use a document camera to project the "Method" section of the draft report from the Investigation Handout (or the investigation log in their workbook) on a screen or board (or take a picture of it and project the picture on a screen or board).
8. Tell the students, "The second part of the report is called the 'Method.' In this section of the report you want to explain to the reader what you did during the investigation, what data you collected and why, and how you went about analyzing your data. All of this information can be found in the 'Plan Your

Investigation' section of your handout [investigation log]. Remember that you all planned and carried out different investigations, so do not assume that the reader will know what you did." Point to the image and say, "There are some sentence starters here to help you begin writing this part of the report." Ask the students, "Do you have any questions about what you need to do?"

9. Answer any questions that come up.
10. Tell the students, "Okay. Let's write."
11. Give the students 10 minutes to write the "Method" section of the report. As they work, move from student to student to check in, ask probing questions, and offer a suggestion if a student gets stuck.
12. If possible, use a document camera to project the "Argument" section of the draft report from the Investigation Handout (or the investigation log in their workbook) on a screen or board (or take a picture of it and project the picture on a screen or board).
13. Tell the students, "The last part of the report is called the 'Argument.' In this section of the report you want to share your claim, evidence, and justification of the evidence with the reader. All of this information can be found on your whiteboard." Point to the image and say, "There are some sentence starters here to help you begin writing this part of the report." Ask the students, "Do you have any questions about what you need to do?"
14. Answer any questions that come up.
15. Tell the students, "Okay. Let's write."
16. Give the students 10 minutes to write the "Argument" section of the report. As they work, move from student to student to check in, ask probing questions, and offer a suggestion if a student gets stuck.

Stage 7: Peer Review (30 minutes)

Your students will use either the Investigation Handout or the investigation log in the student workbook when doing the peer review. When you give the directions shown in quotes in the following steps, substitute "workbook" (as shown in brackets) for "Investigation Handout" if they are using the workbook.

1. Tell the students, "We are now going to review our reports to find ways to make them better. I'm going to come around and collect your Investigation Handout [workbook]. While I do that, please take out a pencil."
2. Collect the Investigation Handouts or workbooks from the students.
3. If possible, use a document camera to project the peer-review guide (PRG; see Appendix 4) on a screen or board (or take a picture of it and project the picture on a screen or board).

4. Tell the students, "We are going to use this peer-review guide to give each other feedback." Point to the image.
5. Give the following instructions:
   - *If using the Investigation Handout,* tell the students, "I'm going to ask you to work with a partner to do this. I'm going to give you and your partner a draft report to read and a peer-review guide to fill out. You two will then read the report together. Once you are done reading the report, I want you to answer each of the questions on the peer-review guide." Point to the review questions on the image of the PRG.
   - *If using the workbook,* tell the students, "I'm going to ask you to work with a partner to do this. I'm going to give you and your partner a draft report to read. You two will then read the report together. Once you are done reading the report, I want you to answer each of the questions on the peer-review guide that is right after the report in the investigation log." Point to the review questions on the image of the PRG.
6. Tell the students, "You can check 'yes,' 'almost,' or 'no' after each question." Point to the checkboxes on the image of the PRG.
7. Tell the students, "This will be your rating for this part of the report. Make sure you agree on the rating you give the author. If you mark 'almost' or 'no,' then you need to tell the author what he or she needs to do to get a 'yes.'" Point to the space for the reviewer feedback on the image of the PRG.
8. Tell the students, "It is really important for you to give the authors feedback that is helpful. That means you need to tell them exactly what they need to do to make their reports better." Ask the students, "Do you have any questions about what you need to do?"
9. Answer any questions that come up.
10. Tell the students, "Please sit with a partner who is not in your current group." Allow the students time to sit with a partner.
11. Give the following instructions:
    - *If using the Investigation Handout,* tell the students, "Okay, I am now going to give you one report to read and one peer-review guide to fill out." Pass out one report to each pair. Make sure that the report you give a pair was not written by one of the students in that pair. Give each pair one PRG to fill out as a team.
    - *If using the workbook,* tell the students, "Okay, I am now going to give you one report to read." Pass out a workbook to each pair. Make sure that the workbook you give a pair is not from one of the students in that pair.
12. Tell the students, "Okay, I'm going to give you 15 minutes to read the report I gave you and to fill out the peer-review guide. Go ahead and get started."

13. Give the students 15 minutes to work. As they work, move around from pair to pair to check in and see how things are going, answer questions, and offer advice.
14. After 15 minutes pass, tell the students, "Okay, time is up." *If using the Investigation Handout,* say, "Please give me the report and the peer-review guide that you filled out." *If using the workbook,* say, "Please give me the workbook that you have."
15. Collect the Investigation Handouts and the PRGs, or collect the workbooks if they are being used. Be sure you keep the handout and the PRG together.
16. Give the following instructions:
    - *If using the Investigation Handout,* tell the students, "Okay, I am now going to give you a different report to read and a new peer-review guide to fill out." Pass out another report to each pair. Make sure that this report was not written by one of the students in that pair. Give each pair a new PRG to fill out as a team.
    - *If using the workbook,* tell the students, "Okay, I am now going to give you a different report to read." Pass out a different workbook to each pair. Make sure that the workbook you give a pair is not from one of the students in that pair.
17. Tell the students, "Okay, I'm going to give you 15 minutes to read this new report and to fill out the peer-review guide. Go ahead and get started."
18. Give the students 15 minutes to work. As they work, move around from pair to pair to check in and see how things are going, answer questions, and offer advice.
19. After 15 minutes pass, tell the students, "Okay, time is up." *If using the Investigation Handout,* say, "Please give me the report and the peer-review guide that you filled out." *If using the workbook,* say, "Please give me the workbook that you have."
20. Collect the Investigation Handouts and the PRGs, or collect the workbooks if they are being used. Be sure you keep the handout and the PRG together.

Stage 8: Revise the Report (30 minutes)

Your students will use either the Investigation Handout or the investigation log in the student workbook when revising the report. Except where noted below, the directions are the same whether using the handout or the log.

1. Give the following instructions:
    - *If using the Investigation Handout,* tell the students, "You are now going to revise your investigation report based on the feedback you get from your classmates. Please take out a pencil while I hand back your draft report and the peer-review guide."

- *If using the investigation log in the student workbook,* tell the students, "You are now going to revise your investigation report based on the feedback you get from your classmates. Please take out a pencil while I hand back your investigation logs."

2. *If using the Investigation Handout,* pass back the handout and the PRG to each student. *If using the investigation log,* pass back the log to each student.
3. Tell the students, "Please take a few minutes to read over the peer-review guide. You should use it to figure out what you need to change in your report and how you will change the report."
4. Allow the students time to read the PRG.
5. *If using the investigation log,* if possible use a document camera to project the "Write Your Final Report" section from the investigation log on a screen or board (or take a picture of it and project the picture on a screen or board).
6. Give the following instructions:
   - *If using the Investigation Handout,* tell the students, "Okay. Let's revise our reports. Please take out a piece of paper. I would like you to rewrite your report. You can use your draft report as a starting point, but use the feedback on the peer-review guide to help make it better."
   - *If using the investigation log,* tell the students, "Okay. Let's revise our reports. I would like you to rewrite your report in the section of the investigation log that says 'Write Your Final Report.'" Point to the image on the screen and tell the students, "You can use your draft report as a starting point, but use the feedback on the peer-review guide to help make your report better."

   Ask the students, "Do you have any questions about what you need to do?"
7. Answer any questions that come up.
8. Tell the students, "Okay. Let's write."
9. Give the students 30 minutes to rewrite their report. As they work, move from student to student to check in, ask probing questions, and offer a suggestion if a student gets stuck.
10. Give the following instructions:
    - *If using the Investigation Handout,* tell the students, "Okay. Time's up. I will now come around and collect your Investigation Handout, the peer-review guide, and your final report."
    - *If using the investigation log,* tell the students, "Okay. Time's up. I will now come around and collect your workbooks."
11. *If using the Investigation Handout,* collect all the Investigation Handouts, PRGs, and final reports. *If using the investigation log,* collect all the workbooks.

12. *If using the Investigation Handout,* use the "Teacher Score" columns in the PRG to grade the final report. *If using the investigation log,* use the "ADI Investigation Report Grading Rubric" in the investigation log to grade the final report. Whether you are using the handout or the log, you can give the students feedback about their writing in the "Teacher Comments" section.

How to Use the Checkout Questions

The Checkout Questions are an optional assessment. We recommend giving them to students one day after they finish stage 8 of the ADI investigation. The Checkout Questions can be used as a formative or summative assessment of student thinking. If you plan to use them as a formative assessment, we recommend that you look over the student answers to determine if you need to reteach the core idea and/or crosscutting concept from the investigation, but do not grade them. If you plan to use them as a summative assessment, we have included a "Teacher Scoring Rubric" at the end of the Checkout Questions that you can use to score a student's ability to apply the core idea in a new scenario and explain their use of a crosscutting concept. The rubric includes a 4-point scale that ranges from 0 (the student cannot apply the core idea correctly in all cases and cannot explain the [crosscutting concept]) to 3 (the student can apply the core idea correctly in all cases and can fully explain the [crosscutting concept]). The Checkout Questions, regardless of how you decide to use them, are a great way to make student thinking visible so you can determine if the students have learned the core idea and the crosscutting concept.

A student who can apply the core idea correctly in all cases and can explain the cause-and-effect relationships would give the following answers for question 1: A, 8–10 wolves; B, 4–6 wolves; C, 4–6 wolves or 1–2 wolves (either answer is correct for C); and D, 1–2 wolves. He or she should then be able to explain that more wolves hunting together can catch and kill larger prey.

Connections to Standards

Table 6.2 (p. 238) highlights how the investigation can be used to address specific performance expectations from the *NGSS, Common Core State Standards (CCSS)* in English language arts (ELA) and in mathematics, and *English Language Proficiency (ELP) Standards.*

TABLE 6.2

Investigation 6 alignment with standards

***NGSS* performance expectation**	Strong alignment • 3-LS2-1: Construct an argument that some animals form groups that help members survive.
***CCSS ELA*—Reading: Informational Text**	Key ideas and details • CCSS.ELA-LITERACY.RI.3.1: Ask and answer questions to demonstrate understanding of a text, referring explicitly to the text as the basis for the answers. • CCSS.ELA-LITERACY.RI.3.2: Determine the main idea of a text; recount the key details and explain how they support the main idea. • CCSS.ELA-LITERACY.RI.3.3: Describe the relationship between a series of historical events, scientific ideas or concepts, or steps in technical procedures in a text, using language that pertains to time, sequence, and cause/effect. Craft and structure • CCSS.ELA-LITERACY.RI.3.4: Determine the meaning of general academic and domain-specific words and phrases in a text relevant to a *grade 3 topic or subject area*. • CCSS.ELA-LITERACY.RI.3.5: Use text features and search tools (e.g., key words, sidebars, hyperlinks) to locate information relevant to a given topic efficiently. • CCSS.ELA-LITERACY.RI.3.6: Distinguish their own point of view from that of the author of a text. Integration of knowledge and ideas • CCSS.ELA-LITERACY.RI.3.7: Use information gained from illustrations (e.g., maps, photographs) and the words in a text to demonstrate understanding of the text (e.g., where, when, why, and how key events occur). • CCSS.ELA-LITERACY.RI.3.8: Describe the logical connection between particular sentences and paragraphs in a text (e.g., comparison, cause/effect, first/second/third in a sequence). • CCSS.ELA-LITERACY.RI.3.9: Compare and contrast the most important points and key details presented in two texts on the same topic. Range of reading and level of text complexity • CCSS.ELA-LITERACY.RI.3.10: By the end of the year, read and comprehend informational texts, including history/social studies, science, and technical texts, at the high end of the grades 2–3 text complexity band independently and proficiently.

Continued

Table 6.2 (*continued*)

***CCSS ELA*—Writing**	Text types and purposes • CCSS.ELA-LITERACY.W.3.1: Write opinion pieces on topics or texts, supporting a point of view with reasons. o CCSS.ELA-LITERACY.W.3.1.A: Introduce the topic or text they are writing about, state an opinion, and create an organizational structure that lists reasons. o CCSS.ELA-LITERACY.W.3.1.B: Provide reasons that support the opinion. o CCSS.ELA-LITERACY.W.3.1.C: Use linking words and phrases (e.g., *because*, *therefore*, *since*, *for example*) to connect opinion and reasons. o CCSS.ELA-LITERACY.W.3.1.D: Provide a concluding statement or section. • CCSS.ELA-LITERACY.W.3.2: Write informative or explanatory texts to examine a topic and convey ideas and information clearly. o CCSS.ELA-LITERACY.W.3.2.A: Introduce a topic and group related information together; include illustrations when useful to aiding comprehension. o CCSS.ELA-LITERACY.W.3.2.B: Develop the topic with facts, definitions, and details. o CCSS.ELA-LITERACY.W.3.2.C: Use linking words and phrases (e.g., *also*, *another*, *and*, *more*, *but*) to connect ideas within categories of information. o CCSS.ELA-LITERACY.W.3.2.D: Provide a concluding statement or section. Production and distribution of writing • CCSS.ELA-LITERACY.W.3.4: With guidance and support from adults, produce writing in which the development and organization are appropriate to task and purpose. • CCSS.ELA-LITERACY.W.3.5: With guidance and support from peers and adults, develop and strengthen writing as needed by planning, revising, and editing. • CCSS.ELA-LITERACY.W.3.6: With guidance and support from adults, use technology to produce and publish writing (using keyboarding skills) as well as to interact and collaborate with others. Research to build and present knowledge • CCSS.ELA-LITERACY.W.3.8: Recall information from experiences or gather information from print and digital sources; take brief notes on sources and sort evidence into provided categories. Range of writing • CCSS.ELA-LITERACY.W.3.10: Write routinely over extended time frames (time for research, reflection, and revision) and shorter time frames (a single sitting or a day or two) for a range of discipline-specific tasks, purposes, and audiences.

Continued

Table 6.2 (*continued*)

<table>
<tr><td>CCSS ELA— Speaking and Listening</td><td>Comprehension and collaboration
• CCSS.ELA-LITERACY.SL.3.1: Engage effectively in a range of collaborative discussions (one-on-one, in groups, and teacher-led) with diverse partners on grade 3 topics and texts, building on others' ideas and expressing their own clearly.
o CCSS.ELA-LITERACY.SL.3.1.A: Come to discussions prepared, having read or studied required material; explicitly draw on that preparation and other information known about the topic to explore ideas under discussion.
o CCSS.ELA-LITERACY.SL.3.1.B: Follow agreed-upon rules for discussions (e.g., gaining the floor in respectful ways, listening to others with care, speaking one at a time about the topics and texts under discussion).
o CCSS.ELA-LITERACY.SL.3.1.C: Ask questions to check understanding of information presented, stay on topic, and link their comments to the remarks of others.
o CCSS.ELA-LITERACY.SL.3.1.D: Explain their own ideas and understanding in light of the discussion.
• CCSS.ELA-LITERACY.SL.3.2: Determine the main ideas and supporting details of a text read aloud or information presented in diverse media and formats, including visually, quantitatively, and orally.
• CCSS.ELA-LITERACY.SL.3.3: Ask and answer questions about information from a speaker, offering appropriate elaboration and detail.
Presentation of knowledge and ideas
• CCSS.ELA-LITERACY.SL.3.4: Report on a topic or text, tell a story, or recount an experience with appropriate facts and relevant, descriptive details, speaking clearly at an understandable pace.
• CCSS.ELA-LITERACY.SL.3.6: Speak in complete sentences when appropriate to task and situation in order to provide requested detail or clarification.</td></tr>
</table>

Continued

Table 6.2 (*continued*)

ELP Standards	Receptive modalities • ELP 1: Construct meaning from oral presentations and literary and informational text through grade-appropriate listening, reading, and viewing. • ELP 8: Determine the meaning of words and phrases in oral presentations and literary and informational text. Productive modalities • ELP 3: Speak and write about grade-appropriate complex literary and informational texts and topics. • ELP 4: Construct grade-appropriate oral and written claims and support them with reasoning and evidence. • ELP 7: Adapt language choices to purpose, task, and audience when speaking and writing. Interactive modalities • ELP 2: Participate in grade-appropriate oral and written exchanges of information, ideas, and analyses, responding to peer, audience, or reader comments and questions. • ELP 5: Conduct research and evaluate and communicate findings to answer questions or solve problems. • ELP 6: Analyze and critique the arguments of others orally and in writing. Linguistic structures of English • ELP 9: Create clear and coherent grade-appropriate speech and text. • ELP 10: Make accurate use of standard English to communicate in grade-appropriate speech and writing.

Investigation 6

Life in Groups: Why Do Wolves Live in Groups?

Introduction

All animals must eat to survive. Some animals eat plants, and some animals eat other animals. The musk ox is an example of an animal that eats plants. The arctic wolf is an example of an animal that eats other animals. Both of these animals live in the Arctic tundra. Arctic wolves often eat musk oxen ("oxen" means more than one ox). Take a few minutes to watch what happens when a group of wolves attacks a group of musk oxen. As you watch the video, keep track of what you observe and what you are wondering about in the boxes below.

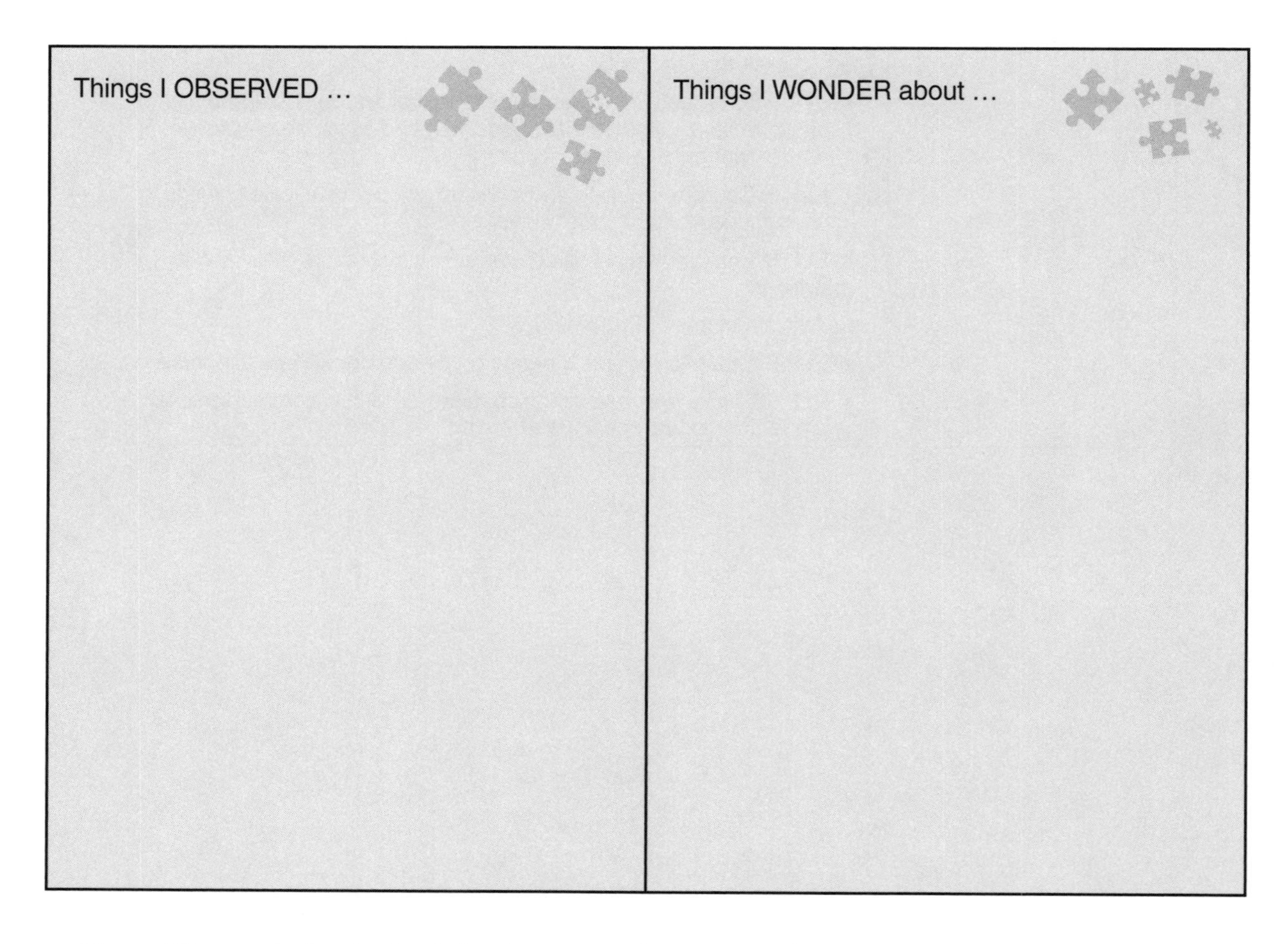

Many different kinds of animals live in groups. Insects often live with other insects in a colony. Fish often travel together in schools. Birds live with other birds in colonies and fly in flocks. Mammals often group together into packs or herds. The size of these groups can range from two or three animals to many thousands of animals.

Wolves are an example of an animal that lives in a group. Scientists often observe 5 to 15 wolves living together for long periods of time. The groups are called wolf packs. There are many potential reasons that may explain why animals, such as wolves, live in a group rather than alone. For example, groups of animals can work together to find food, raise young, or deal with changes in the environment. All of these reasons could make it easier for an animal to survive. Not all animals, however, live in groups. Some animals spend most of their life alone. Therefore, it is important for us to determine why it is a benefit or why it is not a benefit for animals to live as part of a group.

In this investigation you will watch several videos of wolves hunting different types of prey such as caribou, elk, and bison. These three different types of animals are not all the same size. An adult caribou weighs between 200 and 400 pounds, an adult elk weighs between 500 and 700 pounds, and an adult bison weighs between 1,300 and 1,500 pounds. Young caribou, elk, and bison, however, weigh much less.

Your goal in this investigation is to figure out if living in a group (the cause) makes it easier for wolves to get the food they need to survive (the effect). To accomplish this goal, you will need to look for a potential cause-and-effect relationship. Scientists often look for cause-and-effect relationships like this to help explain their observations. You can therefore look for a cause-and-effect relationship to help explain why wolves live in groups.

Things we KNOW from what we read …	What we will NEED to figure out …

Your Task

Use what you know about predators, prey, patterns, and cause-and-effect relationships to design and carry out an investigation to figure out if wolves benefit from hunting in a group.

The *guiding question* of this investigation is, ***Why do wolves live in groups?***

Materials

You will use a computer or tablet with internet access to watch the following videos during your investigation:

- Video showing wolves hunting caribou
- Video showing wolves hunting elk
- Video showing wolves hunting caribou
- Video showing gray wolves chasing down elk
- Video showing baby bison taking on a wolf
- Video showing wolves hunting buffalo
- Video showing wolves taking down elk
- Video showing bison and her calf battling wolves

Safety Rules

Follow all normal safety rules.

Plan Your Investigation

Prepare a plan for your investigation by filling out the chart that follows; this plan is called an *investigation proposal.* Before you start developing your plan, be sure to discuss the following questions with the other members of your group:

- What types of **patterns** might we look for to help answer the guiding question?
- What information do we need to find a **cause-and-effect relationship?**

Our guiding question:

We will collect the following data from the videos:

These are the steps we will follow to collect data as we watch the videos:

I approve of this investigation proposal.

Teacher's signature

Date

Collect Your Data

Keep a record of what you observe as you watch the videos in the space below.

Analyze Your Data

You will need to analyze the data you collected while watching the videos before you can develop an answer to the guiding question. In the space below, you can create a table or graph to show the outcomes of the different hunts.

Draft Argument

Develop an argument on a whiteboard. It should include the following parts:

1. A *claim:* Your answer to the guiding question.
2. *Evidence:* An analysis of the data and an explanation of what the analysis means.
3. A *justification of the evidence:* Why your group thinks the evidence is important.

The Guiding Question:	
Our Claim:	
Our Evidence:	Our Justification of the Evidence:

Argumentation Session

Share your argument with your classmates. Be sure to ask them how to make your draft argument better. Keep track of their suggestions in the space below.

Ways to IMPROVE our argument …

Draft Report

Prepare an *investigation report* to share what you have learned. Use the information in this handout and your group's final argument to write a *draft* of your investigation report.

Introduction

We have been studying ____ in class.

Before we started this investigation, we explored ______________________________

____

____

We noticed ______________________________________

____

____

My goal for this investigation was to figure out ______________________________

____

____

The guiding question was ______________________________

____

____

Method

To gather the data I needed to answer this question, I ______________________________

____

____

____

____

____

I then analyzed the data I collected by ____

____

____

____

____

Argument

My claim is ____

____

____

The ______________________ below shows ____

____

____

____

This evidence is important because ______________________________

____

____

____

____

____

____

____

Review

Your friends need your help! Review the draft of their investigation reports and give them ideas about how to improve. Use the *peer-review guide* when doing your review.

Submit Your Final Report

Once you have received feedback from your friends about your draft report, create your final investigation report and hand it in to your teacher.

Checkout Questions

Investigation 6. Life in Groups: Why Do Wolves Live in Groups?

1. Pictured below are four different animals. Circle the number of wolves that you think would need to hunt together to catch and eat that animal.

A.

Adult moose
1,600–1,800 pounds

1–2 4–6 8–10

B.

Adult caribou
200–400 pounds

1–2 4–6 8–10

C.

Adult white-tailed deer
80–100 pounds

1–2 4–6 8–10

D.

Baby moose
50–80 pounds

1–2 4–6 8–10

Checkout Questions

2. Explain your thinking. What *cause-and-effect relationship* did you use to determine how many wolves would need to hunt together to catch and eat an animal?

____

____

____

____

____

____

____

____

____

Teacher Scoring Rubric for the Checkout Questions

Level	Description
3	The student can apply the core idea correctly in all cases and can fully explain the cause-and-effect relationship.
2	The student can apply the core idea correctly in all cases but cannot fully explain the cause-and-effect relationship.
1	The student cannot apply the core idea correctly in all cases but can fully explain the cause-and-effect relationship.
0	The student cannot apply the core idea correctly in all cases and cannot explain the cause-and-effect relationship.

Section 4

Heredity: Inheritance and Variation of Traits

Investigation 7

Variation Within a Species: How Similar Are Earthworms to Each Other?

Purpose

The purpose of this investigation is to give students an opportunity to use the disciplinary core idea (DCI) of LS3.B: Variation of Traits and the crosscutting concept (CC) of Patterns from *A Framework for K–12 Science Education* (NRC 2012) to figure out which traits of earthworms differ and which do not and how much variation there is in the traits that do differ. Students will also learn about the difference between data and evidence in science during the reflective discussion.

The Disciplinary Core Idea

Students in third grade should understand the following about Variation of Traits and be able to use this DCI to figure out which traits of earthworms differ and which do not and how much variation there is in the traits that do differ:

> *Individuals of the same kind of plant or animal are recognizable as similar but can also vary in many ways. (NRC 2012, p. 160)*

The Crosscutting Concept

Students in third grade should understand the following about the CC Patterns:

> *Noticing patterns is often a first step to organizing and asking scientific questions about why and how the patterns occur. One major use of pattern recognition is in classification, which depends on careful observation of similarities and differences; objects can be classified into groups on the basis of similarities of visible or microscopic features or on the basis of similarities of function. Such classification is useful in codifying relationships and organizing a multitude of objects or processes into a limited number of groups. (NRC 2012, p. 85)*

Students in third grade should also be given opportunities to "investigate the characteristics that allow classification of animal types (e.g., mammals, fish, insects), of plants (e.g., trees, shrubs, grasses), or of materials (e.g., wood, rock, metal, plastic)" (NRC 2012, p. 86).

Students should be encouraged to use their developing understanding of patterns as a tool or a way of thinking about a phenomenon during this investigation to help them figure out which traits of earthworms differ and which do not and how much variation there is in the traits that do differ.

What Students Figure Out

Animals and plants differ in how they look and function. Scientists classify animals and plants based on the traits they do or do not have. Individuals that belong to a specific group share the same traits but can have different versions of a trait. The variation in shared traits is what makes each individual unique within a group. All earthworms, for example, share several traits such as a segmented body, a clitellum, a mouth on one end, and an anus on the other. However, there is variation in these traits among different earthworms. For example, one earthworm might have more segments or a larger clitellum than a different earthworm. These differences in the shared traits of earthworms make each one unique.

Timeline

The time needed to complete this investigation is 270 minutes (4 hours and 30 minutes). The amount of instructional time needed for each stage of the investigation is as follows:

- *Stage 1.* Introduce the task and the guiding question: 35 minutes
- *Stage 2.* Design a method and collect data: 60 minutes
- *Stage 3.* Create a draft argument: 35 minutes
- *Stage 4.* Argumentation session: 30 minutes
- *Stage 5.* Reflective discussion: 15 minutes
- *Stage 6.* Write a draft report: 35 minutes
- *Stage 7.* Peer review: 30 minutes
- *Stage 8.* Revise the report: 30 minutes

This investigation can be completed in one day or over eight days (one day for each stage) during your designated science time in the daily schedule.

Materials and Preparation

The materials needed for this investigation are listed in Table 7.1 (p. 256). The items can be purchased through an online retailer such as Amazon (*www.amazon.com*). The earthworms can also be purchased from a local garden supply store (worms are often used for composting). The materials for this investigation can also be purchased as a complete kit (which includes enough materials for 24 students, or six groups) at *www.argumentdriveninquiry.com*.

TABLE 7.1

Materials for Investigation 7

Item	Quantity
Safety goggles	1 per student
Nitrile gloves	1 pair per student
Earthworms	10 per group
Clear plastic container with lid (snack size)	1 per group
Plastic food tray, 10" × 14"	1 per group
Ruler, 12"	1 per group
Electronic scale	1 per group
Magnifying glass	2 per group
Whiteboard, 2' × 3'*	1 per group
Investigation Handout	1 per student
Peer-review guide and teacher scoring rubric	1 per student
Checkout Questions (optional)	1 per student

* As an alternative, students can use computer and presentation software such as Microsoft PowerPoint or Apple Keynote to create their arguments.

Place one earthworm in each plastic container and close the lid before starting the investigation. You will distribute these containers to the students during stage 1 of the investigation as a way to introduce the task and the guiding question. The other worms will be placed in the trays for the students to study during stage 2 of the investigation.

Be sure to use a set routine for distributing and collecting the materials. One option is to set up the materials for each group in a kit that you can deliver to each group. A second option is to have all the materials on a table or cart at a central location. You can then assign a member of each group to be the "materials manager." This individual is responsible for collecting all the materials his or her group needs from the table or cart during class and for returning all the materials at the end of the class.

Safety Precautions

Remind students to follow all normal safety rules. In addition, tell the students to keep the earthworms healthy by not doing anything to hurt them such as pulling on them, poking them, cutting them, dropping them, or anything else that might hurt them. Also tell the students to take the following safety precautions:

- Wear sanitized indirectly vented chemical-splash goggles and nitrile gloves during setup, investigation activity, and cleanup.
- Wash their hands with soap and water when done collecting data.

Lesson Plan by Stage

Stage 1: Introduce the Task and the Guiding Question (35 minutes)

1. Ask the students to sit in six groups, with three or four students in each group.
2. Ask students to clear off their desks except for a pencil (and their *Student Workbook for Argument-Driven Inquiry in Third-Grade Science* if they have one).
3. Pass out an Investigation Handout to each student (or ask students to turn to Investigation Log 7 in their workbook).
4. Read the first paragraph of the "Introduction" aloud to the class. Ask the students to follow along as you read.
5. Pass out a clear plastic container with one earthworm to each group. As you pass out the container, tell the students not to open it and not to harm the worm in any way.
6. Remind students of the safety rules and explain the safety precautions for this investigation.
7. Tell students to record their observations and questions about the worm in the "OBSERVED/WONDER" chart in the "Introduction" section of their Investigation Handout (or the investigation log in their workbook).
8. Ask students to share *what they observed* about the worm.
9. Ask students to share *what questions they have* about the worm.
10. Collect the containers with the earthworms.
11. Tell the students, "Some of your questions might be answered by reading the rest of the 'Introduction.'"
12. Ask the students to read the rest of the "Introduction" on their own *or* ask them to follow along as you read aloud.
13. Once the students have read rest of the "Introduction," ask them to fill out the "KNOW/NEED" chart on their Investigation Handout (or in their investigation log) as a group.
14. Ask students to share what they learned from the reading. Add these ideas to a class "know / need to figure out" chart.
15. Ask students to share what they think they will need to figure out based on what they read. Add these ideas to the class "know / need to figure out" chart.
16. Tell the students, "It looks like we have something to figure out. Let's see what we will need to do during our investigation."
17. Read the task and the guiding question aloud.
18. Tell the students, "I have lots of different earthworms here that you will be able to study during your investigation."

19. Show the students that they will have 10 earthworms available to each group during the investigation by holding up a tray with 10 earthworms on it. Be sure to tell the students that they need to keep the worms on the tray and then need to be careful with them because they are living things and should not be hurt in any way. You can then show them the other equipment (scale, ruler, magnifying glass) they have available to use by holding each piece of equipment up one at a time and asking them what they might do with it.

Stage 2: Design a Method and Collect Data (60 minutes)

1. Tell the students, "I am now going to give you and the other members of your group about 15 minutes to plan your investigation. Before you begin, I want you all to take a couple of minutes to discuss the following questions with the rest of your group."
2. Show the following questions on the screen or board:
   - What information should we collect so we can *describe* the traits of an earthworm?
   - What types of *patterns* might we look for to help answer the guiding question?
3. Tell the students, "Please take a few minutes to come up with an answer to these questions." Give the students two or three minutes to discuss these two questions.
4. Ask two or three different groups to share their answers. Be sure to highlight or write down any important ideas on the board so students can refer to them later.
5. If possible, use a document camera to project an image of the graphic organizer for this investigation on a screen or board (or take a picture of it and project the picture on a screen or board). Tell the students, "I now want you all to plan out your investigation. To do that, you will need to create an investigation proposal by filling out this graphic organizer."
6. Point to the box labeled "Our guiding question:" and tell the students, "You can put the question we are trying to answer in this box." Then ask, "Where can we find the guiding question?"
7. Wait for a student to answer where to find the guiding question (the answer is "in the handout").
8. Point to the box labeled "We will collect the following data:" and tell the students, "You can list the measurements or observations that you will need to collect during the investigation in this box."
9. Point to the box labeled "These are the steps we will follow to collect data:" and tell the students, "You can list what you are going to do to collect the data

you need and what you will do with your data once you have it. Be sure to give enough detail that I could do your investigation for you."

10. Ask the students, "Do you have any questions about what you need to do?"
11. Answer any questions that come up.
12. Tell the students, "Once you are done, raise your hand and let me know. I'll then come by and look over your proposal and give you some feedback. You may not begin collecting data until I have approved your proposal by signing it. You need to have your proposal done in the next 15 minutes."
13. Give the students 15 minutes to work in their groups on their investigation proposal. As they work, move from group to group to check in, ask probing questions, and offer a suggestion if a group gets stuck.

What should a student-designed investigation look like?

The students' investigation proposal should include the following information:

- The guiding question is "How similar are earthworms to each other?"
- There are a lot of different types of data that students can collect during this investigation. Examples include the following:
  - The length of the earthworm
  - The width of an earthworm
  - The presence of body segments
  - The number of body segments
  - The width of the body segments
  - The presence of a clitellum (a bulge in the body of the earthworm)
  - The size of the clitellum
  - Body color

At a minimum, each group should collect data about two different traits. This investigation works best if each group selects different traits. It also works well if each group picks a trait that does not vary (such as presence of body segments or presence of a clitellum) and a trait that does (such as length of the earthworm, number of segments, or size of clitellum). Be sure to encourage students to collect quantitative (numerical) data when possible.

The steps that the students will follow to collect the data should reflect the traits that they decide to examine. However, a procedure might include the following steps:

> 1. Identify a worm.
> 2. Measure [trait 1].
> 3. Measure [trait 2].
> 4. Repeat steps 1–3 for each worm.
> 5. Create a graph that shows the number of worms with a trait or a version of a trait.
>
> This is just an example of a procedure, and there should be a lot of variation in the student-designed investigations.

14. As each group finishes its investigation proposal, be sure to read it over and determine if it will be productive or not. If you feel the investigation will be productive (not necessarily what you would do or what the other groups are doing), sign your name on the proposal and let the group start collecting data. If the plan needs to be changed, offer some suggestions or ask some probing questions, and have the group make the changes before you approve it.
15. Pass out the materials or have one student from each group collect the materials they need from a central supply table or cart for the groups that have an approved proposal.
16. Remind students of the safety rules and precautions for this investigation.
17. Tell the students to collect their data and record their observations or measurements in the "Collect Your Data" box in their Investigation Handout (or the investigation log in their workbook).
18. Give the students 20 minutes to collect their data.
19. Be sure to collect the materials from each group before asking them to analyze their data.

Stage 3: Create a Draft Argument (35 minutes)

1. Tell the students, "Now that we have all this data, we need to analyze the data so we can figure out an answer to the guiding question."
2. If possible, project an image of the "Analyze Your Data" section for this investigation on a screen or board using a document camera (or take a picture of it and project the picture on a screen or board). Point to the section and tell the students, "You can create a couple of different graphs as a way to analyze your data. You can make your graphs in this section."
3. Ask the students, "What information do we need to include in these graphs?"

4. Tell the students, "Please take a few minutes to discuss this question with your group, and be ready to share."
5. Give the students five minutes to discuss.
6. Ask two or three different groups to share their answers. Be sure to highlight or write down any important ideas on the board so students can refer to them later.
7. Tell the students, "I am now going to give you and the other members of your group about 10 minutes to create your graphs." If the students are having trouble making a graph, you can take a few minutes to provide a mini-lesson about how to create a graph from a bunch of observations or measurements (this strategy is called just-in-time instruction because it is offered only when students get stuck).

What should a graph look like for this investigation?

There are a number of different ways that students can analyze the observations or measurements they collect during this investigation. One of the most straightforward ways is to create two or more scaled bar graphs to represent a data set with several categories. Each bar graph should have the names of the categories for traits on the horizontal or *x*-axis. The traits can be divided into two categories (such as presence or absence of body segments) or three or more categories (such as 100–110 body segments, 111–120 body segments, 121–130 body segments, and 131–140 body segments). The number of worms that belong in a category should be on the *y*-axis. An example of this type of graph can be seen in Figure 7.1 (p. 263). There are other options for analyzing the data they collected. Students often come up with some unique ways of analyzing their data, so be sure to give them some voice and choice during this stage.

8. Give the students 10 minutes to analyze their data. As they work, move from group to group to check in, ask probing questions, and offer suggestions.
9. Tell the students, "I am now going to give you and the other members of your group 15 minutes to create an argument to share what you have learned and convince others that they should believe you. Before you do that, we need to take a few minutes to discuss what you need to include in your argument."
10. If possible, use a document camera to project the "Argument Presentation on a Whiteboard" image from the Investigation Handout (or the investigation log in their workbook) on a screen or board (or take a picture of it and project the picture on a screen or board).

11. Point to the box labeled "The Guiding Question:" and tell the students, "You can put the question we are trying to answer here on your whiteboard."
12. Point to the box labeled "Our Claim:" and tell the students, "You can put your claim here on your whiteboard. The claim is your answer to the guiding question."
13. Point to the box labeled "Our Evidence:" and tell the students, "You can put the evidence that you are using to support your claim here on your whiteboard. Your evidence will need to include the analysis you just did and an explanation of what your analysis means or shows. Scientists always need to support their claims with evidence."
14. Point to the box labeled "Our Justification of the Evidence:" and tell the students, "You can put your justification of your evidence here on your whiteboard. Your justification needs to explain why your evidence is important. Scientists often use core ideas to explain why the evidence they are using matters. Core ideas are important concepts that scientists use to help them make sense of what happens during an investigation."
15. Ask the students, "What are some core ideas that we read about earlier that might help us explain why the evidence we are using is important?"
16. Ask students to share some of the core ideas from the "Introduction" section of the Investigation Handout (or the investigation log in the workbook). List these core ideas on the board.
17. Tell the students, "That is great. I would like to see everyone try to include these core ideas in your justification of the evidence. Your goal is to use these core ideas to help explain why your evidence matters and why the rest of us should pay attention to it."
18. Ask the students, "Do you have any questions about what you need to do?"
19. Answer any questions that come up.
20. Tell the students, "Okay, go ahead and start working on your arguments. You need to have your argument done in the next 15 minutes. It doesn't need to be perfect. We just need something down on the whiteboards so we can share our ideas."
21. Give the students 15 minutes to work in their groups on their arguments. As they work, move from group to group to check in, ask probing questions, and offer a suggestion if a group gets stuck. Figure 7.1 shows an example of an argument created by students for this investigation.

FIGURE 7.1

Example of an argument

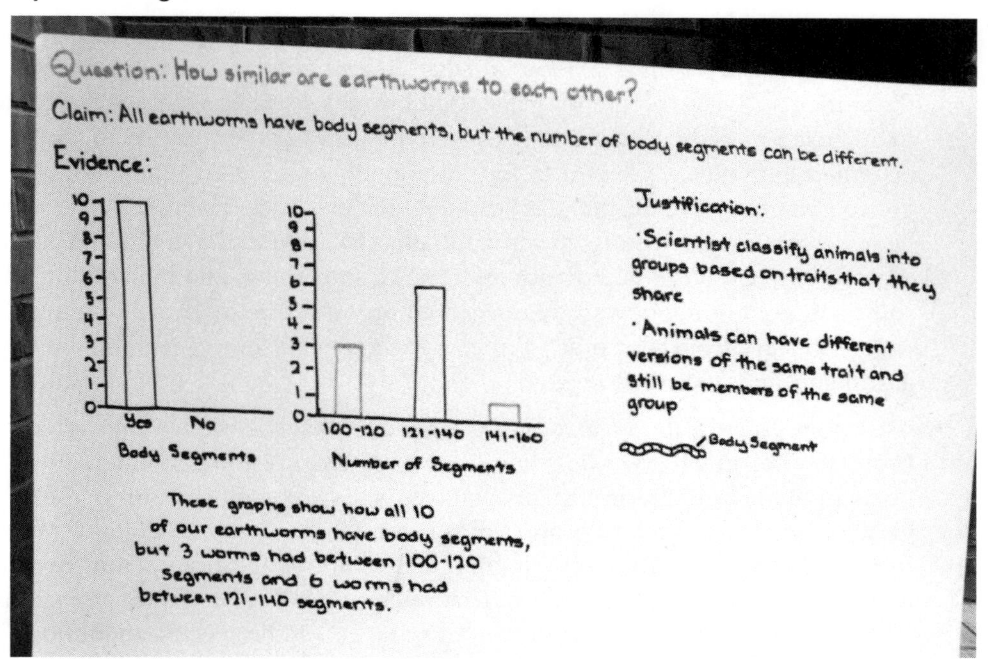

Stage 4: Argumentation Session (30 minutes)

The argumentation session can be conducted in a whole-class presentation format, a gallery walk format, or a modified gallery walk format. We recommend using a whole-class presentation format for the first investigation, but try to transition to either the gallery walk or modified gallery walk format as soon as possible because that will maximize student voice and choice inside the classroom. The following list shows the steps for the three formats; unless otherwise noted, the steps are the same for all three formats.

1. Begin by introducing the use of the whiteboard.
   - *If using the whole-class presentation format,* tell the students, "We are now going to share our arguments. Please set up your whiteboard so everyone can see them."
   - *If using the gallery walk or modified gallery walk format,* tell the students, "We are now going to share our arguments. Please set up your whiteboard so they are facing the walls."
2. Allow the students to set up their whiteboards.
   - *If using the whole-class presentation format,* the whiteboards should be set up on stands or chairs so they are facing toward the center of the room.

- *If using the gallery walk or modified gallery walk format,* the whiteboards should be set up on stands or chairs so they are facing toward the outside of the room.

3. Give the following instructions to the students:
   - *If using the whole-class presentation format or the modified gallery walk format,* tell the students, "Okay, before we get started I want to explain what we are going to do next. I'm going to ask some of you to present your arguments to your classmates. If you are presenting your argument, your job is to share your group's claim, evidence, and justification of the evidence. The rest of you will be reviewers. If you are a reviewer, your job is to listen to the presenters, ask the presenters questions if you do not understand something, and then offer them some suggestions about ways to make their argument better. After we have a chance to learn from each other, I'm going to give you some time to revise your arguments and make them better."
   - *If using the gallery walk format,* tell the students, "Okay, before we get started I want to explain what we are going to do next. You are going to have an opportunity to read the arguments that were created by other groups. Your group will go to a different group's argument. I'll give you a few minutes to read it and review it. Your job is to offer them some suggestions about ways to make their argument better. You can use sticky notes to give them suggestions. Please be specific about what you want to change and be specific about how you think they should change it. After we have a chance to learn from each other, I'm going to give you some time to revise your arguments and make them better."
4. Use a document camera to project the "Ways to IMPROVE our argument …" box from the Investigation Handout (or the investigation log in their workbook) on a screen or board (or take a picture of it and project the picture on a screen or board).
   - *If using the whole-class presentation format or the modified gallery walk format,* point to the box and tell the students, "If you are a presenter, you can write down the suggestions you get from the reviewers here. If you are a reviewer, and you see a good idea from another group, you can write down that idea here. Once we are done with the presentations, I will give you a chance to use these suggestions or ideas to improve your arguments.
   - *If using the gallery walk format,* point to the box and tell the students, "If you see good ideas from another group, you can write them down here. Once we are done reviewing the different arguments, I will give you a chance to use these ideas to improve your own arguments. It is important to share ideas like this."

   Ask the students, "Do you have any questions about what you need to do?"
5. Answer any questions that come up.
6. Give the following instructions:

- *If using the whole-class presentation format,* tell the students, "Okay. Let's get started."
- *If using the gallery walk format,* tell the students, "Okay, I'm now going to tell you which argument to go to and review.
- *If using the modified gallery walk format,* tell the students, "Okay, I'm now going to assign you to be a presenter or a reviewer." Assign one or two students from each group to be presenters and one or two students from each group to be reviewers.

7. Begin the review of the arguments.
   - *If using the whole-class presentation format,* have four or five groups present their argument one at a time. Give each group only two to three minutes to present their argument. Then give the class two to three minutes to ask them questions and offer suggestions. Be sure to encourage as much participation from the students as possible.
   - *If using the gallery walk format,* tell the students, "Okay. Let's get started. Each group, move one argument to the left. Don't move to the next argument until I tell you to move. Once you get there, read the argument and then offer suggestions about how to make it better. I will put some sticky notes next to each argument. You can use the sticky notes to leave your suggestions." Give each group about three to four minutes to read the arguments, talk, and offer suggestions.

     a. Tell the students, "Okay. Let's rotate. Move one group to the left."

     b. Again, give each group three or four minutes to read, talk, and offer suggestions.

     c. Repeat this process for two more rotations.
   - *If using the modified gallery walk format,* tell the students, "Okay. Let's get started. Reviewers, move one group to the left. Don't move to the next group until I tell you to move. Presenters, go ahead and share your argument with the reviewers when they get there." Give each group of presenters and reviewers about three to four minutes to talk.

     a. Tell the students, "Okay. Let's rotate. Reviewers, move one group to the left."

     b. Again, give each group of presenters and reviewers about three or four minutes to talk.

     c. Repeat this process for two more rotations.
8. Tell the students to return to their workstations.
9. Give the following instructions about revising the argument:
   - *If using the whole-class presentation format,* tell the students, "I'm now going to give you about 10 minutes to revise your argument. Take a few minutes to talk in your groups and determine what you want to change to make your

argument better. Once you have decided what to change, go ahead and make the changes to your whiteboard."

- *If using the gallery walk format,* tell the students, "I'm now going to give you about 10 minutes to revise your argument. Take a few minutes to read the suggestions that were left at your argument. Then talk in your groups and determine what you want to change to make your argument better. Once you have decided what to change, go ahead and make the changes to your whiteboard."
- *If using the modified gallery walk format,* "I'm now going to give you about 10 minutes to revise your argument. Please return to your original groups." Wait for the students to move back into their original groups and then tell the students, "Okay, take a few minutes to talk in your groups and determine what you want to change to make your argument better. Once you have decided what to change, go ahead and make the changes to your whiteboard."

Ask the students, "Do you have any questions about what you need to do?"

10. Answer any questions that come up.
11. Tell the students, "Okay. Let's get started."
12. Give the students 10 minutes to work in their groups on their arguments. As they work, be sure to move from group to group to check in, ask probing questions, and offer a suggestion if a group gets stuck.

Stage 5: Reflective Discussion (15 minutes)

1. Tell the students, "We are now going to take a minute to talk about what we did and what we have learned."
2. Show Figure 7.2 on the screen. This image shows several different earthworms in a compost pile.
3. Ask the students, "What do you all see going on here?"
4. Allow students to share their ideas.
5. Ask the students, "What do you think will be similar about all these earthworms?"
6. Allow students to share their ideas.
7. Ask the students, "What do you think will be different about all these earthworms?"
8. Allow students to share their ideas.
9. Show Figure 7.3 on the screen. This image shows several different ladybugs.
10. Ask the students, "What do you all see going on here?"
11. Allow students to share their ideas.

FIGURE 7.2

Earthworms

FIGURE 7.3

A group of ladybugs

Full-color versions of Figures 7.2 and 7.3 are available on the book's Extras page at *www.nsta.org/adi-3rd*.

12. Ask the students, "What do you think will be similar about all these ladybugs?"
13. Allow students to share their ideas. Be sure to point out that they all have six legs, two antennae, two wings, and a wing covering (which is called an elytron [the plural is elytra]) that is red in color, if students do not mention these traits.
14. Ask the students, "What do you think will be different about all these ladybugs?"
15. Allow students to share their ideas. Be sure to point out traits like length, number of spots, and size of spots.
16. Tell the students, "Okay, let's make sure we are on the same page. Animals and plants differ in how they look and function. Scientists classify animals and plants based on the traits they do or do not have. Individuals that belong to a specific group share the same traits but can have different versions of a trait. The differences in the shared traits are what make each individual unique within a group. All earthworms, for example, share several traits such as a segmented body, a clitellum, a mouth on one end, and an anus on the other. However, there is variation in these traits among different earthworms. For example, one earthworm might have more segments or a larger clitellum than a different earthworm. These differences in the shared traits of earthworms make each one unique. All ladybugs also share many of the same traits such as six legs, two antennae, two wings, and a red wing covering. Individual ladybugs are different from each other because they have different numbers of spots on their wing covering or have longer legs. This variation in traits among individuals in a group is a really important core idea in science."
17. Ask the students, "Does anyone have any questions about this core idea?"

18. Answer any questions that come up.
19. Tell the students, "We also looked for patterns during our investigation." Then ask, "Can anyone tell me why we needed to look for patterns?"
20. Allow students to share their ideas.
21. Tell the students, "Patterns are really important in science. Scientists look for patterns all the time. In fact, they even use patterns to help classify different things just like we did."
22. Tell the students, "We are now going take a minute to talk about what went well and what didn't go so well during our investigation. We need to talk about this because you are going to be planning and carrying out your own investigations like this a lot this year, and I want to help you all get better at it."
23. Show an image of the question "What made your investigation scientific?" on the screen. Tell the students, "Take a few minutes to talk about how you would answer this question with the other people in your group. Be ready to share with the rest of the class." Give the students two to three minutes to talk in their group.
24. Ask the students, "What do you all think? Who would like to share an idea?"
25. Allow students to share their ideas. Be sure to expand on their ideas about what makes an investigation scientific.
26. Show an image of the question "What made your investigation not so scientific?" on the screen. Tell the students, "Take a few minutes to talk about how you would answer this question with the other people in your group. Be ready to share with the rest of the class." Give the students two to three minutes to talk in their group.
27. Ask the students, "What do you all think? Who would like to share an idea?"
28. Allow students to share their ideas. Be sure to expand on their ideas about what makes an investigation less scientific.
29. Show an image of the question "What rules can we put into place to help us make sure our next investigation is more scientific?" on the screen. Tell the students, "Take a few minutes to talk about how you would answer this question with the other people in your group. Be ready to share with the rest of the class." Give the students two to three minutes to talk in their group.
30. Ask the students, "What do you all think? Who would like to share an idea?"
31. Allow students to share their ideas. Once they have shared their ideas, offer a suggestion for a possible class rule.
32. Ask the students, "What do you all think? Should we make this a rule?"
33. If the students agree, write the rule on the board or make a class "Rules for Scientific Investigation" chart so you can refer to it during the next investigation.

FIGURE 7.4
An earthworm

34. Tell the students, "We are now going take a minute to talk about what makes science different from other subjects."
35. Show an image of the question "What is the difference between data and evidence in science?" on the screen. Tell the students, "Take a few minutes to talk about how you would answer this question with the other people in your group. Be ready to share with the rest of the class." Give the students two to three minutes to talk in their group.
36. Ask the students, "What do you all think? Who would like to share an idea?"
37. Allow students to share their ideas.
38. Tell the students, "Okay, let's make sure we are all using the same definition. Data is a bunch of observations or measurements. Evidence is an analysis of the data and an interpretation of the analysis."
39. Show Figure 7.4 along with the statement "The earthworm is 26 cm long" on the screen.
40. Ask the students, "Is this statement data or evidence and why?"
41. Allow students to share their ideas.
42. Tell the students, "That statement is data because it is a measurement."
43. Ask the students, "When does data become evidence?"
44. Allow students to share their ideas. Be sure to ask probing questions until students point out that evidence consists of an analysis of multiple measurements or observations and an interpretation of the analysis.
45. Ask the students, "Does anyone have any questions about the difference between data and evidence?"
46. Answer any questions that come up.

Stage 6: Write a Draft Report (35 minutes)

Your students will use either the Investigation Handout or the investigation log in the student workbook when writing the draft report. When you give the directions shown in quotes in the following steps, substitute "investigation log" (as shown in brackets) for "handout" if they are using the workbook.

1. Tell the students, "You are now going to write an investigation report to share what you have learned. Please take out a pencil and turn to the 'Draft Report' section of your handout [investigation log]."
2. If possible, use a document camera to project the "Introduction" section of the draft report from the Investigation Handout (or the investigation log in their workbook) on a screen or board (or take a picture of it and project the picture on a screen or board).
3. Tell the students, "The first part of the report is called the 'Introduction.' In this section of the report you want to explain to the reader what you were investigating, why you were investigating it, and what question you were trying to answer. All of this information can be found in the text at the beginning of your handout [investigation log]." Point to the image and say, "There are some sentence starters here to help you begin writing the report." Ask the students, "Do you have any questions about what you need to do?"
4. Answer any questions that come up.
5. Tell the students, "Okay. Let's write."
6. Give the students 10 minutes to write the "Introduction" section of the report. As they work, move from student to student to check in, ask probing questions, and offer a suggestion if a student gets stuck.
7. If possible, use a document camera to project the "Method" section of the draft report from the Investigation Handout (or the investigation log in their workbook) on a screen or board (or take a picture of it and project the picture on a screen or board).
8. Tell the students, "The second part of the report is called the 'Method.' In this section of the report you want to explain to the reader what you did during the investigation, what data you collected and why, and how you went about analyzing your data. All of this information can be found in the 'Plan Your Investigation' section of your handout [investigation log]. Remember that you all planned and carried out different investigations, so do not assume that the reader will know what you did." Point to the image and say, "There are some sentence starters here to help you begin writing this part of the report." Ask the students, "Do you have any questions about what you need to do?"
9. Answer any questions that come up.
10. Tell the students, "Okay. Let's write."

11. Give the students 10 minutes to write the "Method" section of the report. As they work, move from student to student to check in, ask probing questions, and offer a suggestion if a student gets stuck.
12. If possible, use a document camera to project the "Argument" section of the draft report from the Investigation Handout (or the investigation log in their workbook) on a screen or board (or take a picture of it and project the picture on a screen or board).
13. Tell the students, "The last part of the report is called the 'Argument.' In this section of the report you want to share your claim, evidence, and justification of the evidence with the reader. All of this information can be found on your whiteboard." Point to the image and say, "There are some sentence starters here to help you begin writing this part of the report." Ask the students, "Do you have any questions about what you need to do?"
14. Answer any questions that come up.
15. Tell the students, "Okay. Let's write."
16. Give the students 10 minutes to write the "Argument" section of the report. As they work, move from student to student to check in, ask probing questions, and offer a suggestion if a student gets stuck.

Stage 7: Peer Review (30 minutes)

Your students will use either the Investigation Handout or the investigation log in the student workbook when doing the peer review. When you give the directions shown in quotes in the following steps, substitute "workbook" (as shown in brackets) for "Investigation Handout" if they are using the workbook.

1. Tell the students, "We are now going to review our reports to find ways to make them better. I'm going to come around and collect your Investigation Handout [workbook]. While I do that, please take out a pencil."
2. Collect the Investigation Handouts or workbooks from the students.
3. If possible, use a document camera to project the peer-review guide (PRG; see Appendix 4) on a screen or board (or take a picture of it and project the picture on a screen or board).
4. Tell the students, "We are going to use this peer-review guide to give each other feedback." Point to the image.
5. Give the following instructions:
    - *If using the Investigation Handout,* tell the students, "I'm going to ask you to work with a partner to do this. I'm going to give you and your partner a draft report to read and a peer-review guide to fill out. You two will then read the report together. Once you are done reading the report, I want you to answer each of

the questions on the peer-review guide." Point to the review questions on the image of the PRG.

- *If using the workbook,* tell the students, "I'm going to ask you to work with a partner to do this. I'm going to give you and your partner a draft report to read. You two will then read the report together. Once you are done reading the report, I want you to answer each of the questions on the peer-review guide that is right after the report in the investigation log." Point to the review questions on the image of the PRG.

6. Tell the students, "You can check 'yes,' 'almost,' or 'no' after each question." Point to the checkboxes on the image of the PRG.
7. Tell the students, "This will be your rating for this part of the report. Make sure you agree on the rating you give the author. If you mark 'almost' or 'no,' then you need to tell the author what he or she needs to do to get a 'yes.'" Point to the space for the reviewer feedback on the image of the PRG.
8. Tell the students, "It is really important for you to give the authors feedback that is helpful. That means you need to tell them exactly what they need to do to make their reports better." Ask the students, "Do you have any questions about what you need to do?"
9. Answer any questions that come up.
10. Tell the students, "Please sit with a partner who is not in your current group." Allow the students time to sit with a partner.
11. Give the following instructions:
    - *If using the Investigation Handout,* tell the students, "Okay, I am now going to give you one report to read and one peer-review guide to fill out." Pass out one report to each pair. Make sure that the report you give a pair was not written by one of the students in that pair. Give each pair one PRG to fill out as a team.
    - *If using the workbook,* tell the students, "Okay, I am now going to give you one report to read." Pass out a workbook to each pair. Make sure that the workbook you give a pair is not from one of the students in that pair.
12. Tell the students, "Okay, I'm going to give you 15 minutes to read the report I gave you and to fill out the peer-review guide. Go ahead and get started."
13. Give the students 15 minutes to work. As they work, move around from pair to pair to check in and see how things are going, answer questions, and offer advice.
14. After 15 minutes pass, tell the students, "Okay, time is up." *If using the Investigation Handout,* say, "Please give me the report and the peer-review guide that you filled out." *If using the workbook,* say, "Please give me the workbook that you have."

15. Collect the Investigation Handouts and the PRGs, or collect the workbooks if they are being used. Be sure you keep the handout and the PRG together.
16. Give the following instructions:
    - *If using the Investigation Handout,* tell the students, "Okay, I am now going to give you a different report to read and a new peer-review guide to fill out." Pass out another report to each pair. Make sure that this report was not written by one of the students in that pair. Give each pair a new PRG to fill out as a team.
    - *If using the workbook,* tell the students, "Okay, I am now going to give you a different report to read." Pass out a different workbook to each pair. Make sure that the workbook you give a pair is not from one of the students in that pair.
17. Tell the students, "Okay, I'm going to give you 15 minutes to read this new report and to fill out the peer-review guide. Go ahead and get started."
18. Give the students 15 minutes to work. As they work, move around from pair to pair to check in and see how things are going, answer questions, and offer advice.
19. After 15 minutes pass, tell the students, "Okay, time is up." *If using the Investigation Handout,* say, "Please give me the report and the peer-review guide that you filled out." *If using the workbook,* say, "Please give me the workbook that you have."
20. Collect the Investigation Handouts and the PRGs, or collect the workbooks if they are being used. Be sure you keep the handout and the PRG together.

Stage 8: Revise the Report (30 minutes)

Your students will use either the Investigation Handout or the investigation log in the student workbook when revising the report. Except where noted below, the directions are the same whether using the handout or the log.

1. Give the following instructions:
    - *If using the Investigation Handout,* tell the students, "You are now going to revise your investigation report based on the feedback you get from your classmates. Please take out a pencil while I hand back your draft report and the peer-review guide."
    - *If using the investigation log in the student workbook,* tell the students, "You are now going to revise your investigation report based on the feedback you get from your classmates. Please take out a pencil while I hand back your investigation logs."
2. *If using the Investigation Handout,* pass back the handout and the PRG to each student. *If using the investigation log,* pass back the log to each student.

3. Tell the students, "Please take a few minutes to read over the peer-review guide. You should use it to figure out what you need to change in your report and how you will change the report."
4. Allow the students time to read the PRG.
5. *If using the investigation log,* if possible use a document camera to project the "Write Your Final Report" section from the investigation log on a screen or board (or take a picture of it and project the picture on a screen or board).
6. Give the following instructions:
   - *If using the Investigation Handout,* tell the students, "Okay. Let's revise our reports. Please take out a piece of paper. I would like you to rewrite your report. You can use your draft report as a starting point, but use the feedback on the peer-review guide to help make it better."
   - *If using the investigation log,* tell the students, "Okay. Let's revise our reports. I would like you to rewrite your report in the section of the investigation log that says 'Write Your Final Report.'" Point to the image on the screen and tell the students, "You can use your draft report as a starting point, but use the feedback on the peer-review guide to help make your report better."

   Ask the students, "Do you have any questions about what you need to do?"
7. Answer any questions that come up.
8. Tell the students, "Okay. Let's write."
9. Give the students 30 minutes to rewrite their report. As they work, move from student to student to check in, ask probing questions, and offer a suggestion if a student gets stuck.
10. Give the following instructions:
    - *If using the Investigation Handout,* tell the students, "Okay. Time's up. I will now come around and collect your Investigation Handout, the peer-review guide, and your final report."
    - *If using the investigation log,* tell the students, "Okay. Time's up. I will now come around and collect your workbooks."
11. *If using the Investigation Handout,* collect all the Investigation Handouts, PRGs, and final reports. *If using the investigation log,* collect all the workbooks.
12. *If using the Investigation Handout,* use the "Teacher Score" columns in the PRG to grade the final report. *If using the investigation log,* use the "ADI Investigation Report Grading Rubric" in the investigation log to grade the final report. Whether you are using the handout or the log, you can give the students feedback about their writing in the "Teacher Comments" section.

How to Use the Checkout Questions

The Checkout Questions are an optional assessment. We recommend giving them to students one day after they finish stage 8 of the ADI investigation. The Checkout Questions can be used as a formative or summative assessment of student thinking. If you plan to use them as a formative assessment, we recommend that you look over the student answers to determine if you need to reteach the core idea and/or crosscutting concept from the investigation, but do not grade them. If you plan to use them as a summative assessment, we have included a "Teacher Scoring Rubric" at the end of the Checkout Questions that you can use to score a student's ability to apply the core idea in a new scenario and explain their use of a crosscutting concept. The rubric includes a 4-point scale that ranges from 0 (the student cannot apply the core idea correctly in all cases and cannot explain the [crosscutting concept]) to 3 (the student can apply the core idea correctly in all cases and can fully explain the [crosscutting concept]). The Checkout Questions, regardless of how you decide to use them, are a great way to make student thinking visible so you can determine if the students have learned the core idea and the crosscutting concept.

A student who can apply the core idea correctly in all cases and can explain the pattern would give the following responses to questions 1 and 2:

- For question 1, mark "S" next to "Presence of a clitellum," "Presence of segments," "Presence of a mouth," and "No legs"; and mark "D" next to "Width of clitellum" and "Number of segments."
- For question 2, mark "S" next to "Number of legs," "Number of antennae," "Presence of a head," and "Presence of an elytron"; and mark "D" next to "Number of spots on elytron" and "Length of legs."

He or she should then be able to explain that living things are classified by the traits they share, but these shared traits can differ across the members of a group.

Connections to Standards

Table 7.2 (p. 276) highlights how the investigation can be used to address specific performance expectations from the *NGSS, Common Core State Standards (CCSS)* in English language arts (ELA) and in mathematics, and *English Language Proficiency (ELP) Standards.*

TABLE 7.2

Investigation 7 alignment with standards

***NGSS* performance expectation**	Strong alignment • 3-LS3-1: Analyze and interpret data to provide evidence that plants and animals have traits inherited from parents and that variation of these traits exists in a group of similar organisms.
***CCSS ELA*—Reading: Informational Text**	Key ideas and details • CCSS.ELA-LITERACY.RI.3.1: Ask and answer questions to demonstrate understanding of a text, referring explicitly to the text as the basis for the answers. • CCSS.ELA-LITERACY.RI.3.2: Determine the main idea of a text; recount the key details and explain how they support the main idea. • CCSS.ELA-LITERACY.RI.3.3: Describe the relationship between a series of historical events, scientific ideas or concepts, or steps in technical procedures in a text, using language that pertains to time, sequence, and cause/effect. Craft and structure • CCSS.ELA-LITERACY.RI.3.4: Determine the meaning of general academic and domain-specific words and phrases in a text relevant to a *grade 3 topic or subject area*. • CCSS.ELA-LITERACY.RI.3.5: Use text features and search tools (e.g., key words, sidebars, hyperlinks) to locate information relevant to a given topic efficiently. • CCSS.ELA-LITERACY.RI.3.6: Distinguish their own point of view from that of the author of a text. Integration of knowledge and ideas • CCSS.ELA-LITERACY.RI.3.7: Use information gained from illustrations (e.g., maps, photographs) and the words in a text to demonstrate understanding of the text (e.g., where, when, why, and how key events occur). • CCSS.ELA-LITERACY.RI.3.8: Describe the logical connection between particular sentences and paragraphs in a text (e.g., comparison, cause/effect, first/second/third in a sequence). • CCSS.ELA-LITERACY.RI.3.9: Compare and contrast the most important points and key details presented in two texts on the same topic. Range of reading and level of text complexity • CCSS.ELA-LITERACY.RI.3.10: By the end of the year, read and comprehend informational texts, including history/social studies, science, and technical texts, at the high end of the grades 2–3 text complexity band independently and proficiently.

Continued

Table 7.2 *(continued)*

***CCSS ELA*—Writing**	Text types and purposes • CCSS.ELA-LITERACY.W.3.1: Write opinion pieces on topics or texts, supporting a point of view with reasons. o CCSS.ELA-LITERACY.W.3.1.A: Introduce the topic or text they are writing about, state an opinion, and create an organizational structure that lists reasons. o CCSS.ELA-LITERACY.W.3.1.B: Provide reasons that support the opinion. o CCSS.ELA-LITERACY.W.3.1.C: Use linking words and phrases (e.g., *because*, *therefore*, *since, for example*) to connect opinion and reasons. o CCSS.ELA-LITERACY.W.3.1.D: Provide a concluding statement or section. • CCSS.ELA-LITERACY.W.3.2: Write informative or explanatory texts to examine a topic and convey ideas and information clearly. o CCSS.ELA-LITERACY.W.3.2.A: Introduce a topic and group related information together; include illustrations when useful to aiding comprehension. o CCSS.ELA-LITERACY.W.3.2.B: Develop the topic with facts, definitions, and details. o CCSS.ELA-LITERACY.W.3.2.C: Use linking words and phrases (e.g., *also*, *another*, *and*, *more*, *but*) to connect ideas within categories of information. o CCSS.ELA-LITERACY.W.3.2.D: Provide a concluding statement or section. Production and distribution of writing • CCSS.ELA-LITERACY.W.3.4: With guidance and support from adults, produce writing in which the development and organization are appropriate to task and purpose. • CCSS.ELA-LITERACY.W.3.5: With guidance and support from peers and adults, develop and strengthen writing as needed by planning, revising, and editing. • CCSS.ELA-LITERACY.W.3.6: With guidance and support from adults, use technology to produce and publish writing (using keyboarding skills) as well as to interact and collaborate with others. Research to build and present knowledge • CCSS.ELA-LITERACY.W.3.8: Recall information from experiences or gather information from print and digital sources; take brief notes on sources and sort evidence into provided categories. Range of writing • CCSS.ELA-LITERACY.W.3.10: Write routinely over extended time frames (time for research, reflection, and revision) and shorter time frames (a single sitting or a day or two) for a range of discipline-specific tasks, purposes, and audiences.

Continued

Table 7.2 (*continued*)

***CCSS ELA*—
Speaking and Listening**
	Comprehension and collaboration
	• CCSS.ELA-LITERACY.SL.3.1: Engage effectively in a range of collaborative discussions (one-on-one, in groups, and teacher-led) with diverse partners on *grade 3 topics and texts*, building on others' ideas and expressing their own clearly.
	  o CCSS.ELA-LITERACY.SL.3.1.A: Come to discussions prepared, having read or studied required material; explicitly draw on that preparation and other information known about the topic to explore ideas under discussion.
	  o CCSS.ELA-LITERACY.SL.3.1.B: Follow agreed-upon rules for discussions (e.g., gaining the floor in respectful ways, listening to others with care, speaking one at a time about the topics and texts under discussion).
	  o CCSS.ELA-LITERACY.SL.3.1.C: Ask questions to check understanding of information presented, stay on topic, and link their comments to the remarks of others.
	  o CCSS.ELA-LITERACY.SL.3.1.D: Explain their own ideas and understanding in light of the discussion.
	• CCSS.ELA-LITERACY.SL.3.2: Determine the main ideas and supporting details of a text read aloud or information presented in diverse media and formats, including visually, quantitatively, and orally.
	• CCSS.ELA-LITERACY.SL.3.3: Ask and answer questions about information from a speaker, offering appropriate elaboration and detail.
	Presentation of knowledge and ideas
	• CCSS.ELA-LITERACY.SL.3.4: Report on a topic or text, tell a story, or recount an experience with appropriate facts and relevant, descriptive details, speaking clearly at an understandable pace.
	• CCSS.ELA-LITERACY.SL.3.6: Speak in complete sentences when appropriate to task and situation in order to provide requested detail or clarification.

***CCSS Mathematics*— Operations and Algebraic Thinking**	Solve problems involving the four operations, and identify and explain patterns in arithmetic. • CCSS.MATH.CONTENT.3.OA.D.8: Solve two-step word problems using the four operations. Represent these problems using equations with a letter standing for the unknown quantity. Assess the reasonableness of answers using mental computation and estimation strategies including rounding.
***CCSS Mathematics*— Number and Operations in Base Ten**	Use place value understanding and properties of operations to perform multi-digit arithmetic. • CCSS.MATH.CONTENT.3.NBT.A.2: Fluently add and subtract within 1,000 using strategies and algorithms based on place value, properties of operations, and/or the relationship between addition and subtraction.

Continued

Table 7.2 *(continued)*

***CCSS Mathematics*—Measurement and Data**	Solve problems involving measurement and estimation. • CCSS.MATH.CONTENT.3.MD.A.2: Measure and estimate liquid volumes and masses of objects using standard units of grams (g), kilograms (kg), and liters (l). Add, subtract, multiply, or divide to solve one-step word problems involving masses or volumes that are given in the same units, e.g., by using drawings (such as a beaker with a measurement scale) to represent the problem. Represent and interpret data • CCSS.MATH.CONTENT.3.MD.B.3: Draw a scaled picture graph and a scaled bar graph to represent a data set with several categories. Solve one- and two-step "how many more" and "how many less" problems using information presented in scaled bar graphs. • CCSS.MATH.CONTENT.3.MD.B.4: Generate measurement data by measuring lengths using rulers marked with halves and fourths of an inch. Show the data by making a line plot, where the horizontal scale is marked off in appropriate units—whole numbers, halves, or quarters.
ELP Standards	Receptive modalities • ELP 1: Construct meaning from oral presentations and literary and informational text through grade-appropriate listening, reading, and viewing. • ELP 8: Determine the meaning of words and phrases in oral presentations and literary and informational text. Productive modalities • ELP 3: Speak and write about grade-appropriate complex literary and informational texts and topics. • ELP 4: Construct grade-appropriate oral and written claims and support them with reasoning and evidence. • ELP 7: Adapt language choices to purpose, task, and audience when speaking and writing. Interactive modalities • ELP 2: Participate in grade-appropriate oral and written exchanges of information, ideas, and analyses, responding to peer, audience, or reader comments and questions. • ELP 5: Conduct research and evaluate and communicate findings to answer questions or solve problems. • ELP 6: Analyze and critique the arguments of others orally and in writing. Linguistic structures of English • ELP 9: Create clear and coherent grade-appropriate speech and text. • ELP 10: Make accurate use of standard English to communicate in grade-appropriate speech and writing.

Investigation 7

Variation Within a Species: How Similar Are Earthworms to Each Other?

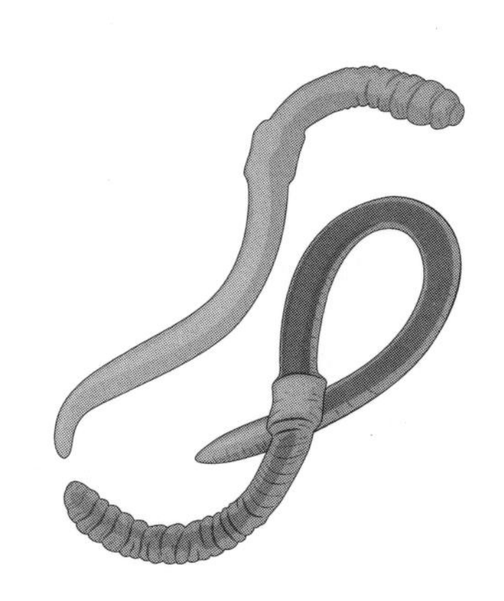

Introduction

When you look at the animals living outside your home or school, you will probably notice a few things. First, there are many different types of animals. Second, you might notice that animals that are the same type share a lot of traits. For example, squirrels have bushy tails, brown eyes, and four legs. Robins have wings, beaks, and two legs. But animals that are the same type of animal can also have some traits that make them look different. For example, not all squirrels are the same size or have the same color fur.

Your teacher will give you two earthworms. Take a moment to observe these two earthworms. Keep track of what you observe and what you are wondering about in the table below.

Things I OBSERVED …	Things I WONDER about …

Scientists classify animals based on the traits they share. For example, dogs have two eyes, four legs, four toes that touch the ground on each foot, a tail, and hair, and they can bark. When an animal has these traits, we call it a dog. There will also be differences in the traits that the animals that belong to a specific group share, because traits come in different versions. For example, an animal can have a tail but the tail can be long or short. The different lengths of tail are different versions of the same trait. No two animals, as a result, will look exactly alike even when they are members of the same group. For example, all dogs have two eyes but some dogs have dark brown eyes and some have blue eyes. All dogs have hair, but the hair can be different colors (such as black, brown, or yellow). Therefore, no two dogs look exactly alike because they each have different versions of the traits that all dogs share.

Your goal in this investigation is to figure out how similar earthworms are to each other. To accomplish this task, you will need to make observations about and take measurements of the traits of earthworms. You will need to compare and contrast at least two different traits to figure out what the worms have in common and what is different about them. Scientists often look for patterns in nature like this and then use these patterns to classify animals into groups. You can therefore use patterns to help determine what is similar about earthworms that make them all the same species of animal, and what is different about individual worms even though they are the same species.

Things we KNOW from what we read …	What we will NEED to figure out …

Your Task

Use what you know about traits and patterns to design and carry out an investigation to compare and contrast at least two different traits of earthworms.

The *guiding question* of this investigation is, ***How similar are earthworms to each other?***

Materials

You may use any of the following materials during your investigation:

- Safety goggles (required)
- Nitrile gloves (required)
- Earthworms
- Ruler
- Electronic scale
- Magnifying glass
- Tray

Safety Rules

Follow all normal safety rules. In addition, be sure to keep the earthworms healthy by not doing anything to hurt them such as pulling on them, poking them, cutting them, or dropping them. You also need to follow these rules:

- Wear sanitized indirectly vented chemical-splash goggles and nitrile gloves during setup, investigation activity, and cleanup.
- Wash your hands with soap and water when done collecting data.

Plan Your Investigation

Prepare a plan for your investigation by filling out the chart that follows; this plan is called an *investigation proposal.* Before you start developing your plan, be sure to discuss the following questions with the other members of your group:

- What information should we collect so we can **describe** the traits of an earthworm?
- What types of **patterns** might we look for to help answer the guiding question?

Our guiding question:

We will collect the following data:

These are the steps we will follow to collect data:

I approve of this investigation proposal.

Teacher's signature

Date

Collect Your Data

Keep a record of what you measure or observe during your investigation in the space below.

Analyze Your Data

You will need to analyze the data you collected before you can develop an answer to the guiding question. To do this, create one or more graphs that show how many earthworms had a specific version of a trait.

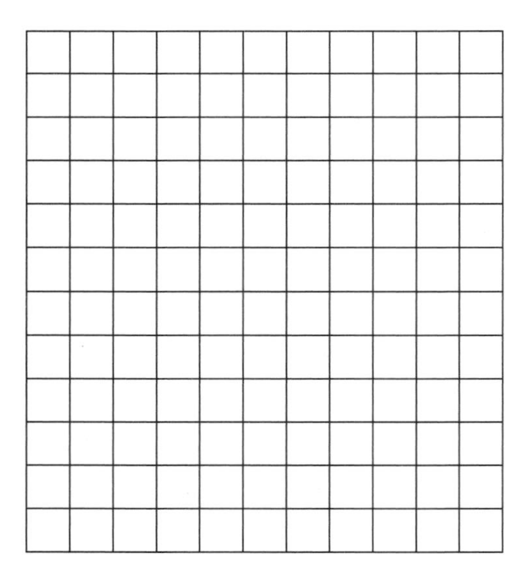

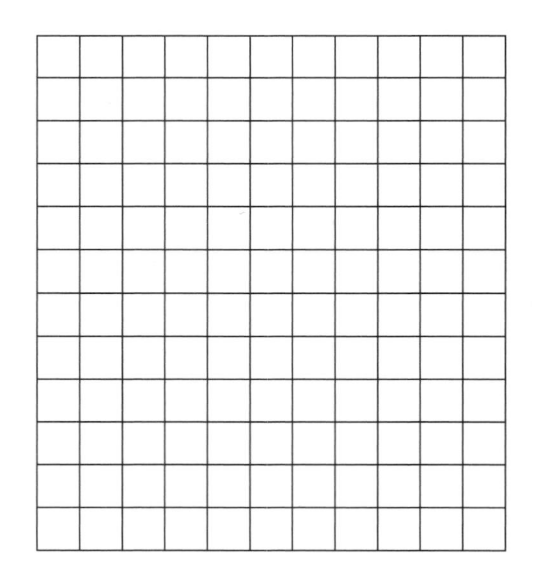

Draft Argument

Develop an argument on a whiteboard. It should include the following parts:

1. A *claim:* Your answer to the guiding question.
2. *Evidence:* An analysis of the data and an explanation of what the analysis means.
3. A *justification of the evidence:* Why your group thinks the evidence is important.

The Guiding Question:	
Our Claim:	
Our Evidence:	Our Justification of the Evidence:

Argumentation Session

Share your argument with your classmates. Be sure to ask them how to make your draft argument better. Keep track of their suggestions in the space below.

Ways to IMPROVE our argument …

Draft Report

Prepare an *investigation report* to share what you have learned. Use the information in this handout and your group's final argument to write a *draft* of your investigation report.

Introduction

We have been studying ______________________________ in class.

Before we started this investigation, we explored ______________________________

____

____

We noticed ______________________________

____

____

My goal for this investigation was to figure out ______________________________

____

____

____

The guiding question was ______________________________

____

____

Method

To gather the data I needed to answer this question, I ______________________________

____

____

____

____

____

I then analyzed the data I collected by ____

____

____

____

Argument

My claim is ____

____

____

____

The graphs below include information about

This analysis of the data I collected shows _____________________________________

____

____

____

____

____

____

This evidence is important because ______________________________

Review

Your friends need your help! Review the draft of their investigation reports and give them ideas about how to improve. Use the *peer-review guide* when doing your review.

Submit Your Final Report

Once you have received feedback from your friends about your draft report, create your final investigation report and hand it in to your teacher.

Checkout Questions

Investigation 7. Variation Within a Species: How Similar Are Earthworms to Each Other?

1. Pictured below is an earthworm. Place an "S" next to the traits that you think will be the same when you compare two different earthworms and a "D" next to the traits that you think will be different.

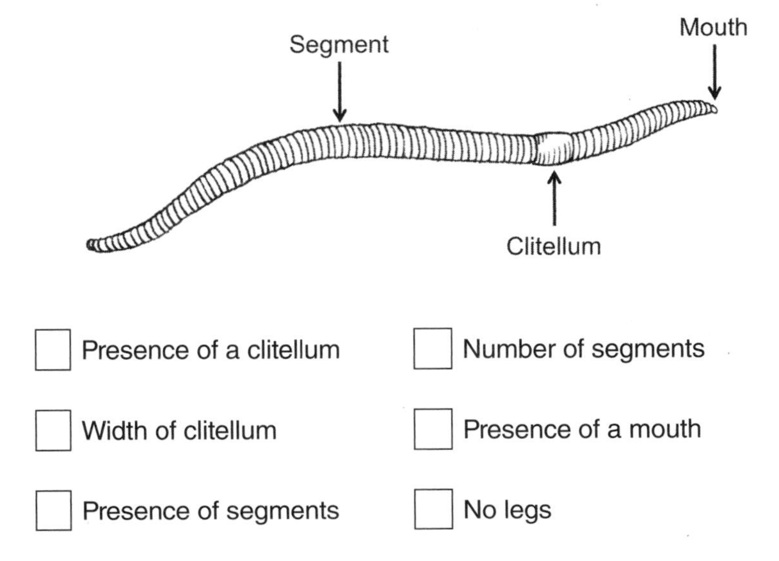

☐ Presence of a clitellum	☐ Number of segments
☐ Width of clitellum	☐ Presence of a mouth
☐ Presence of segments	☐ No legs

2. Pictured below is a ladybug. Place an "S" next to the traits that you think will be the same when you compare two different ladybugs and a "D" next to the traits that you think will be different.

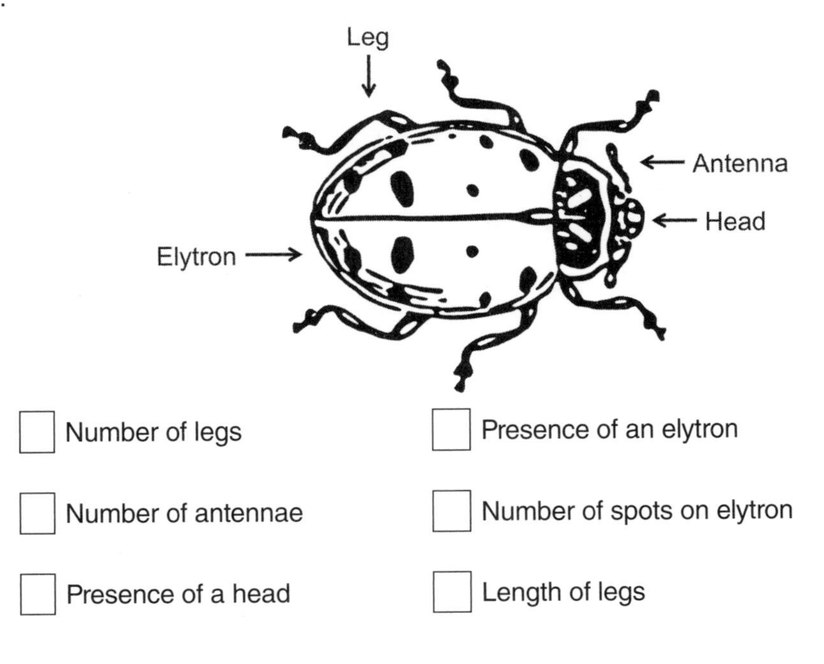

☐ Number of legs	☐ Presence of an elytron
☐ Number of antennae	☐ Number of spots on elytron
☐ Presence of a head	☐ Length of legs

Checkout Questions

3. Explain your thinking. What *pattern* from your investigation did you use to predict if a trait would be the same or different when comparing the traits of two individuals that belong to the same group of organisms?

____

____

____

____

____

____

____

____

Teacher Scoring Rubric for the Checkout Questions

Level	Description
3	The student can apply the core idea correctly in all cases and can fully explain the pattern.
2	The student can apply the core idea correctly in all cases but cannot fully explain the pattern.
1	The student cannot apply the core idea correctly in all cases but can fully explain the pattern.
0	The student cannot apply the core idea correctly in all cases and cannot explain the pattern.

Investigation 8

Inheritance of Traits: How Similar Are Offspring to Their Parents?

Purpose

The purpose of this investigation is to give students an opportunity to use the disciplinary core idea (DCI) of LS3.A: Inheritance of Traits and the crosscutting concept (CC) of Patterns from *A Framework for K–12 Science Education* (NRC 2012) to figure out how similar offspring are to each other and their parents. Students will also learn about the difference between observations and inferences in science during the reflective discussion.

The Disciplinary Core Idea

Students in third grade should understand the following about Inheritance of Traits and be able to use this DCI to figure out how similar offspring are to each other and their parents:

> *Organisms have characteristics that can be similar or different. Young animals are very much, but not exactly, like their parents and also resemble other animals of the same kind. Plants also are very much, but not exactly, like their parents and resemble other plants of the same kind. (NRC 2012, p. 158)*

The Crosscutting Concept

Students in third grade should understand the following about the CC Patterns:

> *Noticing patterns is often a first step to organizing and asking scientific questions about why and how the patterns occur. One major use of pattern recognition is in classification, which depends on careful observation of similarities and differences; objects can be classified into groups on the basis of similarities of visible or microscopic features or on the basis of similarities of function. Such classification is useful in codifying relationships and organizing a multitude of objects or processes into a limited number of groups. (NRC 2012, p. 85)*

Students in third grade should also be given opportunities to "observe and record patterns in the similarities and differences between parents and their offspring" (NRC 2012, p. 86).

Students should be encouraged to use their developing understanding of patterns as a tool or a way of thinking about a phenomenon during this investigation to help them figure out how similar offspring are to each other and their parents.

What Students Figure Out

Offspring acquire a mix of traits from their biological parents. Animals and plants differ in how they look and function because they inherit a different mixture of traits from their biological parents. A biological parent can pass down the exact same version of a trait or a different version of a trait to an offspring. The mixture and variation in traits that are passed down from parent to offspring is what makes each individual unique within a family. Siblings tend to look more alike when the parents have more traits in common.

Timeline

The time needed to complete this investigation is 270 minutes (4 hours and 30 minutes). The amount of instructional time needed for each stage of the investigation is as follows:

- *Stage 1.* Introduce the task and the guiding question: 35 minutes
- *Stage 2.* Design a method and collect data: 60 minutes
- *Stage 3.* Create a draft argument: 35 minutes
- *Stage 4.* Argumentation session: 30 minutes
- *Stage 5.* Reflective discussion: 15 minutes
- *Stage 6.* Write a draft report: 35 minutes
- *Stage 7.* Peer review: 30 minutes
- *Stage 8.* Revise the report: 30 minutes

This investigation can be completed in one day or over eight days (one day for each stage) during your designated science time in the daily schedule.

Materials and Preparation

The materials needed for this investigation are listed in Table 8.1. You can purchase all the sets of pictures as a complete kit (which includes enough materials for 24 students, or six groups) at *www.argumentdriveninquiry.com*.

TABLE 8.1

Materials for Investigation 8

Item	Quantity
Pictures of dog breeds	1 set per group
Pictures of adult dogs and offspring: set A	1 per class
Pictures of adult dogs and offspring: set B	1 per class
Pictures of adult cats and offspring: set A	1 per class
Pictures of adult cats and offspring: set B	1 per class
Pictures of adult guinea pigs and offspring: set A	1 per class
Pictures of adult guinea pigs and offspring: set B	1 per class
Pictures of adult birds and offspring: set A	1 per class
Pictures of adult birds and offspring: set B	1 per class
Pictures of adult snapdragons and offspring	1 per class
Pictures of adult pea plants and offspring	1 per class
Pictures of adult tulips and offspring	1 per class
Whiteboard, 2' × 3'*	1 per group
Investigation Handout	1 per student
Peer-review guide and teacher scoring rubric	1 per student
Checkout Questions (optional)	1 per student

* As an alternative, students can use computer and presentation software such as Microsoft PowerPoint or Apple Keynote to create their arguments.

Safety Precautions

Remind students to follow all normal safety rules.

Lesson Plan by Stage

Stage 1: Introduce the Task and the Guiding Question (35 minutes)

1. Ask the students to sit in six groups, with three or four students in each group.
2. Ask students to clear off their desks except for a pencil (and their *Student Workbook for Argument-Driven Inquiry in Third-Grade Science* if they have one).
3. Pass out an Investigation Handout to each student (or ask students to turn to Investigation Log 8 in their workbook).
4. Read the first paragraph of the "Introduction" aloud to the class. Ask the students to follow along as you read.
5. Pass out a set of the pictures of dog breeds to each group.

6. Tell students to record their observations and questions about the dogs in the "OBSERVED/WONDER" chart in the "Introduction" section of their Investigation Handout (or the investigation log in their workbook).
7. Ask students to share *what they observed* about the dogs.
8. Ask students to share *what questions they have* about the dogs.
9. Tell the students, "Some of your questions might be answered by reading the rest of the 'Introduction.'"
10. Ask the students to read the rest of the "Introduction" on their own *or* ask them to follow along as you read aloud.
11. Once the students have read the rest of the "Introduction," ask them to fill out the "KNOW/NEED" chart on their Investigation Handout (or in their investigation log) as a group.
12. Ask students to share what they learned from the reading. Add these ideas to a class "know / need to figure out" chart.
13. Ask students to share what they think they will need to figure out based on what they read. Add these ideas to the class "know / need to figure out" chart.
14. Tell the students, "It looks like we have something to figure out. Let's see what we will need to do during our investigation."
15. Read the task and the guiding question aloud.
16. Tell the students, "You will need to study at least two different types of living things during the investigation." Hold up each set of pictures of adults and offspring and tell them what type of animal or plant is shown on each picture.

Stage 2: Design a Method and Collect Data (60 minutes)

1. Tell the students, "I am now going to give you and the other members of your group about 15 minutes to plan your investigation. Before you begin, I want you all to take a couple of minutes to discuss the following questions with the rest of your group."
2. Show the following questions on the screen or board:
   - What information should we collect so we can *describe* the traits of an animal?
   - What types of *patterns* might we look for to help answer the guiding question?
3. Tell the students, "Please take a few minutes to come up with an answer to these questions." Give the students two or three minutes to discuss these two questions.
4. Ask two or three different groups to share their answers. Be sure to highlight or write down any important ideas on the board so students can refer to them later.

5. If possible, use a document camera to project an image of the graphic organizer for this investigation on a screen or board (or take a picture of it and project the picture on a screen or board). Tell the students, "I now want you all to plan out your investigation. To do that, you will need to create an investigation proposal by filling out this graphic organizer."
6. Point to the box labeled "Our guiding question:" and tell the students, "You can put the question we are trying to answer in this box." Then ask, "Where can we find the guiding question?"
7. Wait for a student to answer where to find the guiding question (the answer is "in the handout").
8. Point to the box labeled "We will collect the following data:" and tell the students, "You can list the measurements or observations that you will need to collect during the investigation in this box."
9. Point to the box labeled "These are the steps we will follow to collect data:" and tell the students, "You can list what you are going to do to collect the data you need and what you will do with your data once you have it. Be sure to give enough detail that I could do your investigation for you."
10. Ask the students, "Do you have any questions about what you need to do?"
11. Answer any questions that come up.
12. Tell the students, "Once you are done, raise your hand and let me know. I'll then come by and look over your proposal and give you some feedback. You may not begin collecting data until I have approved your proposal by signing it. You need to have your proposal done in the next 15 minutes."
13. Give the students 15 minutes to work in their groups on their investigation proposal. As they work, move from group to group to check in, ask probing questions, and offer a suggestion if a group gets stuck.

What should a student-designed investigation look like?

The students' investigation proposal should include the following information:

- The guiding question is "How similar are offspring to their parents?"
- There are a lot of different types of data that students can collect during this investigation. Examples include (1) hair/feather color, (2) hair texture, (3) eye color, (4) shape of ears, and (5) shape of tail. At a minimum, each group should collect data about two different traits. This investigation works best if each group selects different traits to examine. Students do not need to collect quantitative (numerical) data for this investigation.

The steps that the students will follow to collect the data should reflect the traits that they decide to examine. However, a procedure might include the following steps:

1. Identify the traits to examine for a specific type of animal.
2. Document [trait 1].
3. Document [trait 2].
4. Repeat steps 1–3 for each animal.
5. Create a table that lists (a) the traits the mother and child have in common, (b) the traits the father and child have in common, and (c) the traits the mother, the father, and the child have in common.

This is just an example of a procedure, and there should be a lot of variation in the student-designed investigation.

14. As each group finishes its investigation proposal, be sure to read it over and determine if it will be productive or not. If you feel the investigation will be productive (not necessarily what you would do or what the other groups are doing), sign your name on the proposal and let the group start collecting data. If the plan needs to be changed, offer some suggestions or ask some probing questions, and have the group make the changes before you approve it.
15. Pass out the materials or have one student from each group collect the materials they need from a central supply table or cart for the groups that have an approved proposal.
16. Remind students of the safety rules for this investigation.
17. Tell the students to collect their data and record their observations or measurements in the "Collect Your Data" box in their Investigation Handout (or the investigation log in their workbook).

18. Give the students 20 minutes to collect their data.
19. Be sure to collect the materials from each group before asking them to analyze their data.

Stage 3: Create a Draft Argument (35 minutes)

1. Tell the students, "Now that we have all this data, we need to analyze the data so we can figure out an answer to the guiding question."
2. If possible, project an image of the "Analyze Your Data" text and box for this investigation on a screen or board using a document camera (or take a picture of it and project the picture on a screen or board). Point to the box and tell the students, "You can create a table as a way to analyze your data. You can make your table in this box."
3. Ask the students, "What information do we need to include in this table?"
4. Tell the students, "Please take a few minutes to discuss this question with your group, and be ready to share."
5. Give the students five minutes to discuss.
6. Ask two or three different groups to share their answers. Be sure to highlight or write down any important ideas on the board so students can refer to them later.
7. Tell the groups of students, "I am now going to give you and the other members of your group about 10 minutes to create your tables." If the students are having trouble making a table, you can take a few minutes to provide a mini-lesson about how to create a table from a bunch of observations or measurements (this strategy is called just-in-time instruction because it is offered only when students get stuck).

What should a table look like for this investigation?

There are a number of different ways that students can analyze the observations or measurements they collect during this investigation. One of the most straightforward ways is to create a table with four columns. The first column can be type of animal; the second column can be traits shared by mother and child; the third column can be traits shared by father and child; and the fourth column can be traits shared by the mother, father, and child. An example of this type of table can be seen in Figure 8.1 (p. 299). There are other options for analyzing the data they collected. Students often come up with some unique ways of analyzing their data, so be sure to give them some voice and choice during this stage.

8. Give the students 10 minutes to analyze their data. As they work, move from group to group to check in, ask probing questions, and offer suggestions.
9. Tell the students, "I am now going to give you and the other members of your group 15 minutes to create an argument to share what you have learned and convince others that they should believe you. Before you do that, we need to take a few minutes to discuss what you need to include in your argument."
10. If possible, use a document camera to project the "Argument Presentation on a Whiteboard" image from the Investigation Handout (or the investigation log in their workbook) on a screen or board (or take a picture of it and project the picture on a screen or board).
11. Point to the box labeled "The Guiding Question:" and tell the students, "You can put the question we are trying to answer here on your whiteboard."
12. Point to the box labeled "Our Claim:" and tell the students, "You can put your claim here on your whiteboard. The claim is your answer to the guiding question."
13. Point to the box labeled "Our Evidence:" and tell the students, "You can put the evidence that you are using to support your claim here on your whiteboard. Your evidence will need to include the analysis you just did and an explanation of what your analysis means or shows. Scientists always need to support their claims with evidence."
14. Point to the box labeled "Our Justification of the Evidence:" and tell the students, "You can put your justification of your evidence here on your whiteboard. Your justification needs to explain why your evidence is important. Scientists often use core ideas to explain why the evidence they are using matters. Core ideas are important concepts that scientists use to help them make sense of what happens during an investigation."
15. Ask the students, "What are some core ideas that we read about earlier that might help us explain why the evidence we are using is important?"
16. Ask students to share some of the core ideas from the "Introduction" section of the Investigation Handout (or the investigation log in the workbook). List these core ideas on the board.
17. Tell the students, "That is great. I would like to see everyone try to include these core ideas in your justification of the evidence. Your goal is to use these core ideas to help explain why your evidence matters and why the rest of us should pay attention to it."
18. Ask the students, "Do you have any questions about what you need to do?"
19. Answer any questions that come up.
20. Tell the students, "Okay, go ahead and start working on your arguments. You need to have your argument done in the next 15 minutes. It doesn't need to be

perfect. We just need something down on the whiteboards so we can share our ideas."

21. Give the students 15 minutes to work in their groups on their arguments. As they work, move from group to group to check in, ask probing questions, and offer a suggestion if a group gets stuck. Figure 8.1 shows an example of an argument for this investigation.

FIGURE 8.1

Example of an argument

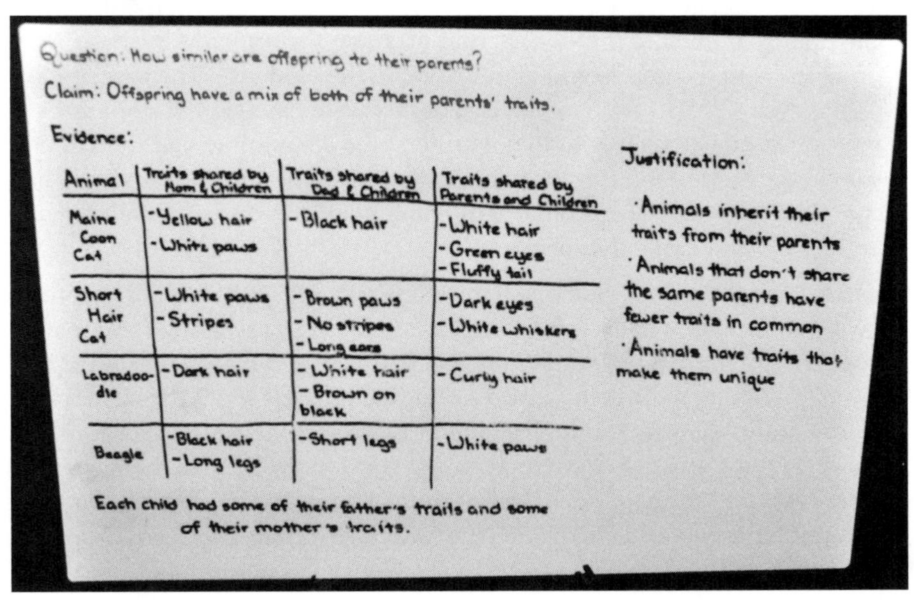

Stage 4: Argumentation Session (30 minutes)

The argumentation session can be conducted in a whole-class presentation format, a gallery walk format, or a modified gallery walk format. We recommend using a whole-class presentation format for the first investigation, but try to transition to either the gallery walk or modified gallery walk format as soon as possible because that will maximize student voice and choice inside the classroom. The following list shows the steps for the three formats; unless otherwise noted, the steps are the same for all three formats.

1. Begin by introducing the use of the whiteboard.
   - *If using the whole-class presentation format,* tell the students, "We are now going to share our arguments. Please set up your whiteboard so everyone can see them."
   - *If using the gallery walk or modified gallery walk format,* tell the students, "We are now going to share our arguments. Please set up your whiteboard so they are facing the walls."

2. Allow the students to set up their whiteboards.

- *If using the whole-class presentation format,* the whiteboards should be set up on stands or chairs so they are facing toward the center of the room.
- *If using the gallery walk or modified gallery walk format,* the whiteboards should be set up on stands or chairs so they are facing toward the outside of the room.

3. Give the following instructions to the students:
   - *If using the whole-class presentation format or the modified gallery walk format,* tell the students, "Okay, before we get started I want to explain what we are going to do next. I'm going to ask some of you to present your arguments to your classmates. If you are presenting your argument, your job is to share your group's claim, evidence, and justification of the evidence. The rest of you will be reviewers. If you are a reviewer, your job is to listen to the presenters, ask the presenters questions if you do not understand something, and then offer them some suggestions about ways to make their argument better. After we have a chance to learn from each other, I'm going to give you some time to revise your arguments and make them better."
   - *If using the gallery walk format,* tell the students, "Okay, before we get started I want to explain what we are going to do next. You are going to have an opportunity to read the arguments that were created by other groups. Your group will go to a different group's argument. I'll give you a few minutes to read it and review it. Your job is to offer them some suggestions about ways to make their argument better. You can use sticky notes to give them suggestions. Please be specific about what you want to change and be specific about how you think they should change it. After we have a chance to learn from each other, I'm going to give you some time to revise your arguments and make them better."

4. Use a document camera to project the "Ways to IMPROVE our argument …" box from the Investigation Handout (or the investigation log in their workbook) on a screen or board (or take a picture of it and project the picture on a screen or board).
   - *If using the whole-class presentation format or the modified gallery walk format,* point to the box and tell the students, "If you are a presenter, you can write down the suggestions you get from the reviewers here. If you are a reviewer, and you see a good idea from another group, you can write down that idea here. Once we are done with the presentations, I will give you a chance to use these suggestions or ideas to improve your arguments.
   - *If using the gallery walk format,* point to the box and tell the students, "If you see good ideas from another group, you can write them down here. Once we are done reviewing the different arguments, I will give you a chance to use these ideas to improve your own arguments. It is important to share ideas like this."

Ask the students, "Do you have any questions about what you need to do?"

5. Answer any questions that come up.
6. Give the following instructions:
   - *If using the whole-class presentation format,* tell the students, "Okay. Let's get started."
   - *If using the gallery walk format,* tell the students, "Okay, I'm now going to tell you which argument to go to and review.
   - *If using the modified gallery walk format,* tell the students, "Okay, I'm now going to assign you to be a presenter or a reviewer." Assign one or two students from each group to be presenters and one or two students from each group to be reviewers.
7. Begin the review of the arguments.
   - *If using the whole-class presentation format,* have four or five groups present their argument one at a time. Give each group only two to three minutes to present their argument. Then give the class two to three minutes to ask them questions and offer suggestions. Be sure to encourage as much participation from the students as possible.
   - *If using the gallery walk format,* tell the students, "Okay. Let's get started. Each group, move one argument to the left. Don't move to the next argument until I tell you to move. Once you get there, read the argument and then offer suggestions about how to make it better. I will put some sticky notes next to each argument. You can use the sticky notes to leave your suggestions." Give each group about three to four minutes to read the arguments, talk, and offer suggestions.
     a. Tell the students, "Okay. Let's rotate. Move one group to the left."
     b. Again, give each group three or four minutes to read, talk, and offer suggestions.
     c. Repeat this process for two more rotations.
   - *If using the modified gallery walk format,* tell the students, "Okay. Let's get started. Reviewers, move one group to the left. Don't move to the next group until I tell you to move. Presenters, go ahead and share your argument with the reviewers when they get there." Give each group of presenters and reviewers about three to four minutes to talk.
     a. Tell the students, "Okay. Let's rotate. Reviewers, move one group to the left."
     b. Again, give each group of presenters and reviewers about three or four minutes to talk.
     c. Repeat this process for two more rotations.
8. Tell the students to return to their workstations.

9. Give the following instructions about revising the argument:
   - *If using the whole-class presentation format,* tell the students, "I'm now going to give you about 10 minutes to revise your argument. Take a few minutes to talk in your groups and determine what you want to change to make your argument better. Once you have decided what to change, go ahead and make the changes to your whiteboard."
   - *If using the gallery walk format,* tell the students, "I'm now going to give you about 10 minutes to revise your argument. Take a few minutes to read the suggestions that were left at your argument. Then talk in your groups and determine what you want to change to make your argument better. Once you have decided what to change, go ahead and make the changes to your whiteboard."
   - *If using the modified gallery walk format,* "I'm now going to give you about 10 minutes to revise your argument. Please return to your original groups." Wait for the students to move back into their original groups and then tell the students, "Okay, take a few minutes to talk in your groups and determine what you want to change to make your argument better. Once you have decided what to change, go ahead and make the changes to your whiteboard."

   Ask the students, "Do you have any questions about what you need to do?"
10. Answer any questions that come up.
11. Tell the students, "Okay. Let's get started."
12. Give the students 10 minutes to work in their groups on their arguments. As they work, move from group to group to check in, ask probing questions, and offer a suggestion if a group gets stuck.

Stage 5: Reflective Discussion (15 minutes)

1. Tell the students, "We are now going to take a minute to talk about what we did and what we have learned."
2. Show Figure 8.2 on the screen. This image shows a litter of puppies and the biological parents of these puppies.
3. Ask the students, "What do you all see going on here?"
4. Allow students to share their ideas.
5. Ask the students, "Why don't all the puppies look the same?"
6. Allow different students to share their ideas.

FIGURE 8.2

A litter of puppies and the biological parents of these puppies

7. Ask the students, "Why doesn't each puppy look exactly like one of its parents?"
8. Allow students to share their ideas.
9. Show Figure 8.3 on the screen. This image shows two Labrador retriever puppies.
10. Ask the students, "What do you all see going on here?"
11. Allow students to share their ideas.
12. Ask the students, "Why do you think these puppies are so similar?"

FIGURE 8.3

Two Labrador retriever puppies

13. Allow students to share their ideas. Be sure to point out that the parents of these puppies likely had the same traits.
14. Tell the students, "Okay, let's make sure we are on the same page. All newborn animals inherit a mix of traits from their biological parents. Animals differ in how they look and function because they inherit a different mixture of traits from their biological parents. A biological parent can pass down the exact same version of a trait or a different version of a trait to a child. The mixture and variation in traits that are passed down from parent to offspring are what make each individual unique within a family. Siblings tend to look more alike when the parents have more traits in common. This description of how traits are passed down from parents to offspring is a really important core idea in science."
15. Ask the students, "Does anyone have any questions about this core idea?"
16. Answer any questions that come up.
17. Tell the students, "We also looked for patterns during our investigation." Then ask, "Can anyone tell me why we needed to look for patterns?"
18. Allow students to share their ideas.
19. Tell the students, "Patterns are really important in science. Scientists look for patterns all the time. In fact, they even use patterns to help understand how traits are passed down from parent to child just like we did."
20. Tell the students, "We are now going take a minute to talk about what went well and what didn't go so well during our investigation. We need to talk about this because you are going to be planning and carrying out your own investigations like this a lot this year, and I want to help you all get better at it."
21. Show an image of the question "What made your investigation scientific?" on the screen. Tell the students, "Take a few minutes to talk about how you would answer this question with the other people in your group. Be ready to share

with the rest of the class." Give the students two to three minutes to talk in their group.

22. Ask the students, "What do you all think? Who would like to share an idea?"
23. Allow students to share their ideas. Be sure to expand on their ideas about what makes an investigation scientific.
24. Show an image of the question "What made your investigation not so scientific?" on the screen. Tell the students, "Take a few minutes to talk about how you would answer this question with the other people in your group. Be ready to share with the rest of the class." Give the students two to three minutes to talk in their group.
25. Ask the students, "What do you all think? Who would like to share an idea?"
26. Allow students to share their ideas. Be sure to expand on their ideas about what makes an investigation less scientific.
27. Show an image of the question "What rules can we put into place to help us make sure our next investigation is more scientific?" on the screen. Tell the students, "Take a few minutes to talk about how you would answer this question with the other people in your group. Be ready to share with the rest of the class." Give the students two to three minutes to talk in their group.
28. Ask the students, "What do you all think? Who would like to share an idea?"
29. Allow students to share their ideas. Once they have shared their ideas, offer a suggestion for a possible class rule.
30. Ask the students, "What do you all think? Should we make this a rule?"
31. If the students agree, write the rule on the board or make a class "Rules for Scientific Investigation" chart so you can refer to it during the next investigation.
32. Tell the students, "We are now going take a minute to talk about what makes science different from other subjects."
33. Show an image of the question "What is the difference between an observation and an inference?" on the screen. Tell the students, "Take a few minutes to talk about how you would answer this question with the other people in your group. Be ready to share with the rest of the class." Give the students two to three minutes to talk in their group.
34. Ask the students, "What do you all think? Who would like to share an idea?"
35. Allow students to share their ideas.
36. Tell the students, "Okay, let's make sure we are all using the same definition. An observation is a descriptive statement about something. An inference is an interpretation of an observation."
37. Show an image of Figure 8.4 along with the statement "These puppies are siblings" on the screen.

FIGURE 8.4

Two puppies

A full-color version of this figure is available on the book's Extras page at *www.nsta.org/adi-3rd*.

38. Ask the students, "Is this statement an observation or an inference and why?"
39. Allow students to share their ideas.
40. Tell the students, "That statement is an inference because it is an interpretation of some observations."
41. Ask the students, "What are some observations we can make about these two dogs?"
42. Allow students to share their ideas.
43. Ask the students, "Does anyone have any questions about the difference between an observation and an inference?"
44. Answer any questions that come up.

Stage 6: Write a Draft Report (35 minutes)

Your students will use either the Investigation Handout or the investigation log in the student workbook when writing the draft report. When you give the directions shown in quotes in the following steps, substitute "investigation log" (as shown in brackets) for "handout" if they are using the workbook.

1. Tell the students, "You are now going to write an investigation report to share what you have learned. Please take out a pencil and turn to the 'Draft Report' section of your handout [investigation log]."
2. If possible, use a document camera to project the "Introduction" section of the draft report from the Investigation Handout (or the investigation log in their

workbook) on a screen or board (or take a picture of it and project the picture on a screen or board).

3. Tell the students, "The first part of the report is called the 'Introduction.' In this section of the report you want to explain to the reader what you were investigating, why you were investigating it, and what question you were trying to answer. All of this information can be found in the text at the beginning of your handout [investigation log]." Point to the image and say, "There are some sentence starters here to help you begin writing the report." Ask the students, "Do you have any questions about what you need to do?"
4. Answer any questions that come up.
5. Tell the students, "Okay. Let's write."
6. Give the students 10 minutes to write the "Introduction" section of the report. As they work, move from student to student to check in, ask probing questions, and offer a suggestion if a student gets stuck.
7. If possible, use a document camera to project the "Method" section of the draft report from the Investigation Handout (or the investigation log in their workbook) on a screen or board (or take a picture of it and project the picture on a screen or board).
8. Tell the students, "The second part of the report is called the 'Method.' In this section of the report you want to explain to the reader what you did during the investigation, what data you collected and why, and how you went about analyzing your data. All of this information can be found in the 'Plan Your Investigation' section of your handout [investigation log]. Remember that you all planned and carried out different investigations, so do not assume that the reader will know what you did." Point to the image and say, "There are some sentence starters here to help you begin writing this part of the report." Ask the students, "Do you have any questions about what you need to do?"
9. Answer any questions that come up.
10. Tell the students, "Okay. Let's write."
11. Give the students 10 minutes to write the "Method" section of the report. As they work, move from student to student to check in, ask probing questions, and offer a suggestion if a student gets stuck.
12. If possible, use a document camera to project the "Argument" section of the draft report from the Investigation Handout (or the investigation log in their workbook) on a screen or board (or take a picture of it and project the picture on a screen or board).
13. Tell the students, "The last part of the report is called the 'Argument.' In this section of the report you want to share your claim, evidence, and justification of the evidence with the reader. All of this information can be found on your whiteboard." Point to the image and say, "There are some sentence starters here

to help you begin writing this part of the report." Ask the students, "Do you have any questions about what you need to do?"

14. Answer any questions that come up.
15. Tell the students, "Okay. Let's write."
16. Give the students 10 minutes to write the "Argument" section of the report. As they work, move from student to student to check in, ask probing questions, and offer a suggestion if a student gets stuck.

Stage 7: Peer Review (30 minutes)

Your students will use either the Investigation Handout or the investigation log in the student workbook when doing the peer review. When you give the directions shown in quotes in the following steps, substitute "workbook" (as shown in brackets) for "Investigation Handout" if they are using the workbook.

1. Tell the students, "We are now going to review our reports to find ways to make them better. I'm going to come around and collect your Investigation Handout [workbook]. While I do that, please take out a pencil."
2. Collect the Investigation Handouts or workbooks from the students.
3. If possible, use a document camera to project the peer-review guide (PRG; see Appendix 4) on a screen or board (or take a picture of it and project the picture on a screen or board).
4. Tell the students, "We are going to use this peer-review guide to give each other feedback." Point to the image.
5. Give the following instructions:
   - *If using the Investigation Handout,* tell the students, "I'm going to ask you to work with a partner to do this. I'm going to give you and your partner a draft report to read and a peer-review guide to fill out. You two will then read the report together. Once you are done reading the report, I want you to answer each of the questions on the peer-review guide." Point to the review questions on the image of the PRG.
   - *If using the workbook,* tell the students, "I'm going to ask you to work with a partner to do this. I'm going to give you and your partner a draft report to read. You two will then read the report together. Once you are done reading the report, I want you to answer each of the questions on the peer-review guide that is right after the report in the investigation log." Point to the review questions on the image of the PRG.
6. Tell the students, "You can check 'yes,' 'almost,' or 'no' after each question." Point to the checkboxes on the image of the PRG.

7. Tell the students, "This will be your rating for this part of the report. Make sure you agree on the rating you give the author. If you mark 'almost' or 'no,' then you need to tell the author what he or she needs to do to get a 'yes.'" Point to the space for the reviewer feedback on the image of the PRG.
8. Tell the students, "It is really important for you to give the authors feedback that is helpful. That means you need to tell them exactly what they need to do to make their reports better." Ask the students, "Do you have any questions about what you need to do?"
9. Answer any questions that come up.
10. Tell the students, "Please sit with a partner who is not in your current group." Allow the students time to sit with a partner.
11. Give the following instructions:
    - *If using the Investigation Handout,* tell the students, "Okay, I am now going to give you one report to read and one peer-review guide to fill out." Pass out one report to each pair. Make sure that the report you give a pair was not written by one of the students in that pair. Give each pair one PRG to fill out as a team.
    - *If using the workbook,* tell the students, "Okay, I am now going to give you one report to read." Pass out a workbook to each pair. Make sure that the workbook you give a pair is not from one of the students in that pair.
12. Tell the students, "Okay, I'm going to give you 15 minutes to read the report I gave you and to fill out the peer-review guide. Go ahead and get started."
13. Give the students 15 minutes to work. As they work, be sure to move around from pair to pair to check in and see how things are going, answer questions, and offer advice.
14. After 15 minutes pass, tell the students, "Okay, time is up." *If using the Investigation Handout,* say, "Please give me the report and the peer-review guide that you filled out." *If using the workbook,* say, "Please give me the workbook that you have."
15. Collect the Investigation Handouts and the PRGs, or collect the workbooks if they are being used. Be sure you keep the handout and the PRG together.
16. Give the following instructions:
    - *If using the Investigation Handout,* tell the students, "Okay, I am now going to give you a different report to read and a new peer-review guide to fill out." Pass out another report to each pair. Make sure that this report was not written by one of the students in that pair. Give each pair a new PRG to fill out as a team.
    - *If using the workbook,* tell the students, "Okay, I am now going to give you a different report to read." Pass out a different workbook to each pair. Make sure that the workbook you give a pair is not from one of the students in that pair.

17. Tell the students, "Okay, I'm going to give you 15 minutes to read this new report and to fill out the peer-review guide. Go ahead and get started."
18. Give the students 15 minutes to work. As they work, be sure to move around from pair to pair to check in and see how things are going, answer questions, and offer advice.
19. After 15 minutes pass, tell the students, "Okay, time is up." *If using the Investigation Handout,* say, "Please give me the report and the peer-review guide that you filled out." *If using the workbook,* say, "Please give me the workbook that you have."
20. Collect the Investigation Handouts and the PRGs, or collect the workbooks if they are being used. Be sure you keep the handout and the PRG together.

Stage 8: Revise the Report (30 minutes)

Your students will use either the Investigation Handout or the investigation log in the student workbook when revising the report. Except where noted below, the directions are the same whether using the handout or the log.

1. Give the following instructions:
    - *If using the Investigation Handout,* tell the students, "You are now going to revise your investigation report based on the feedback you get from your classmates. Please take out a pencil while I hand back your draft report and the peer-review guide."
    - *If using the investigation log in the student workbook,* tell the students, "You are now going to revise your investigation report based on the feedback you get from your classmates. Please take out a pencil while I hand back your investigation logs."
2. *If using the Investigation Handout,* pass back the handout and the PRG to each student. *If using the investigation log,* pass back the log to each student.
3. Tell the students, "Please take a few minutes to read over the peer-review guide. You should use it to figure out what you need to change in your report and how you will change the report."
4. Allow the students time to read the PRG.
5. *If using the investigation log,* if possible use a document camera to project the "Write Your Final Report" section from the investigation log on a screen or board (or take a picture of it and project the picture on a screen or board).
6. Give the following instructions:
    - *If using the Investigation Handout,* tell the students, "Okay. Let's revise our reports. Please take out a piece of paper. I would like you to rewrite your report. You can use your draft report as a starting point, but use the feedback on the peer-review guide to help make it better."

- *If using the investigation log,* tell the students, "Okay. Let's revise our reports. I would like you to rewrite your report in the section of the investigation log that says 'Write Your Final Report.'" Point to the image on the screen and tell the students, "You can use your draft report as a starting point, but use the feedback on the peer-review guide to help make your report better."

Ask the students, "Do you have any questions about what you need to do?"

7. Answer any questions that come up.
8. Tell the students, "Okay. Let's write."
9. Give the students 30 minutes to rewrite their report. As they work, move from student to student to check in, ask probing questions, and offer a suggestion if a student gets stuck.
10. Give the following instructions:
    - *If using the Investigation Handout,* tell the students, "Okay. Time's up. I will now come around and collect your Investigation Handout, the peer-review guide, and your final report."
    - *If using the investigation log,* tell the students, "Okay. Time's up. I will now come around and collect your workbooks."
11. *If using the Investigation Handout,* collect all the Investigation Handouts, PRGs, and final reports. *If using the investigation log,* collect all the workbooks.
12. *If using the Investigation Handout,* use the "Teacher Score" columns in the PRG to grade the final report. *If using the investigation log,* use the "ADI Investigation Report Grading Rubric" in the investigation log to grade the final report. Whether you are using the handout or the log, you can give the students feedback about their writing in the "Teacher Comments" section.

How to Use the Checkout Questions

The Checkout Questions are an optional assessment. We recommend giving them to students one day after they finish stage 8 of the ADI investigation. The Checkout Questions can be used as a formative or summative assessment of student thinking. If you plan to use them as a formative assessment, we recommend that you look over the student answers to determine if you need to reteach the core idea and/or crosscutting concept from the investigation, but do not grade them. If you plan to use them as a summative assessment, we have included a "Teacher Scoring Rubric" at the end of the Checkout Questions that you can use to score a student's ability to apply the core idea in a new scenario and explain their use of a crosscutting concept. The rubric includes a 4-point scale that ranges from 0 (the student cannot apply the core idea correctly in all cases and cannot explain the [crosscutting concept]) to 3 (the student can apply the core idea correctly in all cases and can fully explain the [crosscutting concept]). The Checkout Questions, regardless of how you decide to use them, are a great way to make student thinking visible so you can determine if the students have learned the core idea and the crosscutting concept.

A student who can apply the core idea correctly in all cases and can explain the pattern would choose guinea pig C for the first question. He or she should then be able to explain that living things inherit a mixture of their traits from their biological parents, so the father of the three guinea pigs must have had some black hair.

Connections to Standards

Table 8.2 highlights how the investigation can be used to address specific performance expectations from the *NGSS, Common Core State Standards (CCSS)* in English language arts (ELA) and in mathematics, and *English Language Proficiency (ELP) Standards.*

TABLE 8.2

Investigation 8 alignment with standards

***NGSS* performance expectation**	Strong alignment • 3-LS3-1: Analyze and interpret data to provide evidence that plants and animals have traits inherited from parents and that variation of these traits exists in a group of similar organisms.
***CCSS ELA*—Reading: Informational Text**	Key ideas and details • CCSS.ELA-LITERACY.RI.3.1: Ask and answer questions to demonstrate understanding of a text, referring explicitly to the text as the basis for the answers. • CCSS.ELA-LITERACY.RI.3.2: Determine the main idea of a text; recount the key details and explain how they support the main idea. • CCSS.ELA-LITERACY.RI.3.3: Describe the relationship between a series of historical events, scientific ideas or concepts, or steps in technical procedures in a text, using language that pertains to time, sequence, and cause/effect. Craft and structure • CCSS.ELA-LITERACY.RI.3.4: Determine the meaning of general academic and domain-specific words and phrases in a text relevant to a *grade 3 topic or subject area*. • CCSS.ELA-LITERACY.RI.3.5: Use text features and search tools (e.g., key words, sidebars, hyperlinks) to locate information relevant to a given topic efficiently. • CCSS.ELA-LITERACY.RI.3.6: Distinguish their own point of view from that of the author of a text.

Continued

Table 8.2 (*continued*)

***CCSS ELA*—Reading: Informational Text** (*continued*)	Integration of knowledge and ideas • CCSS.ELA-LITERACY.RI.3.7: Use information gained from illustrations (e.g., maps, photographs) and the words in a text to demonstrate understanding of the text (e.g., where, when, why, and how key events occur). • CCSS.ELA-LITERACY.RI.3.8: Describe the logical connection between particular sentences and paragraphs in a text (e.g., comparison, cause/effect, first/second/third in a sequence). • CCSS.ELA-LITERACY.RI.3.9: Compare and contrast the most important points and key details presented in two texts on the same topic. Range of reading and level of text complexity • CCSS.ELA-LITERACY.RI.3.10: By the end of the year, read and comprehend informational texts, including history/social studies, science, and technical texts, at the high end of the grades 2–3 text complexity band independently and proficiently.
***CCSS ELA*—Writing**	Text types and purposes • CCSS.ELA-LITERACY.W.3.1: Write opinion pieces on topics or texts, supporting a point of view with reasons. o CCSS.ELA-LITERACY.W.3.1.A: Introduce the topic or text they are writing about, state an opinion, and create an organizational structure that lists reasons. o CCSS.ELA-LITERACY.W.3.1.B: Provide reasons that support the opinion. o CCSS.ELA-LITERACY.W.3.1.C: Use linking words and phrases (e.g., *because*, *therefore*, *since, for example*) to connect opinion and reasons. o CCSS.ELA-LITERACY.W.3.1.D: Provide a concluding statement or section. • CCSS.ELA-LITERACY.W.3.2: Write informative or explanatory texts to examine a topic and convey ideas and information clearly. o CCSS.ELA-LITERACY.W.3.2.A: Introduce a topic and group related information together; include illustrations when useful to aiding comprehension. o CCSS.ELA-LITERACY.W.3.2.B: Develop the topic with facts, definitions, and details. o CCSS.ELA-LITERACY.W.3.2.C: Use linking words and phrases (e.g., *also*, *another*, *and*, *more*, *but*) to connect ideas within categories of information. o CCSS.ELA-LITERACY.W.3.2.D: Provide a concluding statement or section.

Continued

Table 8.2 *(continued)*

***CCSS ELA*—Writing** *(continued)*	Production and distribution of writing • CCSS.ELA-LITERACY.W.3.4: With guidance and support from adults, produce writing in which the development and organization are appropriate to task and purpose. • CCSS.ELA-LITERACY.W.3.5: With guidance and support from peers and adults, develop and strengthen writing as needed by planning, revising, and editing. • CCSS.ELA-LITERACY.W.3.6: With guidance and support from adults, use technology to produce and publish writing (using keyboarding skills) as well as to interact and collaborate with others. Research to build and present knowledge • CCSS.ELA-LITERACY.W.3.8: Recall information from experiences or gather information from print and digital sources; take brief notes on sources and sort evidence into provided categories. Range of writing • CCSS.ELA-LITERACY.W.3.10: Write routinely over extended time frames (time for research, reflection, and revision) and shorter time frames (a single sitting or a day or two) for a range of discipline-specific tasks, purposes, and audiences.
***CCSS ELA*—Speaking and Listening**	Comprehension and collaboration • CCSS.ELA-LITERACY.SL.3.1: Engage effectively in a range of collaborative discussions (one-on-one, in groups, and teacher-led) with diverse partners on grade 3 topics and texts, building on others' ideas and expressing their own clearly. o CCSS.ELA-LITERACY.SL.3.1.A: Come to discussions prepared, having read or studied required material; explicitly draw on that preparation and other information known about the topic to explore ideas under discussion. o CCSS.ELA-LITERACY.SL.3.1.B: Follow agreed-upon rules for discussions (e.g., gaining the floor in respectful ways, listening to others with care, speaking one at a time about the topics and texts under discussion). o CCSS.ELA-LITERACY.SL.3.1.C: Ask questions to check understanding of information presented, stay on topic, and link their comments to the remarks of others. o CCSS.ELA-LITERACY.SL.3.1.D: Explain their own ideas and understanding in light of the discussion. • CCSS.ELA-LITERACY.SL.3.2: Determine the main ideas and supporting details of a text read aloud or information presented in diverse media and formats, including visually, quantitatively, and orally. • CCSS.ELA-LITERACY.SL.3.3: Ask and answer questions about information from a speaker, offering appropriate elaboration and detail.

Continued

Table 8.2 *(continued)*

***CCSS ELA—* Speaking and Listening** *(continued)*	Presentation of knowledge and ideas • CCSS.ELA-LITERACY.SL.3.4: Report on a topic or text, tell a story, or recount an experience with appropriate facts and relevant, descriptive details, speaking clearly at an understandable pace. • CCSS.ELA-LITERACY.SL.3.6: Speak in complete sentences when appropriate to task and situation in order to provide requested detail or clarification.
ELP Standards	Receptive modalities • ELP 1: Construct meaning from oral presentations and literary and informational text through grade-appropriate listening, reading, and viewing. • ELP 8: Determine the meaning of words and phrases in oral presentations and literary and informational text. Productive modalities • ELP 3: Speak and write about grade-appropriate complex literary and informational texts and topics. • ELP 4: Construct grade-appropriate oral and written claims and support them with reasoning and evidence. • ELP 7: Adapt language choices to purpose, task, and audience when speaking and writing. Interactive modalities • ELP 2: Participate in grade-appropriate oral and written exchanges of information, ideas, and analyses, responding to peer, audience, or reader comments and questions. • ELP 5: Conduct research and evaluate and communicate findings to answer questions or solve problems. • ELP 6: Analyze and critique the arguments of others orally and in writing. Linguistic structures of English • ELP 9: Create clear and coherent grade-appropriate speech and text. • ELP 10: Make accurate use of standard English to communicate in grade-appropriate speech and writing.

Investigation 8

Inheritance of Traits: How Similar Are Offspring to Their Parents?

Introduction

All living things have some traits in common with other living things and some traits that make them unique. Your teacher will give you some pictures of dogs. Take a minute to look at the pictures. Be sure to write down what you observe and what you are wondering about in the boxes below.

Things I OBSERVED ...	Things I WONDER about ...

The members of any group of living things, such as the group of dogs you just looked at, will often include some individuals that are related to each other and other individuals that are not related. The individuals in the group that are related to each other will look more alike than the individuals in the

group that are not related because animals and plants inherit their traits from their parents. Animals or plants that share the same parents will therefore have many traits in common, and animals or plants that do not share the same parents will have fewer traits in common. You have already seen how some dog traits are the same and others are different. The dogs you examined, however, did not have the same parents. You will now have a chance to examine the traits of parents and children (what scientists call *offspring*).

In this investigation you need to figure out how similar offspring are to their parents. To accomplish this task, you will need to make observations about the traits of parents and offspring for at least two different types of living things. You will need to compare and contrast these traits to figure out which members of the same family share a trait and which members of the same family do not share a trait. You can look for traits that one parent and one offspring have in common; traits that are shared by both parents and an offspring; and traits that are only observed in one or more of the offspring. Your goal is to look for any patterns in the traits that the parents and offspring have in common. Scientists often look for patterns in nature like this and then use these patterns to determine how plants and animals inherit traits.

Things we KNOW from what we read …	What we will NEED to figure out …

Your Task

Use what you know about traits and patterns to design and carry out an investigation to determine which traits parents and offspring have in common and which ones are different.

The *guiding question* of this investigation is, ***How similar are offspring to their parents?***

Materials

You will use at least two of the following sets of pictures of parents and offspring during your investigation:

- Dogs A
- Dogs B
- Cats A
- Cats B
- Guinea pigs A
- Guinea pigs B
- Birds A
- Birds B
- Snapdragons
- Peas
- Tulips

Safety Rules

Follow all normal safety rules.

Plan Your Investigation

Prepare a plan for your investigation by filling out the chart that follows; this plan is called an *investigation proposal.* Before you start developing your plan, be sure to discuss the following questions with the other members of your group:

- What information should we collect so we can **describe** the traits of an animal?
- What types of **patterns** might we look for to help answer the guiding question?

Our guiding question:

We will collect the following data:

These are the steps we will follow to collect data:

I approve of this investigation proposal.

Teacher's signature

Date

Collect Your Data

Keep a record of what you measure or observe during your investigation in the space below.

Analyze Your Data

You will need to analyze the data you collected before you can develop an answer to the guiding question. In the space below, create a table. Your table should include the names of the animals or plants you observed and the traits that the different individuals did or did not have.

Draft Argument

Develop an argument on a whiteboard. It should include the following parts:

1. A *claim:* Your answer to the guiding question.
2. *Evidence:* An analysis of the data and an explanation of what the analysis means.
3. A *justification of the evidence:* Why your group thinks the evidence is important.

The Guiding Question:	
Our Claim:	
Our Evidence:	Our Justification of the Evidence:

Argumentation Session

Share your argument with your classmates. Be sure to ask them how to make your draft argument better. Keep track of their suggestions in the space below.

Ways to IMPROVE our argument …

Draft Report

Prepare an *investigation report* to share what you have learned. Use the information in this handout and your group's final argument to write a draft of your investigation report.

Introduction

We have been studying ______________________________ in class.

Before we started this investigation, we explored ______________________________

____

____

We noticed ______________________________

____

____

My goal for this investigation was to figure out ______________________________

____

____

____

The guiding question was ______________________________

____

____

Method

To gather the data I needed to answer this question, I ______________________________

____

____

____

____

____

I then analyzed the data I collected by ______________________________

____

____

Argument

My claim is ______________________________

____

____

____

The table below shows ______________________________

____

____

____

____

This evidence is important because ____

____

____

____

____

____

____

____

____

____

Review

Your friends need your help! Review the draft of their investigation reports and give them ideas about how to improve. Use the *peer-review guide* when doing your review.

Submit Your Final Report

Once you have received feedback from your friends about your draft report, create your final investigation report and hand it in to your teacher.

Checkout Questions

Investigation 8. Inheritance of Traits: How Similar Are Offspring to Their Parents?

1. Pictured below are a female guinea pig and three of her pups.

Which of the following guinea pigs is most likely the father of the three pups in this litter?

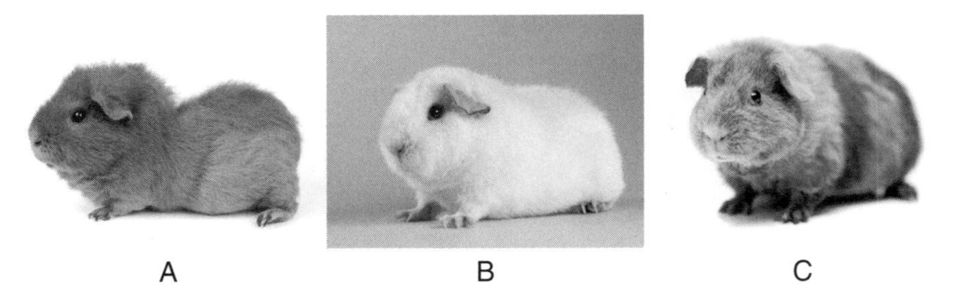

Full-color versions of these images are available on the book's Extras page at *www.nsta.org/adi-3rd*.

2. Explain your thinking. What pattern from your investigation did you use to predict the father of the three guinea pig pups?

____

____

____

____

____

____

Teacher Scoring Rubric for the Checkout Questions

Level	Description
3	The student can apply the core idea correctly in all cases and can fully explain the pattern.
2	The student can apply the core idea correctly in all cases but cannot fully explain the pattern.
1	The student cannot apply the core idea correctly in all cases but can fully explain the pattern.
0	The student cannot apply the core idea correctly in all cases and cannot explain the pattern.

Investigation 9

Traits and the Environment: How Do Differences in Soil Quality Affect the Traits of a Plant?

Purpose

The purpose of this investigation is to give students an opportunity to use the disciplinary core idea (DCI) of LS3.A: Inheritance of Traits and the crosscutting concept (CC) of Cause and Effect from *A Framework for K–12 Science Education* (NRC 2012) to figure out how the environment influences the traits of a plant. Students will also learn about the difference between theories and laws in nature during the reflective discussion.

The Disciplinary Core Idea

Students in third grade should understand the following about Inheritance of Traits and be able to use this DCI to figure out how changes to the environment can affect the traits of a plant:

> *Organisms have characteristics that can be similar or different. Young animals are very much, but not exactly, like their parents and also resemble other animals of the same kind. Plants also are very much, but not exactly, like their parents and resemble other plants of the same kind. (NRC 2012, p. 158)*

The Crosscutting Concept

Students in third grade should understand the following about the CC Cause and Effect:

> *Repeating patterns in nature, or events that occur together with regularity, are clues that scientists can use to start exploring causal, or cause-and-effect, relationships. … Any application of science, or any engineered solution to a problem, is dependent on understanding the cause-and-effect relationships between events; the quality of the application or solution often can be improved as knowledge of the relevant relationships is improved. (NRC, 2012, p. 87)*

Students in third grade should also be given opportunities to begin to "look for and analyze patterns—whether in their observations of the world or in the relationships between different quantities in data (e.g., the sizes of plants over time)"; "they can also begin to consider what might be causing these patterns and relationships and design tests that gather more evidence to support or refute their ideas" (NRC 2012, pp. 88–89).

Students should be encouraged to use their developing understanding of cause-and-effect relationships as a tool or a way of thinking about a phenomenon during this investigation to help them figure out how changes to the environment can affect the traits of a plant.

What Students Figure Out

Plants inherit a mix of traits from their biological parents. Different types of plants differ in how they look and function because they have different inherited information. Individual plants of the same type can also differ in how they look and function because individual plants inherit a different mixture of traits. The environment can also affect the traits of individual plants. An individual plant, as a result, may end up looking or functioning differently than other plants of the same type because of where it germinates and grows over time.

Timeline

The time needed to complete this investigation is 470 minutes (7 hours 50 minutes). The amount of instructional time needed for each stage of the investigation is as follows (note that there are two parts to stage 2):

- *Stage 1.* Introduce the task and the guiding question: 35 minutes
- *Stage 2.* Design a method (and set up the investigation): 60 minutes; collect data (over 10 days): 200 minutes (20 minutes per day)
- *Stage 3.* Create a draft argument: 35 minutes
- *Stage 4.* Argumentation session: 30 minutes
- *Stage 5.* Reflective discussion: 15 minutes
- *Stage 6.* Write a draft report: 35 minutes
- *Stage 7.* Peer review: 30 minutes
- *Stage 8.* Revise the report: 30 minutes

This investigation can be completed in 12 days (1 day for stages 1 and 2, 10 additional days to allow for plant growth and daily observations, and 1 day for stages 3–8) or over 18 days (8 days for each stage of ADI plus 10 additional days to allow for plant growth and daily observations) during your designated science time in the daily schedule.

Materials and Preparation

The materials needed for this investigation are listed in Table 9.1. The Wisconsin Fast Plants seeds can be purchased in packs of 50 or 200 from Carolina Biological Supply Company at *www.carolina.com*. The other items can be purchased from a big-box retail store such as Wal-Mart or Target or through an online retailer such as Amazon. The materials for this investigation can also be purchased as a complete kit (which includes enough materials for 24 students, or six groups) at *www.argumentdriveninquiry.com*.

TABLE 9.1

Materials for Investigation 9

Item	Quantity
Safety goggles	1 per student
Small potted plant	1 per group
Jiffy Windowsill Greenhouse with 12 pellets	3 per class
Osmocote Smart-Release Plant Food Plus Outdoor & Indoor	6 pellets per group
Wisconsin Fast Plant Seeds	12 per group
Ruler, 12"	1 per group
Plant labels	6 per group
Plastic pipette	1 per group
Whiteboard, 2' × 3'*	1 per group
Investigation Handout	1 per student
Peer-review guide and teacher scoring rubric	1 per student
Checkout Questions (optional)	1 per student

* As an alternative, students can use computer and presentation software such as Microsoft PowerPoint or Apple Keynote to create their arguments.

Figure 9.1 illustrates how to set up the materials for this investigation. There are 12 wells in each miniature greenhouse. Two groups can use 6 wells in a greenhouse. The greenhouse comes with 12 sphagnum peat moss pellets (1 for each well). Be sure to soak the pellets and the seeds in water for several hours before you plan to have students plant their seeds. We recommend using the Wisconsin Fast Plants for this investigation because these plants are easy to grow and have a short life cycle. Seeds germinate a day or two after planting, true leaves appear in about 5 days, and flowers start to appear by about day 15 if the plants are kept in fluorescent light 24 hours a day.

FIGURE 9.1

How to set up the materials for Investigation 9

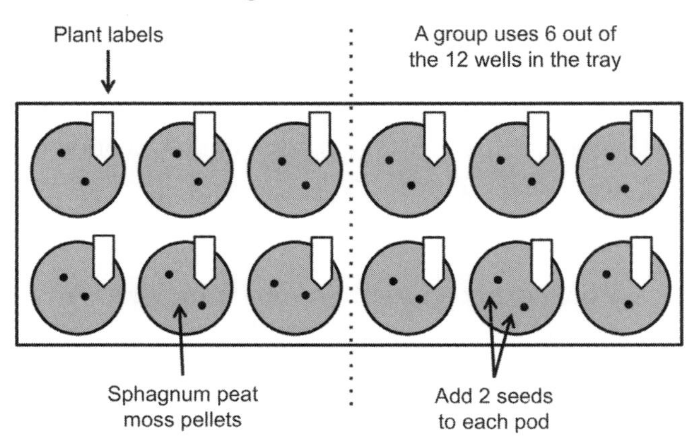

Be sure to use a set routine for distributing and collecting the materials. One option is to set up the materials for each group in a kit that you can deliver to each group. A second option is to have all the materials on a table or cart at a central location. You can then assign a member of each group to be the "materials manager." This individual is responsible for collecting all the materials his or her group needs from the table or cart during class and for returning all the materials at the end of the class.

Safety Precautions

Remind students to follow all normal safety rules. In addition, tell the students to take the following safety precautions:

- Wear sanitized indirectly vented chemical-splash goggles during setup, investigation activity, and cleanup.
- Never eat food or drink when the fertilizer pellets are present.
- Do not throw objects or put any objects in their mouth.
- Wash their hands with soap and water when done collecting the data.

Lesson Plan by Stage

Stage 1: Introduce the Task and the Guiding Question (35 minutes)

1. Ask the students to sit in six groups, with three or four students in each group.
2. Ask students to clear off their desks except for a pencil (and their *Student Workbook for Argument-Driven Inquiry in Third-Grade Science* if they have one).
3. Pass out an Investigation Handout to each student (or ask students to turn to Investigation Log 9 in their workbook).
4. Read the first paragraph of the "Introduction" aloud to the class. Ask the students to follow along as you read.
5. Pass out a potted plant to each group.
6. Remind students of the safety rules and explain the safety precautions for this investigation.
7. Tell students to record their observations and questions about the plant in the "OBSERVED/WONDER" chart in the "Introduction" section of their Investigation Handout (or the investigation log in their workbook).
8. Ask students to share *what they observed* about the plant.
9. Ask students to share *what questions they have* about plants.
10. Tell the students, "Some of your questions might be answered by reading the rest of the 'Introduction.'"

11. Ask the students to read the rest of the "Introduction" on their own *or* ask them to follow along as you read aloud.
12. Once the students have read the rest of the "Introduction," ask them to fill out the "KNOW/NEED" chart on their Investigation Handout (or in their investigation log) as a group.
13. Ask students to share what they learned from the reading. Add these ideas to a class "know / need to figure out" chart.
14. Ask students to share what they think they will need to figure out based on what they read. Add these ideas to the class "know / need to figure out" chart.
15. Tell the students, "It looks like we have something to figure out. Let's see what we will need to do during our investigation."
16. Read the task and the guiding question aloud.
17. Tell the students, "I have lots of materials here that you can use during your investigation."
18. Introduce the students to the materials available for them to use during the investigation by showing them how to plant seeds and the different ways they can change the environment of the plants. Students can set up different environmental conditions by changing the amount of water, fertilizer (which contains the minerals of nitrogen, potassium, and phosphorus), and light in the environment of the plants (see Figure 9.2 for an example).

FIGURE 9.2

An example of an experimental design with four treatment conditions (high and low water, high and low fertilizer)

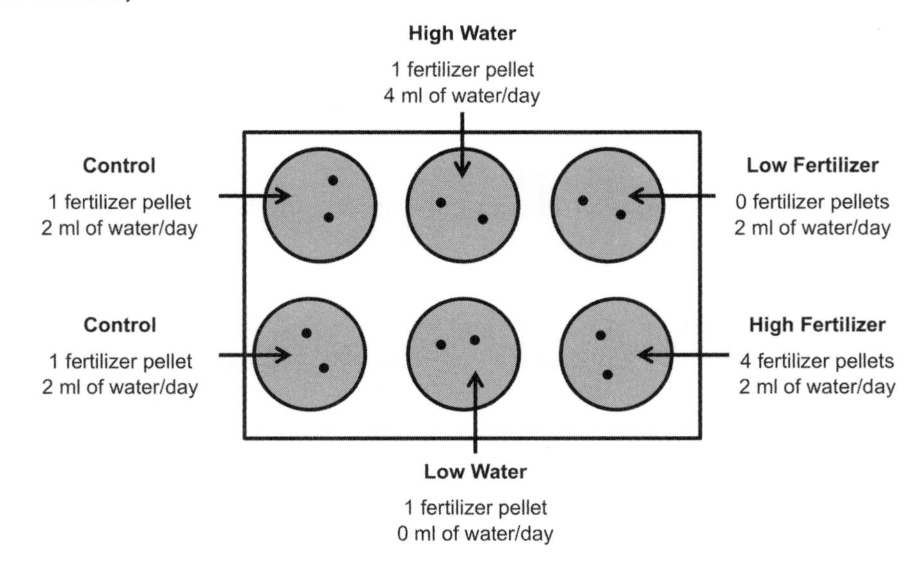

Stage 2: Design a Method and Collect Data (60 minutes for initial setup and 20 minutes a day for 10 additional days to care for and observe the plants as they grow)

1. Tell the students, "I am now going to give you and the other members of your group about 15 minutes to plan your investigation. Before you begin, I want you all to take a couple of minutes to discuss the following questions with the rest of your group."
2. Show the following questions on the screen or board:
   - What information should we collect so we can *describe* the traits of a plant?
   - What information do we need to find a relationship between *a cause and an effect?*
3. Tell the students, "Please take a few minutes to come up with an answer to these questions." Give the students two or three minutes to discuss these two questions.
4. Ask two or three different groups to share their answers. Be sure to highlight or write down any important ideas on the board so students can refer to them later.
5. If possible, use a document camera to project an image of the graphic organizer for this investigation on a screen or board (or take a picture of it and project the picture on a screen or board). Tell the students, "I now want you all to plan out your investigation. To do that, you will need to create an investigation proposal by filling out this graphic organizer."
6. Point to the box labeled "Our guiding question:" and tell the students, "You can put the question we are trying to answer in this box." Then ask, "Where can we find the guiding question?"
7. Wait for a student to answer where to find the guiding question (the answer is "in the handout").
8. Point to the box labeled "We will collect the following data:" and tell the students, "You can list the measurements or observations that you will need to collect during the investigation in this box."
9. Point to the box labeled "This is a picture of how we will set up the equipment:" and tell the students, "You can draw a picture of how you plan to set up the equipment during the investigation so you can collect the data you need in this box."
10. Point to the box labeled "These are the steps we will follow to collect data:" and tell the students, "You can list what you are going to do to collect the data you need and what you will do with your data once you have it. Be sure to give enough detail that I could do your investigation for you."
11. Ask the students, "Do you have any questions about what you need to do?"
12. Answer any questions that come up.

13. Tell the students, "Once you are done, raise your hand and let me know. I'll then come by and look over your proposal and give you some feedback. You may not begin collecting data until I have approved your proposal by signing it. You need to have your proposal done in the next 15 minutes."
14. Give the students 15 minutes to work in their groups on their investigation proposal. As they work, move from group to group to check in, ask probing questions, and offer a suggestion if a group gets stuck.

What should a student-designed investigation look like?

The students' investigation proposal should include the following information:

- The guiding question is "How do differences in soil quality affect the traits of a plant?"
- There are a lot of different types of data that students can collect during this investigation. Examples include (1) height of the plant, (2) leaf size, (3) the number of leaves, and (4) the color of the leaves. At a minimum, each group should collect data about two different traits. This investigation works best if each group selects different traits. Be sure to encourage students to collect quantitative (numerical) data when possible.
- The steps the students will follow will reflect which aspect or aspects of soil quality they decide to investigate and which plant traits that they decide to examine during the investigation. See Figures 9.2–9.4 for examples of different experimental designs that the students could set up for this investigation. In the first experimental design (Figure 9.2, p. 329), the students include a control (seeds grown under normal conditions) and four different treatment conditions (low water, high water, low fertilizer, and high fertilizer). In the second experimental design (Figure 9.3, p. 332), students include a control (seeds grown under normal conditions) and two different treatment conditions (low water and high water). The third experimental design (Figure 9.4, p. 332) also includes a control (seeds grown under normal conditions) and two different treatment conditions (high and low fertilizer). The designs described here are just examples; there should be a lot of variation in the student-designed investigations.

FIGURE 9.3

An example of an experimental design with two treatment conditions (high and low water)

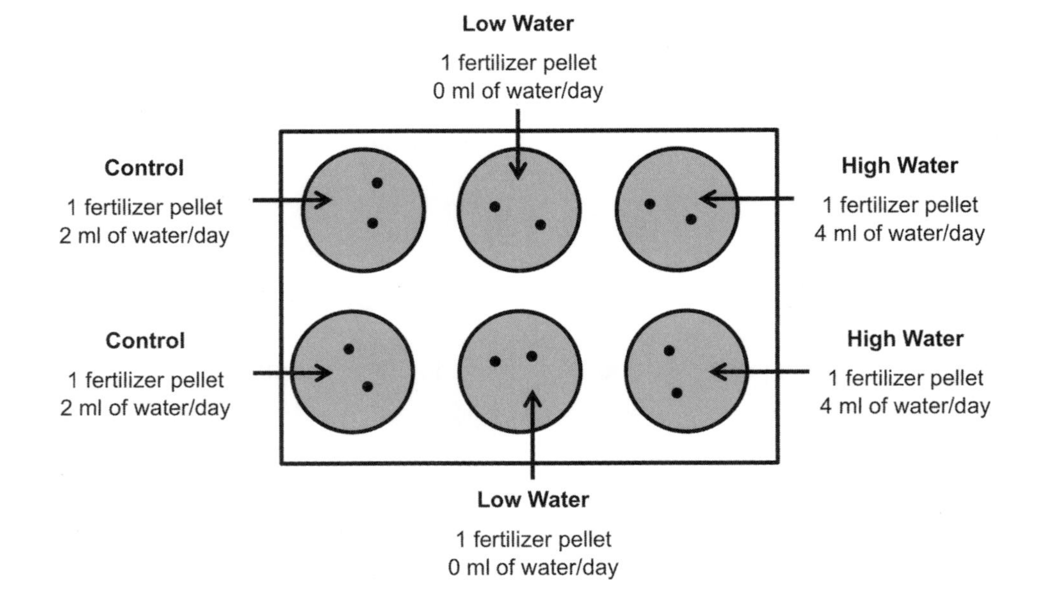

FIGURE 9.4

An example of an experimental design with two treatment conditions (high and low fertilizer)

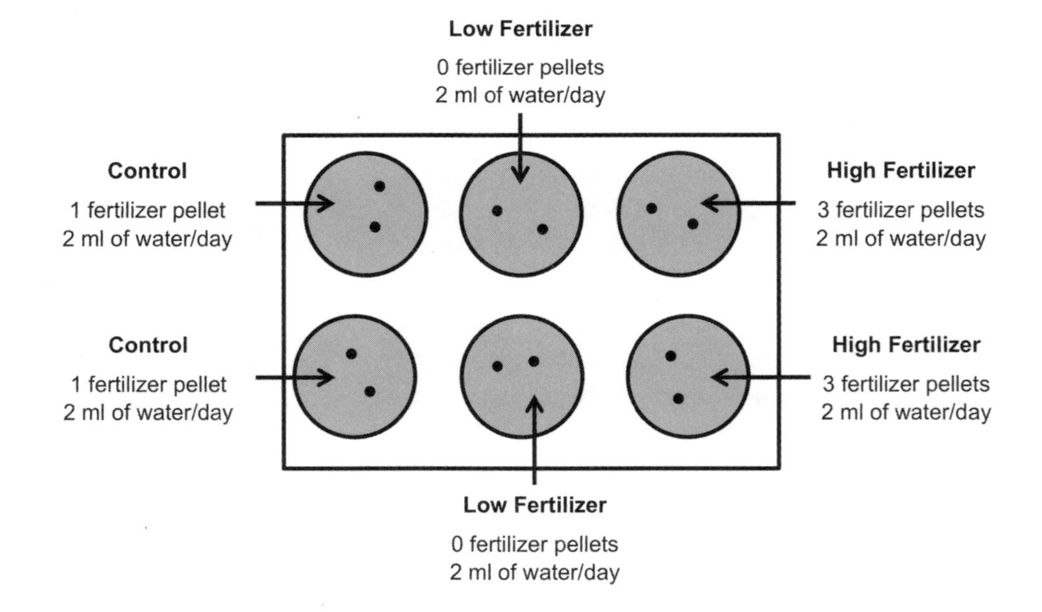

15. As each group finishes its investigation proposal, be sure to read it over and determine if it will be productive or not. If you feel the investigation will be productive (not necessarily what you would do or what the other groups are doing), sign your name on the proposal and let the group start collecting data. If the plan needs to be changed, offer some suggestions or ask some probing questions, and have the group make the changes before you approve it.
16. Pass out the materials or have one student from each group collect the materials they need from a central supply table or cart for the groups that have an approved proposal.
17. Remind students of the safety rules and precautions for this investigation.
18. Give the students 10 minutes to set up their experiments. Be sure to remind students to describe each condition they set up using the plant labels.
19. For the next 10 school days, give the students 20 minutes of class time to care for their plants and make observations and measurements about them. Tell the students to collect their data and record their observations or measurements in the "Collect Your Data" box in their Investigation Handout (or the investigation log in their workbook).

Stage 3: Create a Draft Argument (35 minutes)

1. Tell the students, "Now that we have all this data that we have been collecting over the last 10 days, we need to analyze the data so we can figure out an answer to the guiding question."
2. If possible, project an image of the "Analyze Your Data" section for this investigation on a screen or board using a document camera (or take a picture of it and project the picture on a screen or board). Point to the section and tell the students, "You can create a graph as a way to analyze your data. You can make your graph in this section."
3. Ask the students, "What information do we need to include in the graph?"
4. Tell the students, "Please take a few minutes to discuss this question with your group, and be ready to share."
5. Give the students five minutes to discuss.
6. Ask two or three different groups to share their answers. Be sure to highlight or write down any important ideas on the board so students can refer to them later.
7. Tell the students, "I am now going to give you and the other members of your group about 10 minutes to create your graph." If the students are having trouble making a graph, you can take a few minutes to provide a mini-lesson about how to create a graph from a bunch of observations or measurements (this strategy is called just-in-time instruction because it is offered only when students get stuck).

What should a graph look like for this investigation?

There are a number of different ways that students can analyze the observations or measurements they collect during this investigation. One of the most straightforward ways is to create a scaled bar graph to represent a data set with several categories. This bar graph should have the names of conditions on the horizontal or *x*-axis (e.g., control, low water, and high water). The dependent variable (e.g., height of plant, number of leaves, average size of leaves) should be on the *y*-axis. An example of this type of graph can be seen in Figure 9.5.

FIGURE 9.5

Example of an argument

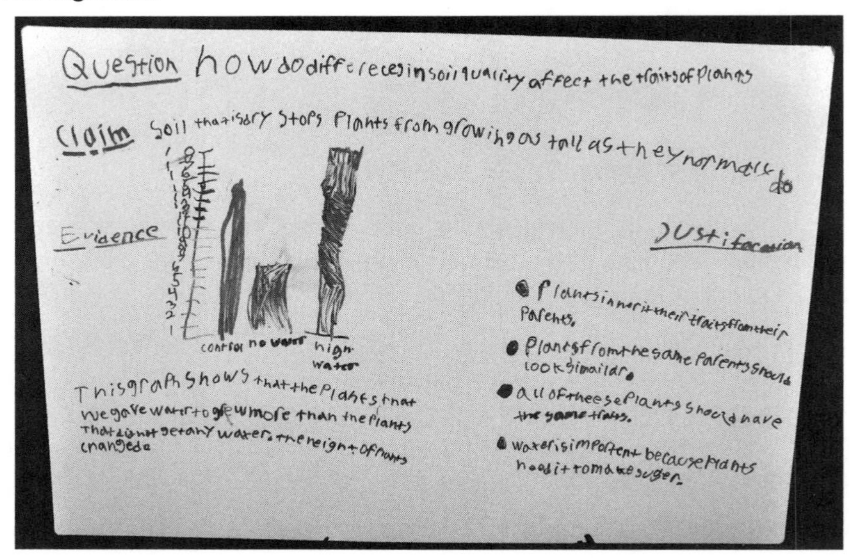

8. Give the students 10 minutes to analyze their data. As they work, move from group to group to check in, ask probing questions, and offer suggestions.
9. Tell the students, "I am now going to give you and the other members of your group 15 minutes to create an argument to share what you have learned and convince others that they should believe you. Before you do that, we need to take a few minutes to discuss what you need to include in your argument."
10. If possible, use a document camera to project the "Argument Presentation on a Whiteboard" image from the Investigation Handout (or the investigation log in their workbook) on a screen or board (or take a picture of it and project the picture on a screen or board).

11. Point to the box labeled "The Guiding Question:" and tell the students, "You can put the question we are trying to answer here on your whiteboard."
12. Point to the box labeled "Our Claim:" and tell the students, "You can put your claim here on your whiteboard. The claim is your answer to the guiding question."
13. Point to the box labeled "Our Evidence:" and tell the students, "You can put the evidence that you are using to support your claim here on your whiteboard. Your evidence will need to include the analysis you just did and an explanation of what your analysis means or shows. Scientists always need to support their claims with evidence."
14. Point to the box labeled "Our Justification of the Evidence:" and tell the students, "You can put your justification of your evidence here on your whiteboard. Your justification needs to explain why your evidence is important. Scientists often use core ideas to explain why the evidence they are using matters. Core ideas are important concepts that scientists use to help them make sense of what happens during an investigation."
15. Ask the students, "What are some core ideas that we read about earlier that might help us explain why the evidence we are using is important?"
16. Ask students to share some of the core ideas from the "Introduction" section of the Investigation Handout (or the investigation log in the workbook). List these core ideas on the board.
17. Tell the students, "That is great. I would like to see everyone try to include these core ideas in your justification of the evidence. Your goal is to use these core ideas to help explain why your evidence matters and why the rest of us should pay attention to it."
18. Ask the students, "Do you have any questions about what you need to do?"
19. Answer any questions that come up.
20. Tell the students, "Okay, go ahead and start working on your arguments. You need to have your argument done in the next 15 minutes. It doesn't need to be perfect. We just need something down on the whiteboards so we can share our ideas."
21. Give the students 15 minutes to work in their groups on their arguments. As they work, move from group to group to check in, ask probing questions, and offer a suggestion if a group gets stuck. Figure 9.5 shows an example of an argument created by students for this investigation.

Stage 4: Argumentation Session (30 minutes)

The argumentation session can be conducted in a whole-class presentation format, a gallery walk format, or a modified gallery walk format. We recommend using a whole-class presentation format for the first investigation, but try to transition to either the gallery walk or modified gallery walk format as soon as possible because that will maximize student voice and choice inside the classroom. The following list shows the steps for the three formats; unless otherwise noted, the steps are the same for all three formats.

1. Begin by introducing the use of the whiteboard.
   - *If using the whole-class presentation format,* tell the students, "We are now going to share our arguments. Please set up your whiteboard so everyone can see them."
   - *If using the gallery walk or modified gallery walk format,* tell the students, "We are now going to share our arguments. Please set up your whiteboard so they are facing the walls."
2. Allow the students to set up their whiteboards.
   - *If using the whole-class presentation format,* the whiteboards should be set up on stands or chairs so they are facing toward the center of the room.
   - *If using the gallery walk or modified gallery walk format,* the whiteboards should be set up on stands or chairs so they are facing toward the outside of the room.
3. Give the following instructions to the students:
   - *If using the whole-class presentation format or the modified gallery walk format,* tell the students, "Okay, before we get started I want to explain what we are going to do next. I'm going to ask some of you to present your arguments to your classmates. If you are presenting your argument, your job is to share your group's claim, evidence, and justification of the evidence. The rest of you will be reviewers. If you are a reviewer, your job is to listen to the presenters, ask the presenters questions if you do not understand something, and then offer them some suggestions about ways to make their argument better. After we have a chance to learn from each other, I'm going to give you some time to revise your arguments and make them better."
   - *If using the gallery walk format,* tell the students, "Okay, before we get started I want to explain what we are going to do next. You are going to have an opportunity to read the arguments that were created by other groups. Your group will go to a different group's argument. I'll give you a few minutes to read it and review it. Your job is to offer them some suggestions about ways to make their argument better. You can use sticky notes to give them suggestions. Please be specific about what you want to change and be specific about how you think they should change it. After we have a chance to learn from each other, I'm going to give you some time to revise your arguments and make them better."
4. Use a document camera to project the "Ways to IMPROVE our argument …" box from the Investigation Handout (or the investigation log in their workbook)

on a screen or board (or take a picture of it and project the picture on a screen or board).

- *If using the whole-class presentation format or the modified gallery walk format,* point to the box and tell the students, "If you are a presenter, you can write down the suggestions you get from the reviewers here. If you are a reviewer, and you see a good idea from another group, you can write down that idea here. Once we are done with the presentations, I will give you a chance to use these suggestions or ideas to improve your arguments.
- *If using the gallery walk format,* point to the box and tell the students, "If you see good ideas from another group, you can write them down here. Once we are done reviewing the different arguments, I will give you a chance to use these ideas to improve your own arguments. It is important to share ideas like this."

Ask the students, "Do you have any questions about what you need to do?"

5. Answer any questions that come up.
6. Give the following instructions:
   - *If using the whole-class presentation format,* tell the students, "Okay. Let's get started."
   - *If using the gallery walk format,* tell the students, "Okay, I'm now going to tell you which argument to go to and review.
   - *If using the modified gallery walk format,* tell the students, "Okay, I'm now going to assign you to be a presenter or a reviewer." Assign one or two students from each group to be presenters and one or two students from each group to be reviewers.
7. Begin the review of the arguments.
   - *If using the whole-class presentation format,* have four or five groups present their argument one at a time. Give each group only two to three minutes to present their argument. Then give the class two to three minutes to ask them questions and offer suggestions. Be sure to encourage as much participation from the students as possible.
   - *If using the gallery walk format,* tell the students, "Okay. Let's get started. Each group, move one argument to the left. Don't move to the next argument until I tell you to move. Once you get there, read the argument and then offer suggestions about how to make it better. I will put some sticky notes next to each argument. You can use the sticky notes to leave your suggestions." Give each group about three to four minutes to read the arguments, talk, and offer suggestions.

     a. Tell the students, "Okay. Let's rotate. Move one group to the left."

     b. Again, give each group three or four minutes to read, talk, and offer suggestions.

c. Repeat this process for two more rotations.

- *If using the modified gallery walk format,* tell the students, "Okay. Let's get started. Reviewers, move one group to the left. Don't move to the next group until I tell you to move. Presenters, go ahead and share your argument with the reviewers when they get there." Give each group of presenters and reviewers about three to four minutes to talk.

  a. Tell the students, "Okay. Let's rotate. Reviewers, move one group to the left."

  b. Again, give each group of presenters and reviewers about three or four minutes to talk.

  c. Repeat this process for two more rotations.

8. Tell the students to return to their workstations.
9. Give the following instructions about revising the argument:
   - *If using the whole-class presentation format,* tell the students, "I'm now going to give you about 10 minutes to revise your argument. Take a few minutes to talk in your groups and determine what you want to change to make your argument better. Once you have decided what to change, go ahead and make the changes to your whiteboard."
   - *If using the gallery walk format,* tell the students, "I'm now going to give you about 10 minutes to revise your argument. Take a few minutes to read the suggestions that were left at your argument. Then talk in your groups and determine what you want to change to make your argument better. Once you have decided what to change, go ahead and make the changes to your whiteboard."
   - *If using the modified gallery walk format,* "I'm now going to give you about 10 minutes to revise your argument. Please return to your original groups." Wait for the students to move back into their original groups and then tell the students, "Okay, take a few minutes to talk in your groups and determine what you want to change to make your argument better. Once you have decided what to change, go ahead and make the changes to your whiteboard."

   Ask the students, "Do you have any questions about what you need to do?"
10. Answer any questions that come up.
11. Tell the students, "Okay. Let's get started."
12. Give the students 10 minutes to work in their groups on their arguments. As they work, move from group to group to check in, ask probing questions, and offer a suggestion if a group gets stuck.

Stage 5: Reflective Discussion (15 minutes)

1. Tell the students, "We are now going to take a minute to talk about what we did and what we have learned."
2. Place several plants that were grown under different conditions on a table (pick a set of plants that were grown by the students that show a great deal of difference based on condition). Ask the students, "What do you all see going on here?"
3. Allow students to share their ideas.
4. Ask the students, "Do you think these plants came from the same biological parents?"
5. Allow students to share their ideas.
6. Ask the students, "Do you think these plants inherited the same traits?"
7. Allow students to share their ideas.
8. Ask the students, "Why don't these plants look the same?"
9. Allow students to share their ideas. Keep probing until someone mentions water or fertilizer.
10. Tell the students, "Okay, let's make sure we are on the same page. Plants inherit a mix of traits from their biological parents. Different types of plants differ in how they look and function because they have different inherited information. The individual plants of the same type can also differ in how they look and function because individual plants inherit a different mixture of traits. The environment can also affect the traits of individual plants. An individual plant can end up looking or functioning differently than other plants of the same type because of where it grows over time. The way living things inherit traits and how the environment can change these traits is a really important core idea in science."
11. Ask the students, "Does anyone have any questions about this core idea?"
12. Answer any questions that come up.
13. Tell the students, "We also looked for a cause-and-effect relationship during our investigation." Then ask, "Can anyone tell me why it is useful to look for cause-and-effect relationships?"
14. Allow students to share their ideas.
15. Tell the students, "Cause-and-effect relationships are important because they allow us to predict what will happen in the future."
16. Ask the students, "What was the cause and what was the effect that we uncovered today?

17. Allow students to share their ideas. Keep probing until someone mentions that the cause was changing the conditions of the soil and the effect was the traits of the plants.
18. Tell the students, "That is great, and if we know that we can predict what will happen the next time we try to grow a plant."
19. Tell the students, "We are now going take a minute to talk about what went well and what didn't go so well during our investigation. We need to talk about this because you are going to be planning and carrying out your own investigations like this a lot this year, and I want to help you all get better at it."
20. Show an image of the question "What made your investigation scientific?" on the screen. Tell the students, "Take a few minutes to talk about how you would answer this question with the other people in your group. Be ready to share with the rest of the class." Give the students two to three minutes to talk in their group.
21. Ask the students, "What do you all think? Who would like to share an idea?"
22. Allow students to share their ideas. Be sure to expand on their ideas about what makes an investigation scientific.
23. Show an image of the question "What made your investigation not so scientific?" on the screen. Tell the students, "Take a few minutes to talk about how you would answer this question with the other people in your group. Be ready to share with the rest of the class." Give the students two to three minutes to talk in their group.
24. Ask the students, "What do you all think? Who would like to share an idea?"
25. Allow students to share their ideas. Be sure to expand on their ideas about what makes an investigation less scientific.
26. Show an image of the question "What rules can we put into place to help us make sure our next investigation is more scientific?" on the screen. Tell the students, "Take a few minutes to talk about how you would answer this question with the other people in your group. Be ready to share with the rest of the class." Give the students two to three minutes to talk in their group.
27. Ask the students, "What do you all think? Who would like to share an idea?"
28. Allow students to share their ideas. Once they have shared their ideas, offer a suggestion for a possible class rule.
29. Ask the students, "What do you all think? Should we make this a rule?"
30. If the students agree, write the rule on the board or make a class "Rules for Scientific Investigation" chart so you can refer to it during the next investigation.
31. Tell the students, "We are now going take a minute to talk about what makes science different from other subjects."

32. Show an image of the question "What is the difference between a theory and a law in science?" on the screen. Tell the students, "Take a few minutes to talk about how you would answer this question with the other people in your group. Be ready to share with the rest of the class." Give the students two to three minutes to talk in their group.
33. Ask the students, "What do you all think? Who would like to share an idea?"
34. Allow students to share their ideas.
35. Tell the students, "Okay, let's make sure we are all using the same definition. Scientists develop or use laws to describe what happens in the world. Scientists develop or use theories to explain why things happen."
36. Show an image of Figure 9.6 along with the statement "Plants grown with less light do not grow as well as plants grown with more light" on the screen.

FIGURE 9.6

Plants grown under different conditions

37. Ask the students, "Is this statement a law or a theory and why?"
38. Allow students to share their ideas.
39. Tell the students, "That statement is a law because it describes what will happen but it does not explain why."
40. Show an image of Figure 9.6 along with the statement "Plants require light to convert carbon dioxide into sugar; without light they cannot make the sugar they need to grow" on the screen.
41. Ask the students, "Is this statement a law or a theory and why?"
42. Allow students to share their ideas.
43. Tell the students, "That statement is theory because it explains why plants do not grow as well when they do not have enough light."

44. Ask the students, "Does anyone have any questions about the difference between a law and a theory in science?"
45. Answer any questions that come up.

Stage 6: Write a Draft Report (35 minutes)

Your students will use either the Investigation Handout or the investigation log in the student workbook when writing the draft report. When you give the directions shown in quotes in the following steps, substitute "investigation log" (as shown in brackets) for "handout" if they are using the workbook.

1. Tell the students, "You are now going to write an investigation report to share what you have learned. Please take out a pencil and turn to the 'Draft Report' section of your handout [investigation log]."
2. If possible, use a document camera to project the "Introduction" section of the draft report from the Investigation Handout (or the investigation log in their workbook) on a screen or board (or take a picture of it and project the picture on a screen or board).
3. Tell the students, "The first part of the report is called the 'Introduction.' In this section of the report you want to explain to the reader what you were investigating, why you were investigating it, and what question you were trying to answer. All of this information can be found in the text at the beginning of your handout [investigation log]." Point to the image and say, "There are some sentence starters here to help you begin writing the report." Ask the students, "Do you have any questions about what you need to do?"
4. Answer any questions that come up.
5. Tell the students, "Okay. Let's write."
6. Give the students 10 minutes to write the "Introduction" section of the report. As they work, move from student to student to check in, ask probing questions, and offer a suggestion if a student gets stuck.
7. If possible, use a document camera to project the "Method" section of the draft report from the Investigation Handout (or the investigation log in their workbook) on a screen or board (or take a picture of it and project the picture on a screen or board).
8. Tell the students, "The second part of the report is called the 'Method.' In this section of the report you want to explain to the reader what you did during the investigation, what data you collected and why, and how you went about analyzing your data. All of this information can be found in the 'Plan Your Investigation' section of your handout [investigation log]. Remember that you all planned and carried out different investigations, so do not assume that the reader will know what you did." Point to the image and say, "There are some

sentence starters here to help you begin writing this part of the report." Ask the students, "Do you have any questions about what you need to do?"

9. Answer any questions that come up.
10. Tell the students, "Okay. Let's write."
11. Give the students 10 minutes to write the "Method" section of the report. As they work, move from student to student to check in, ask probing questions, and offer a suggestion if a student gets stuck.
12. If possible, use a document camera to project the "Argument" section of the draft report from the Investigation Handout (or the investigation log in their workbook) on a screen or board (or take a picture of it and project the picture on a screen or board).
13. Tell the students, "The last part of the report is called the 'Argument.' In this section of the report you want to share your claim, evidence, and justification of the evidence with the reader. All of this information can be found on your whiteboard." Point to the image and say, "There are some sentence starters here to help you begin writing this part of the report." Ask the students, "Do you have any questions about what you need to do?"
14. Answer any questions that come up.
15. Tell the students, "Okay. Let's write."
16. Give the students 10 minutes to write the "Argument" section of the report. As they work, move from student to student to check in, ask probing questions, and offer a suggestion if a student gets stuck.

Stage 7: Peer Review (30 minutes)

Your students will use either the Investigation Handout or the investigation log in the student workbook when doing the peer review. When you give the directions shown in quotes in the following steps, substitute "workbook" (as shown in brackets) for "Investigation Handout" if they are using the workbook.

1. Tell the students, "We are now going to review our reports to find ways to make them better. I'm going to come around and collect your Investigation Handout [workbook]. While I do that, please take out a pencil."
2. Collect the Investigation Handouts or workbooks from the students.
3. If possible, use a document camera to project the peer-review guide (PRG; see Appendix 4) on a screen or board (or take a picture of it and project the picture on a screen or board).
4. Tell the students, "We are going to use this peer-review guide to give each other feedback." Point to the image.
5. Give the following instructions:

- *If using the Investigation Handout,* tell the students, "I'm going to ask you to work with a partner to do this. I'm going to give you and your partner a draft report to read and a peer-review guide to fill out. You two will then read the report together. Once you are done reading the report, I want you to answer each of the questions on the peer-review guide." Point to the review questions on the image of the PRG.
- *If using the workbook,* tell the students, "I'm going to ask you to work with a partner to do this. I'm going to give you and your partner a draft report to read. You two will then read the report together. Once you are done reading the report, I want you to answer each of the questions on the peer-review guide that is right after the report in the investigation log." Point to the review questions on the image of the PRG.

6. Tell the students, "You can check 'yes,' 'almost,' or 'no' after each question." Point to the checkboxes on the image of the PRG.
7. Tell the students, "This will be your rating for this part of the report. Make sure you agree on the rating you give the author. If you mark 'almost' or 'no,' then you need to tell the author what he or she needs to do to get a 'yes.'" Point to the space for the reviewer feedback on the image of the PRG.
8. Tell the students, "It is really important for you to give the authors feedback that is helpful. That means you need to tell them exactly what they need to do to make their reports better." Ask the students, "Do you have any questions about what you need to do?"
9. Answer any questions that come up.
10. Tell the students, "Please sit with a partner who is not in your current group." Allow the students time to sit with a partner.
11. Give the following instructions:
    - *If using the Investigation Handout,* tell the students, "Okay, I am now going to give you one report to read and one peer-review guide to fill out." Pass out one report to each pair. Make sure that the report you give a pair was not written by one of the students in that pair. Give each pair one PRG to fill out as a team.
    - *If using the workbook,* tell the students, "Okay, I am now going to give you one report to read." Pass out a workbook to each pair. Make sure that the workbook you give a pair is not from one of the students in that pair.
12. Tell the students, "Okay, I'm going to give you 15 minutes to read the report I gave you and to fill out the peer-review guide. Go ahead and get started."
13. Give the students 15 minutes to work. As they work, move around from pair to pair to check in and see how things are going, answer questions, and offer advice.

14. After 15 minutes pass, tell the students, "Okay, time is up." *If using the Investigation Handout,* say, "Please give me the report and the peer-review guide that you filled out." *If using the workbook,* say, "Please give me the workbook that you have."
15. Collect the Investigation Handouts and the PRGs, or collect the workbooks if they are being used. Be sure you keep the handout and the PRG together.
16. Give the following instructions:
    - *If using the Investigation Handout,* tell the students, "Okay, I am now going to give you a different report to read and a new peer-review guide to fill out." Pass out another report to each pair. Make sure that this report was not written by one of the students in that pair. Give each pair a new PRG to fill out as a team.
    - *If using the workbook,* tell the students, "Okay, I am now going to give you a different report to read." Pass out a different workbook to each pair. Make sure that the workbook you give a pair is not from one of the students in that pair.
17. Tell the students, "Okay, I'm going to give you 15 minutes to read this new report and to fill out the peer-review guide. Go ahead and get started."
18. Give the students 15 minutes to work. As they work, move around from pair to pair to check in and see how things are going, answer questions, and offer advice.
19. After 15 minutes pass, tell the students, "Okay, time is up." *If using the Investigation Handout,* say, "Please give me the report and the peer-review guide that you filled out." *If using the workbook,* say, "Please give me the workbook that you have."
20. Collect the Investigation Handouts and the PRGs, or collect the workbooks if they are being used. Be sure you keep the handout and the PRG together.

Stage 8: Revise the Report (30 minutes)

Your students will use either the Investigation Handout or the investigation log in the student workbook when revising the report. Except where noted below, the directions are the same whether using the handout or the log.

1. Give the following instructions:
    - *If using the Investigation Handout,* tell the students, "You are now going to revise your investigation report based on the feedback you get from your classmates. Please take out a pencil while I hand back your draft report and the peer-review guide."
    - *If using the investigation log in the student workbook,* tell the students, "You are now going to revise your investigation report based on the feedback you

get from your classmates. Please take out a pencil while I hand back your investigation logs."

2. *If using the Investigation Handout,* pass back the handout and the PRG to each student. *If using the investigation log,* pass back the log to each student.
3. Tell the students, "Please take a few minutes to read over the peer-review guide. You should use it to figure out what you need to change in your report and how you will change the report."
4. Allow the students time to read the PRG.
5. *If using the investigation log,* if possible use a document camera to project the "Write Your Final Report" section from the investigation log on a screen or board (or take a picture of it and project the picture on a screen or board).
6. Give the following instructions:
   - *If using the Investigation Handout,* tell the students, "Okay. Let's revise our reports. Please take out a piece of paper. I would like you to rewrite your report. You can use your draft report as a starting point, but use the feedback on the peer-review guide to help make it better."
   - *If using the investigation log,* tell the students, "Okay. Let's revise our reports. I would like you to rewrite your report in the section of the investigation log that says 'Write Your Final Report.'" Point to the image on the screen and tell the students, "You can use your draft report as a starting point, but use the feedback on the peer-review guide to help make your report better."

   Ask the students, "Do you have any questions about what you need to do?"
7. Answer any questions that come up.
8. Tell the students, "Okay. Let's write."
9. Give the students 30 minutes to rewrite their report. As they work, move from student to student to check in, ask probing questions, and offer a suggestion if a student gets stuck.
10. Give the following instructions:
    - *If using the Investigation Handout,* tell the students, "Okay. Time's up. I will now come around and collect your Investigation Handout, the peer-review guide, and your final report."
    - *If using the investigation log,* tell the students, "Okay. Time's up. I will now come around and collect your workbooks."
11. *If using the Investigation Handout,* collect all the Investigation Handouts, PRGs, and final reports. *If using the investigation log,* collect all the workbooks.
12. *If using the Investigation Handout,* use the "Teacher Score" columns in the PRG to grade the final report. *If using the investigation log,* use the "ADI Investigation Report Grading Rubric" in the investigation log to grade the final report.

Whether you are using the handout or the log, you can give the students feedback about their writing in the "Teacher Comments" section.

How to Use the Checkout Questions

The Checkout Questions are an optional assessment. We recommend giving them to students one day after they finish stage 8 of the ADI investigation. The Checkout Questions can be used as a formative or summative assessment of student thinking. If you plan to use them as a formative assessment, we recommend that you look over the student answers to determine if you need to reteach the core idea and/or crosscutting concept from the investigation, but do not grade them. If you plan to use them as a summative assessment, we have included a "Teacher Scoring Rubric" at the end of the Checkout Questions that you can use to score a student's ability to apply the core idea in a new scenario and explain their use of a crosscutting concept. The rubric includes a 4-point scale that ranges from 0 (the student cannot apply the core idea correctly in all cases and cannot explain the [crosscutting concept]) to 3 (the student can apply the core idea correctly in all cases and can fully explain the [crosscutting concept]). The Checkout Questions, regardless of how you decide to use them, are a great way to make student thinking visible so you can determine if the students have learned the core idea and the crosscutting concept.

A student who can apply the core idea correctly in all cases and can explain the cause-and-effect relationship would select A for question 1, C for question 2, and A for question 3. He or she should then be able to explain that plants inherit traits from their biological parents but the environment can also affect the traits of a plant. An individual plant can end up looking different from other plants because of a lack of minerals or water in the soil.

Connections to Standards

Table 9.2 highlights how the investigation can be used to address specific performance expectations from the *NGSS, Common Core State Standards (CCSS)* in English language arts (ELA) and in mathematics, and *English Language Proficiency (ELP) Standards.*

TABLE 9.2

Investigation 9 alignment with standards

***NGSS* performance expectations**	Strong alignment • 3-LS3-2. Use evidence to support the explanation that traits can be influenced by the environment. Moderate alignment (this investigation can be used to build toward this performance expectation) • 3-LS3-1: Analyze and interpret data to provide evidence that plants and animals have traits inherited from parents and that variation of these traits exists in a group of similar organisms.

Continued

Table 9.2 (*continued*)

***CCSS ELA*—Reading: Informational Text**	Key ideas and details • CCSS.ELA-LITERACY.RI.3.1: Ask and answer questions to demonstrate understanding of a text, referring explicitly to the text as the basis for the answers. • CCSS.ELA-LITERACY.RI.3.2: Determine the main idea of a text; recount the key details and explain how they support the main idea. • CCSS.ELA-LITERACY.RI.3.3: Describe the relationship between a series of historical events, scientific ideas or concepts, or steps in technical procedures in a text, using language that pertains to time, sequence, and cause/effect. Craft and structure • CCSS.ELA-LITERACY.RI.3.4: Determine the meaning of general academic and domain-specific words and phrases in a text relevant to a *grade 3 topic or subject area*. • CCSS.ELA-LITERACY.RI.3.5: Use text features and search tools (e.g., key words, sidebars, hyperlinks) to locate information relevant to a given topic efficiently. • CCSS.ELA-LITERACY.RI.3.6: Distinguish their own point of view from that of the author of a text. Integration of knowledge and ideas • CCSS.ELA-LITERACY.RI.3.7: Use information gained from illustrations (e.g., maps, photographs) and the words in a text to demonstrate understanding of the text (e.g., where, when, why, and how key events occur). • CCSS.ELA-LITERACY.RI.3.8: Describe the logical connection between particular sentences and paragraphs in a text (e.g., comparison, cause/effect, first/second/third in a sequence). • CCSS.ELA-LITERACY.RI.3.9: Compare and contrast the most important points and key details presented in two texts on the same topic. Range of reading and level of text complexity • CCSS.ELA-LITERACY.RI.3.10: By the end of the year, read and comprehend informational texts, including history/social studies, science, and technical texts, at the high end of the grades 2–3 text complexity band independently and proficiently.

Continued

Table 9.2 (*continued*)

***CCSS ELA*—Writing**	Text types and purposes • CCSS.ELA-LITERACY.W.3.1: Write opinion pieces on topics or texts, supporting a point of view with reasons. o CCSS.ELA-LITERACY.W.3.1.A: Introduce the topic or text they are writing about, state an opinion, and create an organizational structure that lists reasons. o CCSS.ELA-LITERACY.W.3.1.B: Provide reasons that support the opinion. o CCSS.ELA-LITERACY.W.3.1.C: Use linking words and phrases (e.g., *because*, *therefore*, *since*, *for example*) to connect opinion and reasons. o CCSS.ELA-LITERACY.W.3.1.D: Provide a concluding statement or section. • CCSS.ELA-LITERACY.W.3.2: Write informative or explanatory texts to examine a topic and convey ideas and information clearly. o CCSS.ELA-LITERACY.W.3.2.A: Introduce a topic and group related information together; include illustrations when useful to aiding comprehension. o CCSS.ELA-LITERACY.W.3.2.B: Develop the topic with facts, definitions, and details. o CCSS.ELA-LITERACY.W.3.2.C: Use linking words and phrases (e.g., *also*, *another*, *and*, *more*, *but*) to connect ideas within categories of information. o CCSS.ELA-LITERACY.W.3.2.D: Provide a concluding statement or section. Production and distribution of writing • CCSS.ELA-LITERACY.W.3.4: With guidance and support from adults, produce writing in which the development and organization are appropriate to task and purpose. • CCSS.ELA-LITERACY.W.3.5: With guidance and support from peers and adults, develop and strengthen writing as needed by planning, revising, and editing. • CCSS.ELA-LITERACY.W.3.6: With guidance and support from adults, use technology to produce and publish writing (using keyboarding skills) as well as to interact and collaborate with others. Research to build and present knowledge • CCSS.ELA-LITERACY.W.3.8: Recall information from experiences or gather information from print and digital sources; take brief notes on sources and sort evidence into provided categories. Range of writing • CCSS.ELA-LITERACY.W.3.10: Write routinely over extended time frames (time for research, reflection, and revision) and shorter time frames (a single sitting or a day or two) for a range of discipline-specific tasks, purposes, and audiences.

Continued

Table 9.2 (*continued*)

***CCSS ELA*—Speaking and Listening**	Comprehension and collaboration • CCSS.ELA-LITERACY.SL.3.1: Engage effectively in a range of collaborative discussions (one-on-one, in groups, and teacher-led) with diverse partners on *grade 3 topics and texts*, building on others' ideas and expressing their own clearly. o CCSS.ELA-LITERACY.SL.3.1.A: Come to discussions prepared, having read or studied required material; explicitly draw on that preparation and other information known about the topic to explore ideas under discussion. o CCSS.ELA-LITERACY.SL.3.1.B: Follow agreed-upon rules for discussions (e.g., gaining the floor in respectful ways, listening to others with care, speaking one at a time about the topics and texts under discussion). o CCSS.ELA-LITERACY.SL.3.1.C: Ask questions to check understanding of information presented, stay on topic, and link their comments to the remarks of others. o CCSS.ELA-LITERACY.SL.3.1.D: Explain their own ideas and understanding in light of the discussion. • CCSS.ELA-LITERACY.SL.3.2: Determine the main ideas and supporting details of a text read aloud or information presented in diverse media and formats, including visually, quantitatively, and orally. • CCSS.ELA-LITERACY.SL.3.3: Ask and answer questions about information from a speaker, offering appropriate elaboration and detail. Presentation of knowledge and ideas • CCSS.ELA-LITERACY.SL.3.4: Report on a topic or text, tell a story, or recount an experience with appropriate facts and relevant, descriptive details, speaking clearly at an understandable pace. • CCSS.ELA-LITERACY.SL.3.6: Speak in complete sentences when appropriate to task and situation in order to provide requested detail or clarification.
***CCSS Mathematics*—Operations and Algebraic Thinking**	Solve problems involving the four operations, and identify and explain patterns in arithmetic. • CCSS.MATH.CONTENT.3.OA.D.8: Solve two-step word problems using the four operations. Represent these problems using equations with a letter standing for the unknown quantity. Assess the reasonableness of answers using mental computation and estimation strategies including rounding.
***CCSS Mathematics*—Number and Operations in Base Ten**	Use place value understanding and properties of operations to perform multi-digit arithmetic. • CCSS.MATH.CONTENT.3.NBT.A.2: Fluently add and subtract within 1,000 using strategies and algorithms based on place value, properties of operations, and/or the relationship between addition and subtraction.

Continued

Table 9.2 (*continued*)

***CCSS Mathematics*— Measurement and Data**	Solve problems involving measurement and estimation. • CCSS.MATH.CONTENT.3.MD.A.2: Measure and estimate liquid volumes and masses of objects using standard units of grams (g), kilograms (kg), and liters (l). Add, subtract, multiply, or divide to solve one-step word problems involving masses or volumes that are given in the same units, e.g., by using drawings (such as a beaker with a measurement scale) to represent the problem. Represent and interpret data. • CCSS.MATH.CONTENT.3.MD.B.3: Draw a scaled picture graph and a scaled bar graph to represent a data set with several categories. Solve one- and two-step "how many more" and "how many less" problems using information presented in scaled bar graphs. • CCSS.MATH.CONTENT.3.MD.B.4: Generate measurement data by measuring lengths using rulers marked with halves and fourths of an inch. Show the data by making a line plot, where the horizontal scale is marked off in appropriate units—whole numbers, halves, or quarters.
ELP Standards	Receptive modalities • ELP 1: Construct meaning from oral presentations and literary and informational text through grade-appropriate listening, reading, and viewing. • ELP 8: Determine the meaning of words and phrases in oral presentations and literary and informational text. Productive modalities • ELP 3: Speak and write about grade-appropriate complex literary and informational texts and topics. • ELP 4: Construct grade-appropriate oral and written claims and support them with reasoning and evidence. • ELP 7: Adapt language choices to purpose, task, and audience when speaking and writing. Interactive modalities • ELP 2: Participate in grade-appropriate oral and written exchanges of information, ideas, and analyses, responding to peer, audience, or reader comments and questions. • ELP 5: Conduct research and evaluate and communicate findings to answer questions or solve problems. • ELP 6: Analyze and critique the arguments of others orally and in writing. Linguistic structures of English • ELP 9: Create clear and coherent grade-appropriate speech and text. • ELP 10: Make accurate use of standard English to communicate in grade-appropriate speech and writing.

Investigation 9

Traits and the Environment: How Do Differences in Soil Quality Affect the Traits of a Plant?

Introduction

Plants have many unique traits that animals do not have. These traits enable plants to turn carbon dioxide into sugar, get water from the soil, and reproduce. Take a moment to examine a plant. Be sure to keep track of what you observe and what you are wondering about in the boxes below.

Things I OBSERVED ...	Things I WONDER about ...

The plant you observed is called a flowering plant. Flowering plants have several leaves, a stem, and one or more flowers. The leaves of plants are very important. Plants use carbon dioxide from the air and water from the soil to make sugar inside their leaves though a process called *photosynthesis.* Plants then use this sugar as a source of energy. Plants are able to move water from the soil to the leaves inside their stems. The flowers are used in reproduction.

Flowering plants produce pollen inside the flowers. This pollen is then spread to other plants by the wind or by sticking to animals such as bees and birds. Once a plant is fertilized by pollen, it will produce seeds that can grow and develop into a new plant.

The leaves, stem, and flowers of flowering plants make them different from other living things, such as animals or fungi.

Flowering plants can have very different-looking leaves, stems, and flowers. The leaves and stems of a plant can be big and wide or small and narrow. Flowers can also come in a wide range of shapes, sizes, and colors. The size, shape, and color of leaves, stems, and flowers are all traits that are passed down from parent to offspring. This is one reason that there are so many different kinds of plants. Adult plants produce offspring with traits that are similar to the ones they have. The environment, however, can also change the traits of a plant as it grows and develops.

In this investigation, your goal is to figure out how a change in a characteristic of soil (a cause) affects one or more different plant traits (the effect). You will be able to change one of two different characteristics of soil during your investigation: either the moisture of the soil or the amount of minerals in the soil. These minerals include the elements of nitrogen, phosphorus, and potassium. You can change the amount of minerals found in the soil by adding one or more fertilizer pellets to it. Plants use water to produce sugar for energy and minerals to create more plant parts (such as leaves and roots).

There are many different plant traits that you can study during your investigation. You can look at the height of the plant, leaf size, the number of leaves, or the color of the leaves. You can also examine a combination of these four different plant traits.

Things we KNOW from what we read ...	What we will NEED to figure out ...

Your Task

Use what you know about cause and effect to plan and carry out an investigation to determine how a change in amount of water or minerals in the soil will or will not change the traits of a plant. Be sure to set up a fair test so you can determine if there is a cause-and-effect relationship or not. To do that, you will need to grow the same type of plant under at least three different conditions.

The *guiding question* of this investigation is, ***How do differences in soil quality affect the traits of a plant?***

Materials

You may use any of the following materials during your investigation:

Equipment

- Windowsill greenhouse
- Plant labels
- Ruler
- Plastic pipette

Consumables

- Potting pellets
- Seeds
- Fertilizer pellets

Safety Rules

Follow all normal safety rules. In addition, be sure to follow these rules:

- Wear sanitized indirectly vented chemical-splash goggles during setup, investigation activity, and cleanup.
- Never eat food or drink during the activity using fertilizer pellets.
- Wash your hands with soap and water when done collecting the data.

Plan Your Investigation

Prepare a plan for your investigation by filling out the chart that follows; this plan is called an *investigation proposal.* Before you start developing your plan, be sure to discuss the following questions with the other members of your group:

- What information should we collect so we can **describe** the traits of a plant?
- What information do we need to find a relationship between **a cause and an effect?**

Our guiding question:

This is a picture of how we will set up the equipment:

We will collect the following data:

These are the steps we will follow to collect data:

I approve of this investigation proposal.

Teacher's signature

Date

Collect Your Data

Keep a record of what you measure or observe during your investigation in the space below.

Analyze Your Data

You will need to analyze the data you collected before you can develop an answer to the guiding question. To do this, create a graph that shows the relationship between the cause (change in soil quality) and the effect (traits of the plant).

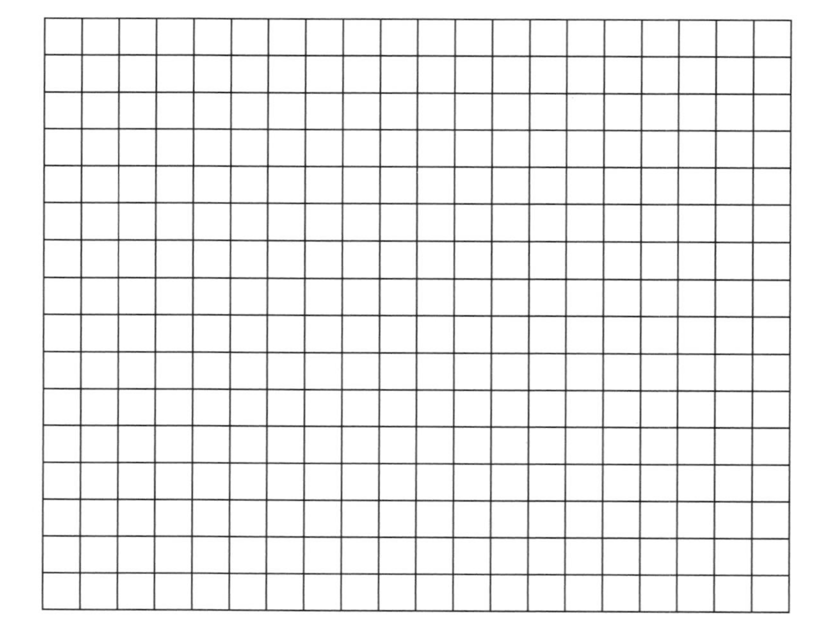

Draft Argument

Develop an argument on a whiteboard. It should include the following parts:

1. A *claim:* Your answer to the guiding question.
2. *Evidence:* An analysis of the data and an explanation of what the analysis means.
3. A *justification of the evidence:* Why your group thinks the evidence is important.

The Guiding Question:	
Our Claim:	
Our Evidence:	Our Justification of the Evidence:

Argumentation Session

Share your argument with your classmates. Be sure to ask them how to make your draft argument better. Keep track of their suggestions in the space below.

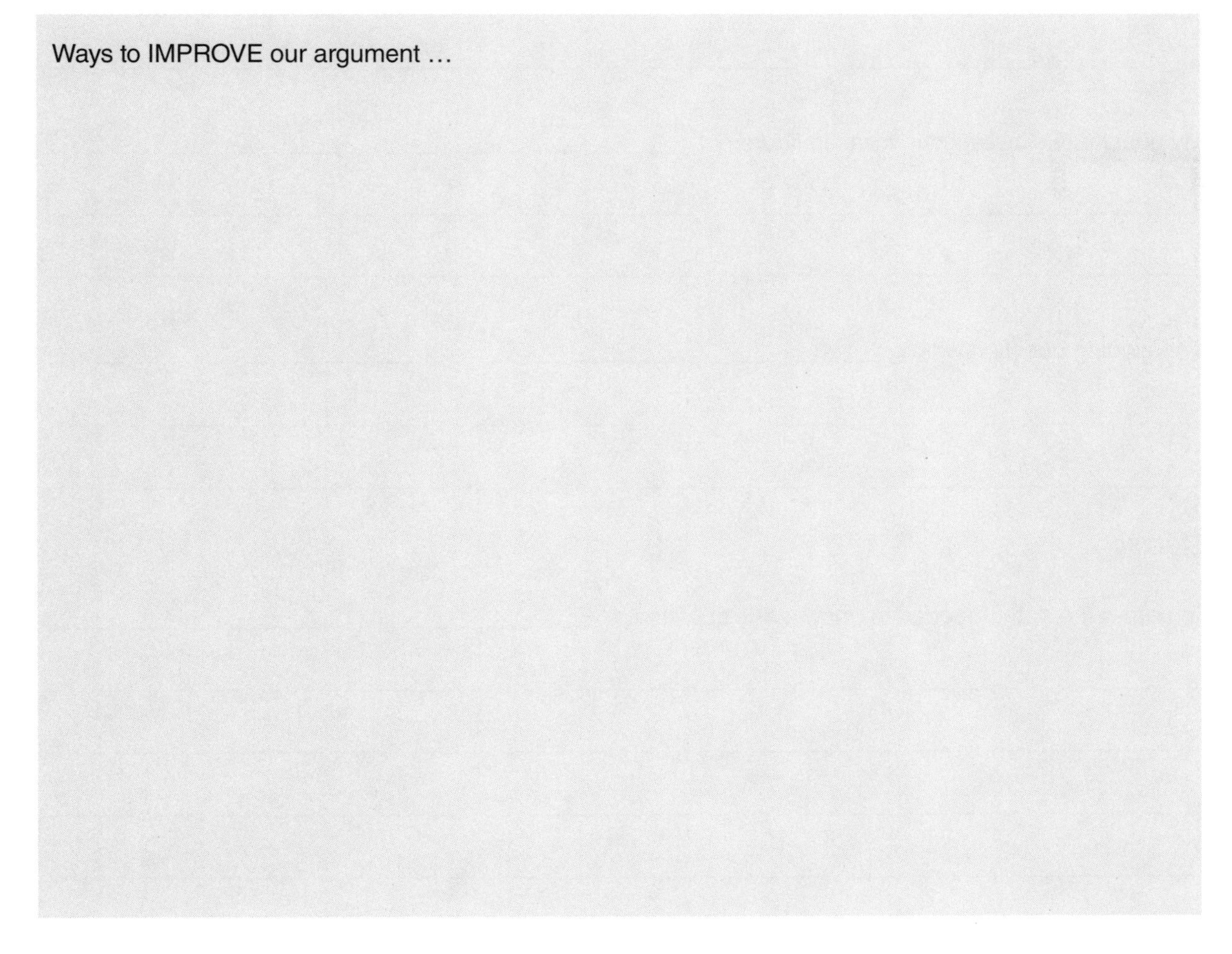

Draft Report

Prepare an *investigation report* to share what you have learned. Use the information in this handout and your group's final argument to write a *draft* of your investigation report.

Introduction

We have been studying ________________________________ in class.

Before we started this investigation, we explored ________________________________

____

____

____

We noticed ________________________________

____

____

My goal for this investigation was to figure out ________________________________

____

____

The guiding question was ________________________________

____

____

Method

To gather the data I needed to answer this question, I ________________________________

____

____

____

____

____

I then analyzed the data I collected by ____

____

____

Argument

My claim is ____

____

____

____

The graph below shows ____

____

____

____

____

____

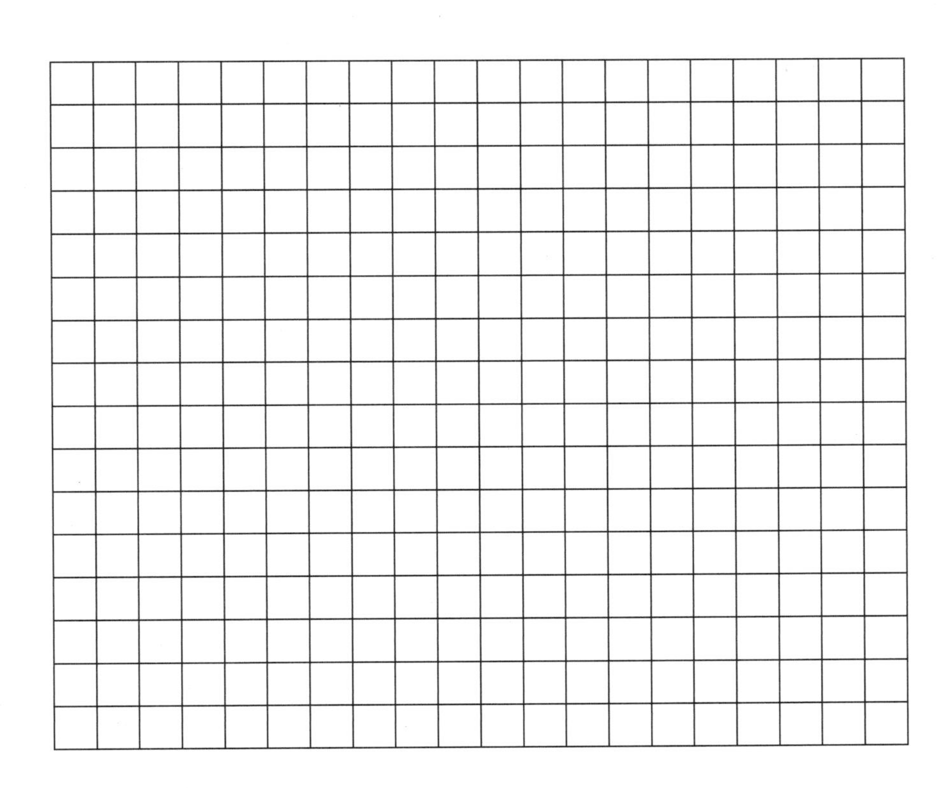

This evidence is important because

____

____

____

____

____

____

____

____

Review

Your friends need your help! Review the draft of their investigation reports and give them ideas about how to improve. Use the *peer-review guide* when doing your review.

Submit Your Final Report

Once you have received feedback from your friends about your draft report, create your final investigation report and hand it in to your teacher.

Checkout Questions

Investigation 9. Traits and the Environment: How Do Differences in Soil Quality Affect the Traits of a Plant?

Imagine that you plant five seeds in some potting soil. All the seeds are from the same type of plant. Each day you add 2 ml of water to the soil. After a few weeks you have five plants.

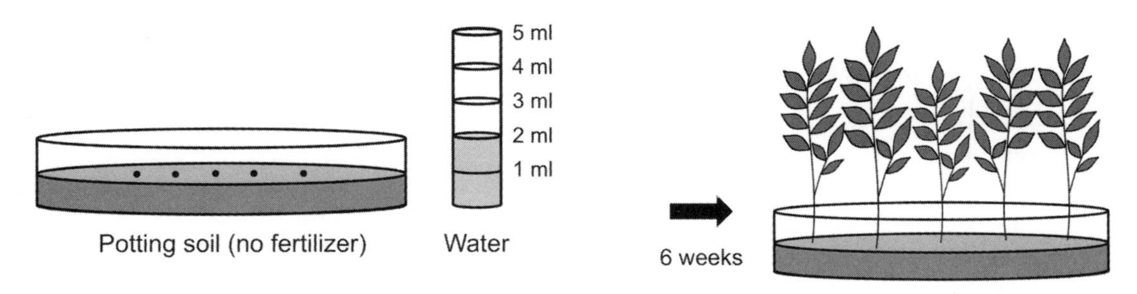

1. What would the plants look like if you decided to add fertilizer to the soil? Circle the letter (A, B, or C) under the picture that matches what they would look like.

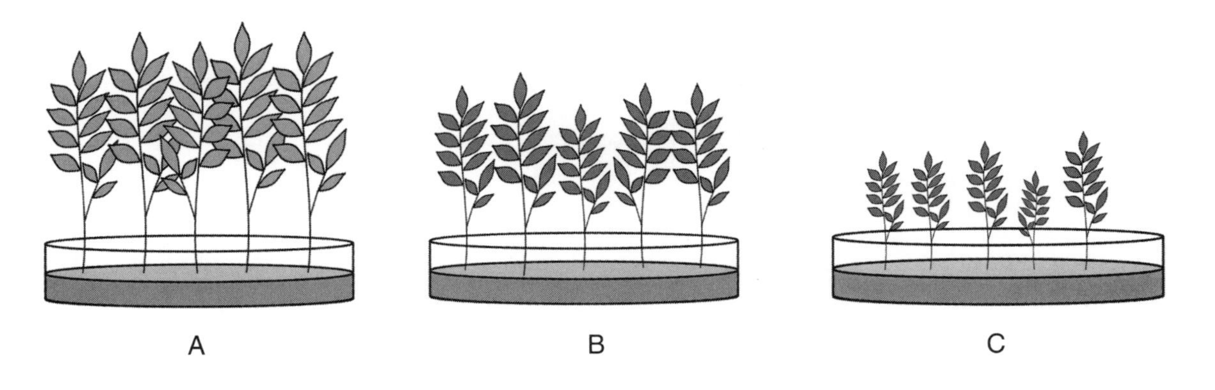

2. What would the plants look like if you decided to add less water to the soil? Circle the letter (A, B, or C) under the picture that matches what they would look like.

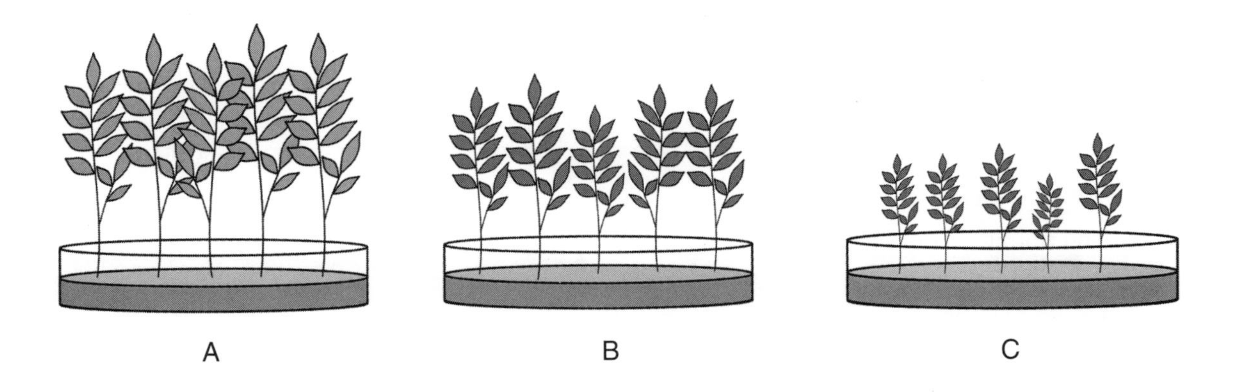

Checkout Questions

3. What would the plants look like if you decided to add more water to the soil? Circle the letter (A, B, or C) under the picture that matches what they would look like.

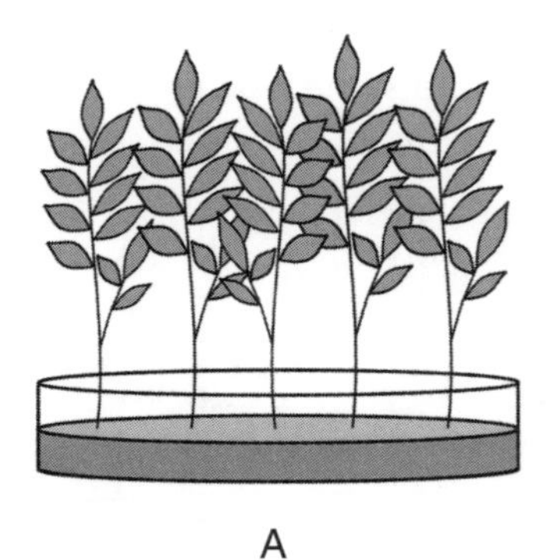

A

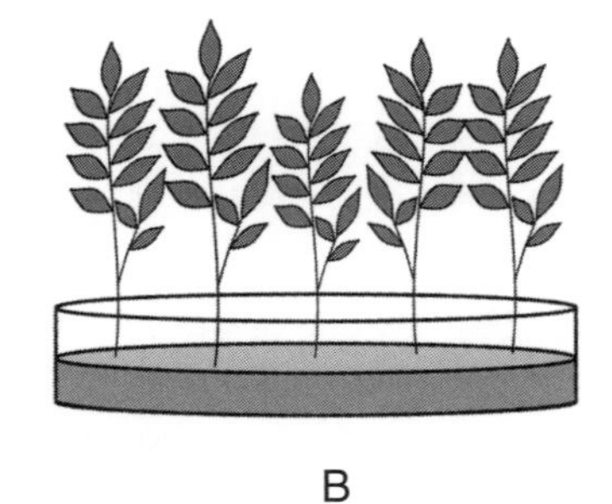

B

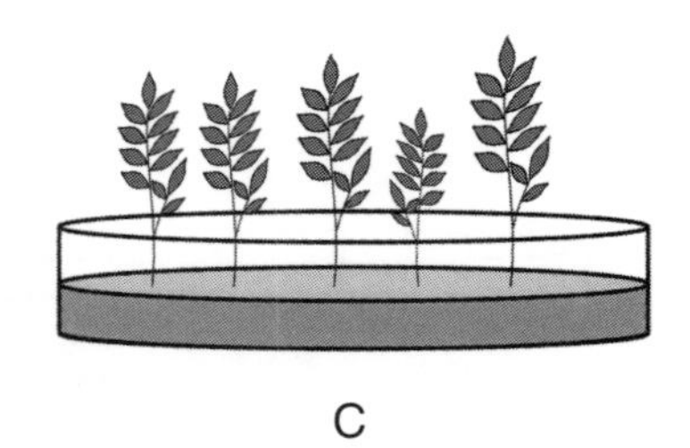

C

4. Explain your thinking. What *cause-and-effect relationship* did you use to determine how the plants would look under each condition?

____

____

____

____

____

____

____

____

____

____

____

Teacher Scoring Rubric for the Checkout Questions

Level	Description
3	The student can apply the core idea correctly in all cases and can fully explain the cause-and-effect relationship.
2	The student can apply the core idea correctly in all cases but cannot fully explain the cause-and-effect relationship.
1	The student cannot apply the core idea correctly in all cases but can fully explain the cause-and-effect relationship.
0	The student cannot apply the core idea correctly in all cases and cannot explain the cause-and-effect relationship.

Section 5

Biological Evolution: Unity and Diversity

Investigation 10

Fossils: What Was the Ecosystem at Darmstadt Like 49 Million Years Ago?

Purpose

The purpose of this investigation is to give students an opportunity to use the disciplinary core idea (DCI) of LS4.A: Evidence of Common Ancestry and Diversity along with the crosscutting concept (CC) of Structure and Function from *A Framework for K–12 Science Education* (NRC 2012) to figure out how fossils can be used as clues to determine what an ecosystem was like in the past. Students will also learn about how scientists assume that there is order and consistency in nature during the reflective discussion.

The Disciplinary Core Idea

Students in third grade should understand the following about Evidence of Common Ancestry and Diversity and be able to use this DCI to figure out how fossils can be used as clues to determine what an ecosystem was like in the past:

> *Some kinds of plants and animals that once lived on Earth (e.g., dinosaurs) are no longer found anywhere, although others now living (e.g., lizards) resemble them in some ways. (NRC 2012, p. 162)*

The Crosscutting Concept

Students in third grade should understand the following about the CC Structure and Function:

> *The functioning of natural and built systems alike depends on the shapes and relationships of certain key parts as well as on the properties of the materials from which they are made. (NRC 2012, p. 96)*

Students in third grade should also be given opportunities to "explore how shape and stability are related for a variety of structures (e.g., a bridge's diagonal brace) or purposes (e.g., different animals get their food using different parts of their bodies). As children move through the elementary grades, they progress to understanding the relationships of structure and mechanical function" (NRC 2012, p. 97).

Students should be encouraged to use their developing understanding of structure and function as a tool or a way of thinking about a phenomenon during this investigation to help them figure out how fossils can be used as clues to determine what an ecosystem was like in the past.

What Students Figure Out

Fossils can be compared with one another and with living organisms to provide evidence about the types of organisms that lived at different locations on Earth long ago. The similarities and differences in the fossils that are found at a specific location can provide information about the what the environment was like at that location in the past, because plants and animals have structures (e.g., legs, fins, wings) that serve specific functions (e.g. walking, swimming, flying) that allow them to live in a specific type of environment (e.g., grassland, ocean, forest).

Timeline

The time needed to complete this investigation is 270 minutes (4 hours and 30 minutes). The amount of instructional time needed for each stage of the investigation is as follows:

- *Stage 1.* Introduce the task and the guiding question: 35 minutes
- *Stage 2.* Design a method and collect data: 60 minutes
- *Stage 3.* Create a draft argument: 35 minutes
- *Stage 4.* Argumentation session: 30 minutes
- *Stage 5.* Reflective discussion: 15 minutes
- *Stage 6.* Write a draft report: 35 minutes
- *Stage 7.* Peer review: 30 minutes
- *Stage 8.* Revise the report: 30 minutes

This investigation can be completed in one day or over eight days (one day for each stage of ADI) during your designated science time in the daily schedule.

Materials and Preparation

The materials needed for this investigation are listed in Table 10.1 (p. 366). The fossils can be purchased from Prehistoric Planet at *www.prehistoricstore.com*. The materials for this investigation can also be purchased as a complete kit (which includes enough materials for 24 students, or six groups) at *www.argumentdriveninquiry.com*.

All of these fossils were found in the Messel Pit near Darmstadt, Germany. To learn more about the Messel Pit and why it is such a remarkable excavation site for fossils, go to *www.youtube.com/watch?v=9oHY9weNJVM.*

TABLE 10.1

Materials for Investigation 10

Item	Quantity
Fossil A: Messel frog (item 1114)	1 per class
Fossil B: Messel bat (item 835)	1 per class
Fossil C: Messel bat (item 3213)	1 per class
Fossil D: Messel garpike fish (item 418)	1 per class
Fossil E: Messel turtle (item 1379)	1 per class
Fossil F: Messel Pit snake (item 1380)	1 per class
Whiteboard, 2' × 3'*	1 per group
Investigation Handout	1 per student
Peer-review guide and teacher scoring rubric	1 per student
Checkout Questions (optional)	1 per student

* As an alternative, students can use computer and presentation software such as Microsoft PowerPoint or Apple Keynote to create their arguments.

Be sure to label each fossil with a letter (A–F) and not by name of the animal (frog, bat, garpike fish, etc.). Students must make inferences about what type of animal died at Darmstadt 49 million years ago based on the structure preserved in each fossil during this investigation, so it is important that the students do not know the names of the animal that they examine.

Safety Precautions

Remind students to follow all safety rules.

Lesson Plan by Stage

Stage 1: Introduce the Task and the Guiding Question (35 minutes)

1. Ask the students to sit in six groups, with three or four students in each group.
2. Ask students to clear off their desks except for a pencil (and their *Student Workbook for Argument-Driven Inquiry in Third-Grade Science* if they have one).
3. Pass out an Investigation Handout to each student (or ask students to turn to Investigation Log 10 in their workbook).
4. Read the first paragraph of the "Introduction" aloud to the class. Ask the students to follow along as you read.
5. Pass out one of the six available fossils to each group.

6. Tell students to record their observations and questions about the fossil in the "OBSERVED/WONDER" chart in the "Introduction" section of their Investigation Handout (or the investigation log in their workbook).
7. Ask students to share *what they observed* about the fossil.
8. Ask students to share *what questions they have* about fossils.
9. Tell the students, "Some of your questions might be answered by reading the rest of the 'Introduction.'"
10. Ask the students to read the rest of the "Introduction" on their own *or* ask them to follow along as you read aloud.
11. Once the students have read the rest of the "Introduction," ask them to fill out the "KNOW/NEED" chart on their Investigation Handout (or in their investigation log) as a group.
12. Ask students to share what they learned from the reading. Add these ideas to a class "know / need to figure out" chart.
13. Ask students to share what they think they will need to figure out based on what they read. Add these ideas to the class "know / need to figure out" chart.
14. Tell the students, "It looks like we have something to figure out. Let's see what we will need to do during our investigation."
15. Read the task and the guiding question aloud.
16. Tell the students, "I have six different fossils that were found at Darmstadt for you to use. These fossils are 49 million years old."
17. Introduce the students to the fossils by holding each fossil up.

Stage 2: Design a Method and Collect Data (60 minutes)

1. Tell the students, "I am now going to give you and the other members of your group about 15 minutes to plan your investigation. Before you begin, I want you all to take a couple of minutes to discuss the following questions with the rest of your group."
2. Show the following questions on the screen or board:
   - How might the *structure* of what you are studying relate to its *function?*
   - What types of *patterns* might we look for to help answer the guiding question?
3. Tell the students, "Please take a few minutes to come up with an answer to these questions." Give the students two or three minutes to discuss these two questions.
4. Ask two or three different groups to share their answers. Be sure to highlight or write down any important ideas on the board so students can refer to them later.

5. If possible, use a document camera to project an image of the graphic organizer for this investigation on a screen or board (or take a picture of it and project the picture on a screen or board). Tell the students, "I now want you all to plan out your investigation. To do that, you will need to create an investigation proposal by filling out this graphic organizer."
6. Point to the box labeled "Our guiding question:" and tell the students, "You can put the question we are trying to answer in this box." Then ask, "Where can we find the guiding question?"
7. Wait for a student to answer where to find the guiding question (the answer is "in the handout").
8. Point to the box labeled "We will collect the following data:" and tell the students, "You can list the measurements or observations that you will need to collect during the investigation in this box."
9. Point to the box labeled "These are the steps we will follow to collect data:" and tell the students, "You can list what you are going to do to collect the data you need and what you will do with your data once you have it. Be sure to give enough detail that I could do your investigation for you."
10. Ask the students, "Do you have any questions about what you need to do?"
11. Answer any questions that come up.
12. Tell the students, "Once you are done, raise your hand and let me know. I'll then come by and look over your proposal and give you some feedback. You may not begin collecting data until I have approved your proposal by signing it. You need to have your proposal done in the next 15 minutes."
13. Give the students 15 minutes to work in their groups on their investigation proposal. As they work, move from group to group to check in, ask probing questions, and offer a suggestion if a group gets stuck.

What should a student-designed investigation look like?

The students' investigation proposal should include the following information:

- The guiding question is "What was the ecosystem at Darmstadt like 49 million years ago?"
- There are a lot of different types of data that students can collect during this investigation. Students can collect the following data for each fossil: (1) presence or absence of legs, (2) presence or absence of wings, (3) presence or absence of tail, and (4) presence or absence of fins (other traits are possible). This investigation works best if each group selects different traits.

There are other options in addition to the traits listed here.

- The steps that the students will follow to collect the data should reflect the traits that they decide to examine. However, a procedure might include the following steps:
  1. Identify the traits to examine.
  2. Document [trait 1].
  3. Document [trait 2].
  4. Repeat steps 1–3 for each fossil.
  5. Create a table or graph that highlights the characteristics of the fossils that are similar to the traits of animals that live in aquatic ecosystems or similar to the traits of animals that live in terrestrial ecosystems.

This is just an example of a procedure, and there should be a lot of variation in the student-designed investigations. At a minimum, each group should collect data about two different traits.

14. As each group finishes its investigation proposal, be sure to read it over and determine if it will be productive or not. If you feel the investigation will be productive (not necessarily what you would do or what the other groups are doing), sign your name on the proposal and let the group start collecting data. If the plan needs to be changed, offer some suggestions or ask some probing questions, and have the group make the changes before you approve it.
15. Pass out the materials or have one student from each group collect the materials they need from a central supply table or cart for the groups that have an approved proposal.
16. Remind students of the safety rules and precautions for this investigation.
17. Give the students about 5 minutes to collect data on each fossil (30 minutes total).
18. Tell the students to collect their data and record their observations or measurements in the "Collect Your Data" box in their Investigation Handout (or the investigation log in their workbook).

Stage 3: Create a Draft Argument (35 minutes)

1. Tell the students, "Now that we have all this data, we need to analyze the data so we can figure out an answer to the guiding question."
2. If possible, project an image of the "Analyze Your Data" text and box for this investigation on a screen or board using a document camera (or take a picture

of it and project the picture on a screen or board). Point to the box and tell the students, "You can create a table or graph as a way to analyze your data. You can make your table or graph in this box."

3. Ask the students, "What information do we need to include in a table or graph?"
4. Tell the students, "Please take a few minutes to discuss this question with your group, and be ready to share."
5. Give the students five minutes to discuss.
6. Ask two or three different groups to share their answers. Be sure to highlight or write down any important ideas on the board so students can refer to them later.
7. Tell the groups of students, "I am now going to give you and the other members of your group about 10 minutes to create your table or graph." If the students are having trouble making a table or graph, you can take a few minutes to provide a mini-lesson about how to create a table or graph from a bunch of observations or measurements (this strategy is called just-in-time instruction because it is offered only when students get stuck).

What should a table or graph look like for this investigation?

There are a number of different ways that students can analyze the observations or measurements they collect during this investigation. One of the most straightforward ways is to create a scaled bar graph to represent a data set with several categories. This bar graph should have the three categories related to the characteristics of animals (traits of animals that live on land, traits of animals that live in the water, traits of animals that fly) on the horizontal or *x*-axis. The dependent variable (number of fossils that have each type of trait) should be on the *y*-axis. An example of this type of graph can be seen in Figure 10.1 (p. 372).

8. Give the students 10 minutes to analyze their data. As they work, move from group to group to check in, ask probing questions, and offer suggestions.
9. Tell the students, "I am now going to give you and the other members of your group 15 minutes to create an argument to share what you have learned and convince others that they should believe you. Before you do that, we need to take a few minutes to discuss what you need to include in your argument."
10. If possible, use a document camera to project the "Argument Presentation on a Whiteboard" image from the Investigation Handout (or the investigation log

in their workbook) on a screen or board (or take a picture of it and project the picture on a screen or board).

11. Point to the box labeled "The Guiding Question:" and tell the students, "You can put the question we are trying to answer here on your whiteboard."
12. Point to the box labeled "Our Claim:" and tell the students, "You can put your claim here on your whiteboard. The claim is your answer to the guiding question."
13. Point to the box labeled "Our Evidence:" and tell the students, "You can put the evidence that you are using to support your claim here on your whiteboard. Your evidence will need to include the analysis you just did and an explanation of what your analysis means or shows. Scientists always need to support their claims with evidence."
14. Point to the box labeled "Our Justification of the Evidence:" and tell the students, "You can put your justification of your evidence here on your whiteboard. Your justification needs to explain why your evidence is important. Scientists often use core ideas to explain why the evidence they are using matters. Core ideas are important concepts that scientists use to help them make sense of what happens during an investigation."
15. Ask the students, "What are some core ideas that we read about earlier that might help us explain why the evidence we are using is important?"
16. Ask students to share some of the core ideas from the "Introduction" section of the Investigation Handout (or the investigation log in the workbook). List these core ideas on the board.
17. Tell the students, "That is great. I would like to see everyone try to include these core ideas in your justification of the evidence. Your goal is to use these core ideas to help explain why your evidence matters and why the rest of us should pay attention to it."
18. Ask the students, "Do you have any questions about what you need to do?"
19. Answer any questions that come up.
20. Tell the students, "Okay, go ahead and start working on your arguments. You need to have your argument done in the next 15 minutes. It doesn't need to be perfect. We just need something down on the whiteboards so we can share our ideas."
21. Give the students 15 minutes to work in their groups on their arguments. As they work, move from group to group to check in, ask probing questions, and offer a suggestion if a group gets stuck. Figure 10.1 (p. 372) shows an example of an argument created by students for this investigation.

FIGURE 10.1

Example of an argument

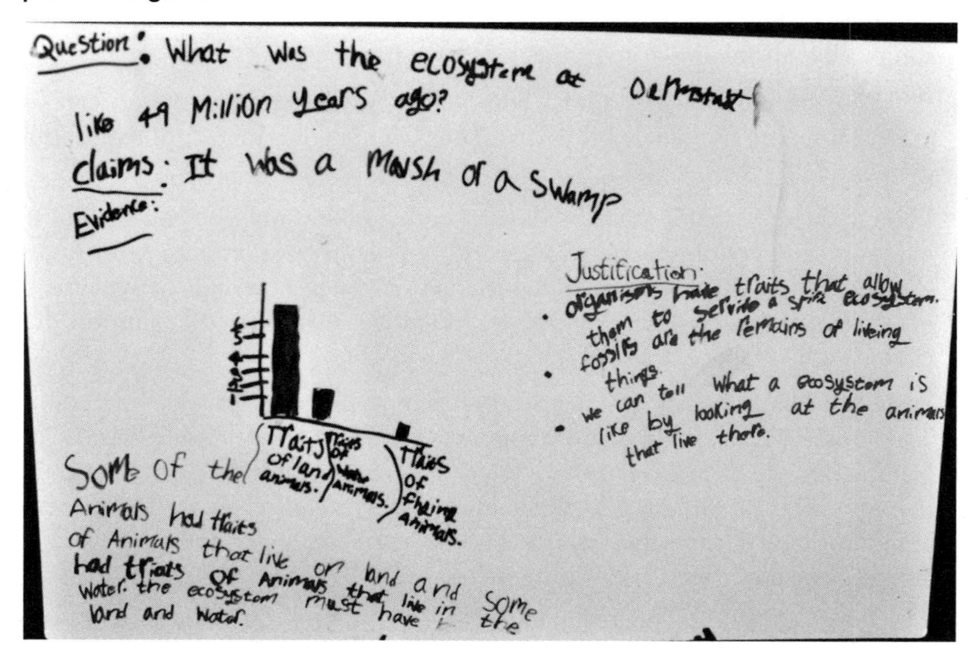

Stage 4: Argumentation Session (30 minutes)

The argumentation session can be conducted in a whole-class presentation format, a gallery walk format, or a modified gallery walk format. We recommend using a whole-class presentation format for the first investigation, but try to transition to either the gallery walk or modified gallery walk format as soon as possible because that will maximize student voice and choice inside the classroom. The following list shows the steps for the three formats; unless otherwise noted, the steps are the same for all three formats.

1. Begin by introducing the use of the whiteboard.
   - *If using the whole-class presentation format,* tell the students, "We are now going to share our arguments. Please set up your whiteboard so everyone can see them."
   - *If using the gallery walk or modified gallery walk format,* tell the students, "We are now going to share our arguments. Please set up your whiteboard so they are facing the walls."
2. Allow the students to set up their whiteboards.
   - *If using the whole-class presentation format,* the whiteboards should be set up on stands or chairs so they are facing toward the center of the room.
   - *If using the gallery walk or modified gallery walk format,* the whiteboards should be set up on stands or chairs so they are facing toward the outside of the room.

3. Give the following instructions to the students:
   - *If using the whole-class presentation format or the modified gallery walk format,* tell the students, "Okay, before we get started I want to explain what we are going to do next. I'm going to ask some of you to present your arguments to your classmates. If you are presenting your argument, your job is to share your group's claim, evidence, and justification of the evidence. The rest of you will be reviewers. If you are a reviewer, your job is to listen to the presenters, ask the presenters questions if you do not understand something, and then offer them some suggestions about ways to make their argument better. After we have a chance to learn from each other, I'm going to give you some time to revise your arguments and make them better."
   - *If using the gallery walk format,* tell the students, "Okay, before we get started I want to explain what we are going to do next. You are going to have an opportunity to read the arguments that were created by other groups. Your group will go to a different group's argument. I'll give you a few minutes to read it and review it. Your job is to offer them some suggestions about ways to make their argument better. You can use sticky notes to give them suggestions. Please be specific about what you want to change and be specific about how you think they should change it. After we have a chance to learn from each other, I'm going to give you some time to revise your arguments and make them better."
4. Use a document camera to project the "Ways to IMPROVE our argument …" box from the Investigation Handout (or the investigation log in their workbook) on a screen or board (or take a picture of it and project the picture on a screen or board).
   - *If using the whole-class presentation format or the modified gallery walk format,* point to the box and tell the students, "If you are a presenter, you can write down the suggestions you get from the reviewers here. If you are a reviewer, and you see a good idea from another group, you can write down that idea here. Once we are done with the presentations, I will give you a chance to use these suggestions or ideas to improve your arguments.
   - *If using the gallery walk format,* point to the box and tell the students, "If you see good ideas from another group, you can write them down here. Once we are done reviewing the different arguments, I will give you a chance to use these ideas to improve your own arguments. It is important to share ideas like this."

   Ask the students, "Do you have any questions about what you need to do?"
5. Answer any questions that come up.
6. Give the following instructions:
   - *If using the whole-class presentation format,* tell the students, "Okay. Let's get started."

- *If using the gallery walk format,* tell the students, "Okay, I'm now going to tell you which argument to go to and review.
- *If using the modified gallery walk format,* tell the students, "Okay, I'm now going to assign you to be a presenter or a reviewer." Assign one or two students from each group to be presenters and one or two students from each group to be reviewers.

7. Begin the review of the arguments.
   - *If using the whole-class presentation format,* have four or five groups present their argument one at a time. Give each group only two to three minutes to present their argument. Then give the class two to three minutes to ask them questions and offer suggestions. Be sure to encourage as much participation from the students as possible.
   - *If using the gallery walk format,* tell the students, "Okay. Let's get started. Each group, move one argument to the left. Don't move to the next argument until I tell you to move. Once you get there, read the argument and then offer suggestions about how to make it better. I will put some sticky notes next to each argument. You can use the sticky notes to leave your suggestions." Give each group about three to four minutes to read the arguments, talk, and offer suggestions.
     a. Tell the students, "Okay. Let's rotate. Move one group to the left."
     b. Again, give each group three or four minutes to read, talk, and offer suggestions.
     c. Repeat this process for two more rotations.
   - *If using the modified gallery walk format,* tell the students, "Okay. Let's get started. Reviewers, move one group to the left. Don't move to the next group until I tell you to move. Presenters, go ahead and share your argument with the reviewers when they get there." Give each group of presenters and reviewers about three to four minutes to talk.
     a. Tell the students, "Okay. Let's rotate. Reviewers, move one group to the left."
     b. Again, give each group of presenters and reviewers about three or four minutes to talk.
     c. Repeat this process for two more rotations.
8. Tell the students to return to their workstations.
9. Give the following instructions about revising the argument:
   - *If using the whole-class presentation format,* tell the students, "I'm now going to give you about 10 minutes to revise your argument. Take a few minutes to talk in your groups and determine what you want to change to make your argument better. Once you have decided what to change, go ahead and make the changes to your whiteboard."

- *If using the gallery walk format,* tell the students, "I'm now going to give you about 10 minutes to revise your argument. Take a few minutes to read the suggestions that were left at your argument. Then talk in your groups and determine what you want to change to make your argument better. Once you have decided what to change, go ahead and make the changes to your whiteboard."
- *If using the modified gallery walk format,* "I'm now going to give you about 10 minutes to revise your argument. Please return to your original groups." Wait for the students to move back into their original groups and then tell the students, "Okay, take a few minutes to talk in your groups and determine what you want to change to make your argument better. Once you have decided what to change, go ahead and make the changes to your whiteboard."

Ask the students, "Do you have any questions about what you need to do?"

10. Answer any questions that come up.
11. Tell the students, "Okay. Let's get started."
12. Give the students 10 minutes to work in their groups on their arguments. As they work, move from group to group to check in, ask probing questions, and offer a suggestion if a group gets stuck.

Stage 5: Reflective Discussion (15 minutes)

1. Tell the students, "We are now going to take a minute to talk about what we did and what we have learned."
2. Show Figure 10.2 on the screen. This image shows a fossil fish that was found at Fossil Butte National Monument in Wyoming.
3. Ask the students, "What do you all see going on here?"
4. Allow students to share their ideas.
5. Ask the students, "What type of ecosystem do you think this organism lived in and how do you know?"
6. Allow students to share their ideas.
7. Show Figure 10.3 on the screen. This image shows a fossil horse that was found at Fossil Butte National Monument.
8. Ask the students, "What do you all see going on here?"
9. Allow students to share their ideas.

FIGURE 10.2

Fossil fish

FIGURE 10.3

Fossil horse

10. Ask the students, "What type of ecosystem do you think this organism lived in and how do you know?"
11. Allow students to share their ideas.
12. Tell the students, "Okay, let's make sure we are on the same page. Fossils are the preserved remains of a living thing. Fossils allow us to investigate what life was like in the past. Some kinds of plants and animals that once lived on Earth are no longer found anywhere, but we know about them because we have found fossils of them. We can compare fossils with living things to figure out the types of plants or animals that lived at different locations on Earth long ago. The similarities and differences in the fossils that are found at a specific location can also provide us with information about what the environment was like at a location in the past. Our ability to use fossils to learn about the types of organisms that lived long ago and also the nature of their environments is a really important core idea in science."
13. Ask the students, "Does anyone have any questions about this core idea?"
14. Answer any questions that come up.
15. Tell the students, "We also needed to think about how structure and function are related to each other during our investigation." Then ask, "Can anyone tell me why it is useful to think about the relationship between structure and function in nature?"
16. Allow students to share their ideas.
17. Tell the students, "Plants and animals have structures that have a specific function that allow them to live in a specific type of environment. For example, fish have fins that allow them to swim. Horses have legs that allow them to walk."
18. Ask the students, "What were some other structures of animals that we looked at today, and how did these structures allow us to answer our guiding question?"
19. Allow students to share their ideas. Keep probing until someone mentions that the animals have structures that allow them to live a specific type of environment.
20. Tell the students, "That is great, and if we know the structures of a living thing we can figure out where it lived because of how that structure functions. For example, if we see a fossil with fins, we can assume that it lived in the water because animals use fins to swim."
21. Ask the students, "Can anyone give me another example?"
22. Allow students to share their ideas.
23. Tell the students, "We are now going take a minute to talk about what went well and what didn't go so well during our investigation. We need to talk about this because you are going to be planning and carrying out your own investigations like this a lot this year, and I want to help you all get better at it."

24. Show an image of the question "What made your investigation scientific?" on the screen. Tell the students, "Take a few minutes to talk about how you would answer this question with the other people in your group. Be ready to share with the rest of the class." Give the students two to three minutes to talk in their group.
25. Ask the students, "What do you all think? Who would like to share an idea?"
26. Allow students to share their ideas. Be sure to expand on their ideas about what makes an investigation scientific.
27. Show an image of the question "What made your investigation not so scientific?" on the screen. Tell the students, "Take a few minutes to talk about how you would answer this question with the other people in your group. Be ready to share with the rest of the class." Give the students two to three minutes to talk in their group.
28. Ask the students, "What do you all think? Who would like to share an idea?"
29. Allow students to share their ideas. Be sure to expand on their ideas about what makes an investigation less scientific.
30. Show an image of the question "What rules can we put into place to help us make sure our next investigation is more scientific?" on the screen. Tell the students, "Take a few minutes to talk about how you would answer this question with the other people in your group. Be ready to share with the rest of the class." Give the students two to three minutes to talk in their group.
31. Ask the students, "What do you all think? Who would like to share an idea?"
32. Allow students to share their ideas. Once they have shared their ideas, offer a suggestion for a possible class rule.
33. Ask the students, "What do you all think? Should we make this a rule?"
34. If the students agree, write the rule on the board or make a class "Rules for Scientific Investigation" chart so you can refer to it during the next investigation.
35. Tell the students, "We are now going take a minute to talk about how scientists think about the world."
36. Show an image of the question "How can scientists study what happened in the past?" on the screen. Tell the students, "Take a few minutes to talk about how you would answer this question with the other people in your group. Be ready to share with the rest of the class." Give the students two to three minutes to talk in their group.
37. Ask the students, "What do you all think? Who would like to share an idea?"
38. Allow students to share their ideas.

39. Tell the students, "Okay, let's make sure we are all on the same page. Scientists can study the past because they assume that the basic laws of nature are the same now as they were in the past."
40. Ask the students, "For example, what does an animal that lives in the water need to be able to do to survive.
41. Allow students to share their ideas.
42. Ask the students, "Do you think it was any different for animals that lived in the water a long, long time ago?
43. Allow students to share their ideas.
44. Tell the students, "The laws of nature don't change. So, if it is a law now, it was a law in the past, and it will continue to be a law in the future. This is an important idea in science. This is one of things scientists try to keep in mind when they are trying to figure something out."
45. Ask the students, "Does anyone have any questions about how scientists think about the world"?
46. Answer any questions that come up.

Stage 6: Write a Draft Report (35 minutes)

Your students will use either the Investigation Handout or the investigation log in the student workbook when writing the draft report. When you give the directions shown in quotes in the following steps, substitute "investigation log" (as shown in brackets) for "handout" if they are using the workbook.

1. Tell the students, "You are now going to write an investigation report to share what you have learned. Please take out a pencil and turn to the 'Draft Report' section of your handout [investigation log]."
2. If possible, use a document camera to project the "Introduction" section of the draft report from the Investigation Handout (or the investigation log in their workbook) on a screen or board (or take a picture of it and project the picture on a screen or board).
3. Tell the students, "The first part of the report is called the 'Introduction.' In this section of the report you want to explain to the reader what you were investigating, why you were investigating it, and what question you were trying to answer. All of this information can be found in the text at the beginning of your handout [investigation log]." Point to the image and say, "There are some sentence starters here to help you begin writing the report." Ask the students, "Do you have any questions about what you need to do?"
4. Answer any questions that come up.
5. Tell the students, "Okay. Let's write."

6. Give the students 10 minutes to write the "Introduction" section of the report. As they work, move from student to student to check in, ask probing questions, and offer a suggestion if a student gets stuck.
7. If possible, use a document camera to project the "Method" section of the draft report from the Investigation Handout (or the investigation log in their workbook) on a screen or board (or take a picture of it and project the picture on a screen or board).
8. Tell the students, "The second part of the report is called the 'Method.' In this section of the report you want to explain to the reader what you did during the investigation, what data you collected and why, and how you went about analyzing your data. All of this information can be found in the 'Plan Your Investigation' section of your handout [investigation log]. Remember that you all planned and carried out different investigations, so do not assume that the reader will know what you did." Point to the image and say, "There are some sentence starters here to help you begin writing this part of the report." Ask the students, "Do you have any questions about what you need to do?"
9. Answer any questions that come up.
10. Tell the students, "Okay. Let's write."
11. Give the students 10 minutes to write the "Method" section of the report. As they work, move from student to student to check in, ask probing questions, and offer a suggestion if a student gets stuck.
12. If possible, use a document camera to project the "Argument" section of the draft report from the Investigation Handout (or the investigation log in their workbook) on a screen or board (or take a picture of it and project the picture on a screen or board).
13. Tell the students, "The last part of the report is called the 'Argument.' In this section of the report you want to share your claim, evidence, and justification of the evidence with the reader. All of this information can be found on your whiteboard." Point to the image and say, "There are some sentence starters here to help you begin writing this part of the report." Ask the students, "Do you have any questions about what you need to do?"
14. Answer any questions that come up.
15. Tell the students, "Okay. Let's write."
16. Give the students 10 minutes to write the "Argument" section of the report. As they work, move from student to student to check in, ask probing questions, and offer a suggestion if a student gets stuck.

Stage 7: Peer Review (30 minutes)

Your students will use either the Investigation Handout or the investigation log in the student workbook when doing the peer review. When you give the directions shown in quotes in the following steps, substitute "workbook" (as shown in brackets) for "Investigation Handout" if they are using the workbook.

1. Tell the students, "We are now going to review our reports to find ways to make them better. I'm going to come around and collect your Investigation Handout [workbook]. While I do that, please take out a pencil."
2. Collect the Investigation Handouts or workbooks from the students.
3. If possible, use a document camera to project the peer-review guide (PRG; see Appendix 4) on a screen or board (or take a picture of it and project the picture on a screen or board).
4. Tell the students, "We are going to use this peer-review guide to give each other feedback." Point to the image.
5. Give the following instructions:
   - *If using the Investigation Handout,* tell the students, "I'm going to ask you to work with a partner to do this. I'm going to give you and your partner a draft report to read and a peer-review guide to fill out. You two will then read the report together. Once you are done reading the report, I want you to answer each of the questions on the peer-review guide." Point to the review questions on the image of the PRG.
   - *If using the workbook,* tell the students, "I'm going to ask you to work with a partner to do this. I'm going to give you and your partner a draft report to read. You two will then read the report together. Once you are done reading the report, I want you to answer each of the questions on the peer-review guide that is right after the report in the investigation log." Point to the review questions on the image of the PRG.
6. Tell the students, "You can check 'yes,' 'almost,' or 'no' after each question." Point to the checkboxes on the image of the PRG.
7. Tell the students, "This will be your rating for this part of the report. Make sure you agree on the rating you give the author. If you mark 'almost' or 'no,' then you need to tell the author what he or she needs to do to get a 'yes.'" Point to the space for the reviewer feedback on the image of the PRG.
8. Tell the students, "It is really important for you to give the authors feedback that is helpful. That means you need to tell them exactly what they need to do to make their reports better." Ask the students, "Do you have any questions about what you need to do?"
9. Answer any questions that come up.

10. Tell the students, "Please sit with a partner who is not in your current group." Allow the students time to sit with a partner.
11. Give the following instructions:
    - *If using the Investigation Handout,* tell the students, "Okay, I am now going to give you one report to read and one peer-review guide to fill out." Pass out one report to each pair. Make sure that the report you give a pair was not written by one of the students in that pair. Give each pair one PRG to fill out as a team.
    - *If using the workbook,* tell the students, "Okay, I am now going to give you one report to read." Pass out a workbook to each pair. Make sure that the workbook you give a pair is not from one of the students in that pair.
12. Tell the students, "Okay, I'm going to give you 15 minutes to read the report I gave you and to fill out the peer-review guide. Go ahead and get started."
13. Give the students 15 minutes to work. As they work, move around from pair to pair to check in and see how things are going, answer questions, and offer advice.
14. After 15 minutes pass, tell the students, "Okay, time is up." *If using the Investigation Handout,* say, "Please give me the report and the peer-review guide that you filled out." *If using the workbook,* say, "Please give me the workbook that you have."
15. Collect the Investigation Handouts and the PRGs, or collect the workbooks if they are being used. Be sure you keep the handout and the PRG together.
16. Give the following instructions:
    - *If using the Investigation Handout,* tell the students, "Okay, I am now going to give you a different report to read and a new peer-review guide to fill out." Pass out another report to each pair. Make sure that this report was not written by one of the students in that pair. Give each pair a new PRG to fill out as a team.
    - *If using the workbook,* tell the students, "Okay, I am now going to give you a different report to read." Pass out a different workbook to each pair. Make sure that the workbook you give a pair is not from one of the students in that pair.
17. Tell the students, "Okay, I'm going to give you 15 minutes to read this new report and to fill out the peer-review guide. Go ahead and get started."
18. Give the students 15 minutes to work. As they work, move around from pair to pair to check in and see how things are going, answer questions, and offer advice.
19. After 15 minutes pass, tell the students, "Okay, time is up." *If using the Investigation Handout,* say, "Please give me the report and the peer-review guide that you filled out." *If using the workbook,* say, "Please give me the workbook that you have."

20. Collect the Investigation Handouts and the PRGs, or collect the workbooks if they are being used. Be sure you keep the handout and the PRG together.

Stage 8: Revise the Report (30 minutes)

Your students will use either the Investigation Handout or the investigation log in the student workbook when revising the report. Except where noted below, the directions are the same whether using the handout or the log.

1. Give the following instructions:
    - *If using the Investigation Handout,* tell the students, "You are now going to revise your investigation report based on the feedback you get from your classmates. Please take out a pencil while I hand back your draft report and the peer-review guide."
    - *If using the investigation log in the student workbook,* tell the students, "You are now going to revise your investigation report based on the feedback you get from your classmates. Please take out a pencil while I hand back your investigation logs."
2. *If using the Investigation Handout,* pass back the handout and the PRG to each student. *If using the investigation log,* pass back the log to each student.
3. Tell the students, "Please take a few minutes to read over the peer-review guide. You should use it to figure out what you need to change in your report and how you will change the report."
4. Allow the students time to read the PRG.
5. *If using the investigation log,* if possible use a document camera to project the "Write Your Final Report" section from the investigation log on a screen or board (or take a picture of it and project the picture on a screen or board).
6. Give the following instructions:
    - *If using the Investigation Handout,* tell the students, "Okay. Let's revise our reports. Please take out a piece of paper. I would like you to rewrite your report. You can use your draft report as a starting point, but use the feedback on the peer-review guide to help make it better."
    - *If using the investigation log,* tell the students, "Okay. Let's revise our reports. I would like you to rewrite your report in the section of the investigation log that says 'Write Your Final Report.'" Point to the image on the screen and tell the students, "You can use your draft report as a starting point, but use the feedback on the peer-review guide to help make your report better."

    Ask the students, "Do you have any questions about what you need to do?"
7. Answer any questions that come up.
8. Tell the students, "Okay. Let's write."

9. Give the students 30 minutes to rewrite their report. As they work, move from student to student to check in, ask probing questions, and offer a suggestion if a student gets stuck.
10. Give the following instructions:
    - *If using the Investigation Handout,* tell the students, "Okay. Time's up. I will now come around and collect your Investigation Handout, the peer-review guide, and your final report."
    - *If using the investigation log,* tell the students, "Okay. Time's up. I will now come around and collect your workbooks."
11. *If using the Investigation Handout,* collect all the Investigation Handouts, PRGs, and final reports. *If using the investigation log,* collect all the workbooks.
12. *If using the Investigation Handout,* use the "Teacher Score" columns in the PRG to grade the final report. *If using the investigation log,* use the "ADI Investigation Report Grading Rubric" in the investigation log to grade the final report. Whether you are using the handout or the log, you can give the students feedback about their writing in the "Teacher Comments" section.

How to Use the Checkout Questions

The Checkout Questions are an optional assessment. We recommend giving them to students one day after they finish stage 8 of the ADI investigation. The Checkout Questions can be used as a formative or summative assessment of student thinking. If you plan to use them as a formative assessment, we recommend that you look over the student answers to determine if you need to reteach the core idea and/or crosscutting concept from the investigation, but do not grade them. If you plan to use them as a summative assessment, we have included a "Teacher Scoring Rubric" at the end of the Checkout Questions that you can use to score a student's ability to apply the core idea in a new scenario and explain their use of a crosscutting concept. The rubric includes a 4-point scale that ranges from 0 (the student cannot apply the core idea correctly in all cases and cannot explain the [crosscutting concept]) to 3 (the student can apply the core idea correctly in all cases and can fully explain the [crosscutting concept]). The Checkout Questions, regardless of how you decide to use them, are a great way to make student thinking visible so you can determine if the students have learned the core idea and the crosscutting concept.

A student who can apply the core idea correctly in all cases and can explain the relationship between structure and function would select freshwater aquatic and/or terrestrial for question 1, freshwater aquatic and marine aquatic for question 2, and terrestrial for question 3. He or she should then be able to explain that animals have specific structures that allow them to live (or function) in specific environments.

Connections to Standards

Table 10.2 highlights how the investigation can be used to address specific performance expectations from the *NGSS, Common Core State Standards (CCSS)* in English language arts (ELA) and in mathematics, and *English Language Proficiency (ELP) Standards.*

TABLE 10.2

Investigation 10 alignment with standards

***NGSS* performance expectations**	Strong alignment • 3-LS4-1: Analyze and interpret data from fossils to provide evidence of the organisms and the environments in which they lived long ago.
***CCSS ELA*—Reading: Informational Text**	Key ideas and details • CCSS.ELA-LITERACY.RI.3.1: Ask and answer questions to demonstrate understanding of a text, referring explicitly to the text as the basis for the answers. • CCSS.ELA-LITERACY.RI.3.2: Determine the main idea of a text; recount the key details and explain how they support the main idea. • CCSS.ELA-LITERACY.RI.3.3: Describe the relationship between a series of historical events, scientific ideas or concepts, or steps in technical procedures in a text, using language that pertains to time, sequence, and cause/effect. Craft and structure • CCSS.ELA-LITERACY.RI.3.4: Determine the meaning of general academic and domain-specific words and phrases in a text relevant to a *grade 3 topic or subject area*. • CCSS.ELA-LITERACY.RI.3.5: Use text features and search tools (e.g., key words, sidebars, hyperlinks) to locate information relevant to a given topic efficiently. • CCSS.ELA-LITERACY.RI.3.6: Distinguish their own point of view from that of the author of a text. Integration of knowledge and ideas • CCSS.ELA-LITERACY.RI.3.7: Use information gained from illustrations (e.g., maps, photographs) and the words in a text to demonstrate understanding of the text (e.g., where, when, why, and how key events occur). • CCSS.ELA-LITERACY.RI.3.8: Describe the logical connection between particular sentences and paragraphs in a text (e.g., comparison, cause/effect, first/second/third in a sequence). • CCSS.ELA-LITERACY.RI.3.9: Compare and contrast the most important points and key details presented in two texts on the same topic. Range of reading and level of text complexity • CCSS.ELA-LITERACY.RI.3.10: By the end of the year, read and comprehend informational texts, including history/social studies, science, and technical texts, at the high end of the grades 2–3 text complexity band independently and proficiently.

Continued

Table 10.2 (*continued*)

***CCSS ELA*—Writing**	Text types and purposes • CCSS.ELA-LITERACY.W.3.1: Write opinion pieces on topics or texts, supporting a point of view with reasons. o CCSS.ELA-LITERACY.W.3.1.A: Introduce the topic or text they are writing about, state an opinion, and create an organizational structure that lists reasons. o CCSS.ELA-LITERACY.W.3.1.B: Provide reasons that support the opinion. o CCSS.ELA-LITERACY.W.3.1.C: Use linking words and phrases (e.g., *because*, *therefore*, *since, for example*) to connect opinion and reasons. o CCSS.ELA-LITERACY.W.3.1.D: Provide a concluding statement or section. • CCSS.ELA-LITERACY.W.3.2: Write informative or explanatory texts to examine a topic and convey ideas and information clearly. o CCSS.ELA-LITERACY.W.3.2.A: Introduce a topic and group related information together; include illustrations when useful to aiding comprehension. o CCSS.ELA-LITERACY.W.3.2.B: Develop the topic with facts, definitions, and details. o CCSS.ELA-LITERACY.W.3.2.C: Use linking words and phrases (e.g., *also*, *another*, *and*, *more*, *but*) to connect ideas within categories of information. o CCSS.ELA-LITERACY.W.3.2.D: Provide a concluding statement or section. Production and distribution of writing • CCSS.ELA-LITERACY.W.3.4: With guidance and support from adults, produce writing in which the development and organization are appropriate to task and purpose. • CCSS.ELA-LITERACY.W.3.5: With guidance and support from peers and adults, develop and strengthen writing as needed by planning, revising, and editing. • CCSS.ELA-LITERACY.W.3.6: With guidance and support from adults, use technology to produce and publish writing (using keyboarding skills) as well as to interact and collaborate with others. Research to build and present knowledge • CCSS.ELA-LITERACY.W.3.8: Recall information from experiences or gather information from print and digital sources; take brief notes on sources and sort evidence into provided categories. Range of writing • CCSS.ELA-LITERACY.W.3.10: Write routinely over extended time frames (time for research, reflection, and revision) and shorter time frames (a single sitting or a day or two) for a range of discipline-specific tasks, purposes, and audiences.

Continued

Table 10.2 (*continued*)

***CCSS ELA—* Speaking and Listening**	Comprehension and collaboration • CCSS.ELA-LITERACY.SL.3.1: Engage effectively in a range of collaborative discussions (one-on-one, in groups, and teacher-led) with diverse partners on *grade 3 topics and texts*, building on others' ideas and expressing their own clearly. o CCSS.ELA-LITERACY.SL.3.1.A: Come to discussions prepared, having read or studied required material; explicitly draw on that preparation and other information known about the topic to explore ideas under discussion. o CCSS.ELA-LITERACY.SL.3.1.B: Follow agreed-upon rules for discussions (e.g., gaining the floor in respectful ways, listening to others with care, speaking one at a time about the topics and texts under discussion). o CCSS.ELA-LITERACY.SL.3.1.C: Ask questions to check understanding of information presented, stay on topic, and link their comments to the remarks of others. o CCSS.ELA-LITERACY.SL.3.1.D: Explain their own ideas and understanding in light of the discussion. • CCSS.ELA-LITERACY.SL.3.2: Determine the main ideas and supporting details of a text read aloud or information presented in diverse media and formats, including visually, quantitatively, and orally. • CCSS.ELA-LITERACY.SL.3.3: Ask and answer questions about information from a speaker, offering appropriate elaboration and detail. Presentation of knowledge and ideas • CCSS.ELA-LITERACY.SL.3.4: Report on a topic or text, tell a story, or recount an experience with appropriate facts and relevant, descriptive details, speaking clearly at an understandable pace. • CCSS.ELA-LITERACY.SL.3.6: Speak in complete sentences when appropriate to task and situation in order to provide requested detail or clarification.
***CCSS Mathematics—* Measurement and Data**	Represent and interpret data. • CCSS.MATH.CONTENT.3.MD.B.3: Draw a scaled picture graph and a scaled bar graph to represent a data set with several categories. Solve one- and two-step "how many more" and "how many less" problems using information presented in scaled bar graphs.

Continued

Table 10.2 *(continued)*

ELP Standards	Receptive modalities • ELP 1: Construct meaning from oral presentations and literary and informational text through grade-appropriate listening, reading, and viewing. • ELP 8: Determine the meaning of words and phrases in oral presentations and literary and informational text. Productive modalities • ELP 3: Speak and write about grade-appropriate complex literary and informational texts and topics. • ELP 4: Construct grade-appropriate oral and written claims and support them with reasoning and evidence. • ELP 7: Adapt language choices to purpose, task, and audience when speaking and writing. Interactive modalities • ELP 2: Participate in grade-appropriate oral and written exchanges of information, ideas, and analyses, responding to peer, audience, or reader comments and questions. • ELP 5: Conduct research and evaluate and communicate findings to answer questions or solve problems. • ELP 6: Analyze and critique the arguments of others orally and in writing. Linguistic structures of English • ELP 9: Create clear and coherent grade-appropriate speech and text. • ELP 10: Make accurate use of standard English to communicate in grade-appropriate speech and writing.

Investigation 10
Fossils: What Was the Ecosystem at Darmstadt Like 49 Million Years Ago?

Introduction

A *fossil* is the preserved remains of a plant or animal. Fossils provide information about organisms that lived long ago and the environments where these organisms lived. Take a minute to examine a fossil. Keep track of what you observe and what you are wondering about in the boxes below.

Fossils provide clues about the traits of organisms that lived on Earth a long time ago. There are two main types of fossils. The first type of fossil is called a *body fossil*. Body fossils are the preserved remains of a plant or animal. The second type of fossil is called a *trace fossil*. Trace fossils are the remains of the activity of an animal. Trace fossils include footprints, imprints of shells or body parts, and nests. Fossils can be found high on mountains, underwater, in the desert, on beaches, or underground. Scientists can use fossils to learn about different organisms that are no longer found on Earth and how the traits of different types of organisms have changed over time.

Fossils can also provide clues about what an ecosystem at a specific location was like in the past. An ecosystem includes all of the living things and non-living things in a given area. There are two main types of ecosystems. The first main type of ecosystem is called an *aquatic ecosystem.* An aquatic ecosystem is found in a body of water. Aquatic ecosystems are classified as either marine (ocean or sea) or freshwater (lake, marsh, river, or swamp). The second main type of ecosystem is called a *terrestrial ecosystem.* A terrestrial ecosystem is found on land. Terrestrial ecosystems are classified as forest, desert, grassland, or mountain.

Scientists can use fossils to learn about what an ecosystem was like in the past because organisms have specific traits that allow them to survive in a specific ecosystem. For example, fish, clams, and seaweed can survive in an aquatic ecosystem but not in a terrestrial one. Therefore, if a scientist finds a fossil fish in a desert, he or she can assume that there was once an ocean or lake at that location.

In this investigation you will examine several different fossils that were found near Darmstadt, Germany. These fossils are about 49 million years old. Your goal is to figure out what the ecosystem near Darmstadt was like 49 million years ago. To accomplish this task, you will need to make observations about these fossils to make inferences about the traits of the organisms that created those fossils and the type of environment in which they lived. Scientists can determine where an animal lived based on its traits because the structure of an organism determines how it functions and places limits on what it can and cannot do. You can therefore use the relationship between structure and function in animal bodies to determine what the ecosystem near Darmstadt was like in the past.

Things we KNOW from what we read …	What we will NEED to figure out …

Your Task

Use what you know about plants, animals, and the relationship between structure and function in nature to plan and carry out an investigation to determine what types of organisms lived near Darmstadt 49 million years ago and what the environment was like at that time.

The *guiding question* of this investigation is, ***What was the ecosystem at Darmstadt like 49 million years ago?***

Materials

You may use any of the following materials during your investigation:

- Fossil A
- Fossil B
- Fossil C
- Fossil D
- Fossil E
- Fossil F

Safety Rules

Follow all normal safety rules.

Plan Your Investigation

Prepare a plan for your investigation by filling out the chart that follows; this plan is called an *investigation proposal.* Before you start developing your plan, be sure to discuss the following questions with the other members of your group:

- How might the **structure** of what you are studying relate to its **function?**
- What types of **patterns** might we look for to help answer the guiding question?

Our guiding question:

We will collect the following data:

These are the steps we will follow to collect data:

I approve of this investigation proposal.

______________________________ ______________________________

Teacher's signature Date

Collect Your Data

Keep a record of what you measure or observe during your investigation in the space below.

Analyze Your Data

You will need to analyze the data you collected before you can develop an answer to the guiding question. In the space below, you can create a table or a graph or use pictures to show the traits or structures of the organisms found in the fossils.

Draft Argument

Develop an argument on a whiteboard. It should include the following parts:

1. A *claim:* Your answer to the guiding question.
2. *Evidence:* An analysis of the data and an explanation of what the analysis means.
3. A *justification of the evidence:* Why your group thinks the evidence is important.

The Guiding Question:	
Our Claim:	
Our Evidence:	Our Justification of the Evidence:

Argumentation Session

Share your argument with your classmates. Be sure to ask them how to make your draft argument better. Keep track of their suggestions in the space below.

Ways to IMPROVE our argument …

Draft Report

Prepare an *investigation report* to share what you have learned. Use the information in this handout and your group's final argument to write a *draft* of your investigation report.

Investigation Handout

Introduction

We have been studying ______________________________ in class. Before we started

this investigation, we explored ______________________________

____

____

We noticed ______________________________

____

____

My goal for this investigation was to figure out ______________________________

____

____

The guiding question was ______________________________

____

____

Method

To gather the data I needed to answer this question, I ______________________________

____

____

____

____

I then analyzed the data I collected by ______________________________

____

____

____

Argument

My claim is ____

____

____

____

____

The ____________________ below shows ____________________________

____

____

____

This evidence is important because ____

____

____

____

____

____

____

____

____

____

____

____

____

Review

Your friends need your help! Review the draft of their investigation reports and give them ideas about how to improve. Use the *peer-review guide* when doing your review.

Submit Your Final Report

Once you have received feedback from your friends about your draft report, create your final investigation report and hand it in to your teacher.

Checkout Questions

Investigation 10. Fossils: What Was the Ecosystem at Darmstadt Like 49 Million Years Ago?

1. Pictured below is the skeleton of an animal. What type of ecosystem do you think this animal lived in while it was alive? You may choose more than one ecosystem.

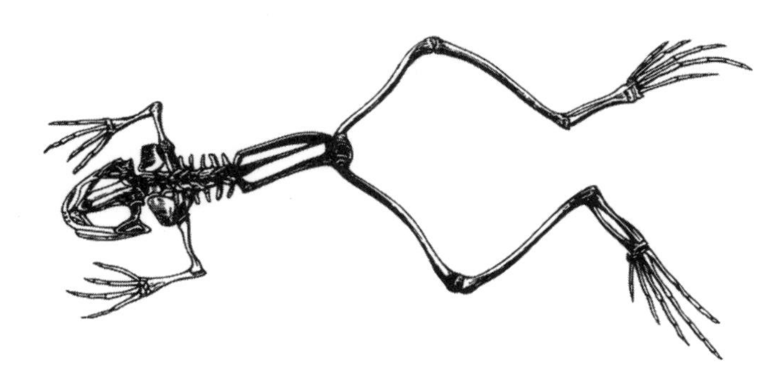

- [ ] Freshwater aquatic (lake, marsh, river, or swamp)
- [ ] Marine aquatic (ocean or sea)
- [ ] Terrestrial (forest, desert, grassland, or mountain)

2. Pictured below is the skeleton of an animal. What type of ecosystem do you think this animal lived in while it was alive? You may choose more than one ecosystem.

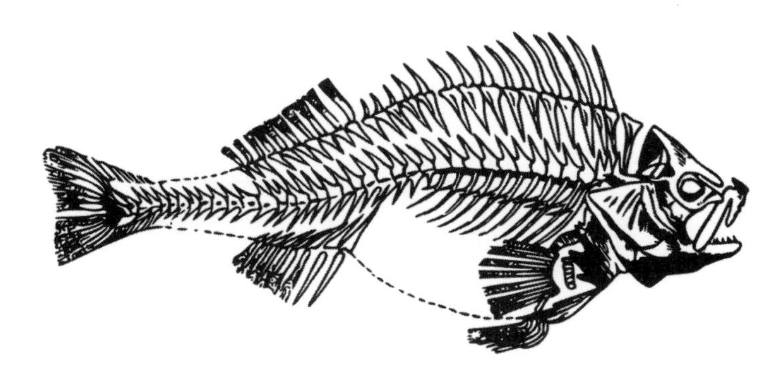

- [ ] Freshwater aquatic (lake, marsh, river, or swamp)
- [ ] Marine aquatic (ocean or sea)
- [ ] Terrestrial (forest, desert, grassland, or mountain)

Checkout Questions

3. Pictured below is the skeleton of an animal. What type of ecosystem do you think this animal lived in while it was alive? You may choose more than one ecosystem.

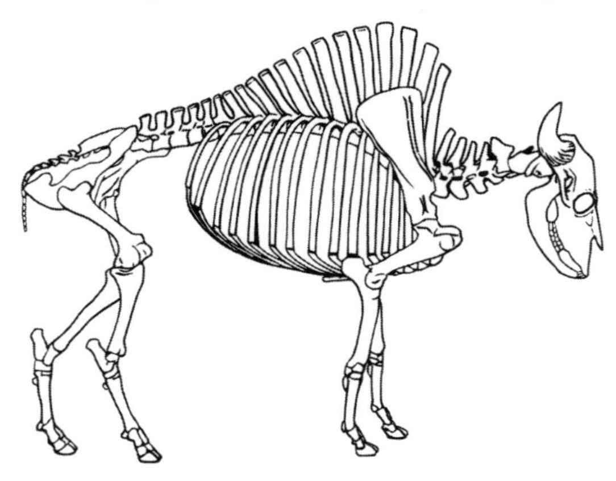

☐ Freshwater aquatic (lake, marsh, river, or swamp)

☐ Marine aquatic (ocean or sea)

☐ Terrestrial (forest, desert, grassland, or mountain)

4. Explain your thinking. How did the *structure* of the skeletons of these animals allow you to determine where they might have lived when they were alive?

____

____

____

____

____

____

Teacher Scoring Rubric for the Checkout Questions

Level	Description
3	The student can apply the core idea correctly in all cases and can fully explain the relationship between structure and function.
2	The student can apply the core idea correctly in all cases but cannot fully explain the relationship between structure and function.
1	The student cannot apply the core idea correctly in all cases but can fully explain the relationship between structure and function.
0	The student cannot apply the core idea correctly in all cases and cannot explain the relationship between structure and function.

Investigation 11

Differences in Traits: How Does Fur Color Affect the Likelihood That a Rabbit Will Survive?

Purpose

The purpose of this investigation is to give students an opportunity to use the disciplinary core idea (DCI) of LS4.B: Natural Selection and the crosscutting concept (CC) of Cause and Effect from *A Framework for K–12 Science Education* (NRC 2012) to figure out how variations in traits that organisms possess may provide advantages in survival. Students will also learn about how scientists use different methods to answer different types of questions during the reflective discussion.

The Disciplinary Core Idea

Students in third grade should understand the following about Natural Selection and be able to use this DCI to figure out how variations in traits that organisms possess may provide advantages in survival:

> *Sometimes the differences in characteristics between individuals of the same species provide advantages in surviving, finding mates, and reproducing. (NRC 2012, p. 164)*

The Crosscutting Concept

Students in third grade should understand the following about the CC Cause and Effect:

> *Repeating patterns in nature, or events that occur together with regularity, are clues that scientists can use to start exploring causal, or cause-and-effect, relationships. … Any application of science, or any engineered solution to a problem, is dependent on understanding the cause-and-effect relationships between events; the quality of the application or solution often can be improved as knowledge of the relevant relationships is improved. (NRC 2012, p. 87)*

Students in third grade should also be given opportunities to begin to "look for and analyze patterns—whether in their observations of the world or in the relationships between different quantities in data (e.g., the sizes of plants over time)"; "they can also begin to consider what might be causing these patterns and relationships and design tests that gather more evidence to support or refute their ideas" (NRC 2012, pp. 88–89).

Students should be encouraged to use their developing understanding of cause-and-effect relationships as a tool or a way of thinking about a phenomenon during this investigation to help them figure out how traits of an animal can affect the likelihood of that animal's survival.

What Students Figure Out

Traits possessed by organisms can be advantageous (helping the organism survive and/or reproduce), disadvantageous (preventing the organism from surviving and/or reproducing), or neutral (producing no discernible effects on an organism's ability to survive or reproduce) in a given environment. Organisms with advantageous traits tend to survive and reproduce more than those without those traits for the environment in which they live. Organisms with advantageous traits also tend to survive and reproduce more than those with disadvantageous traits for the environment in which they live. Therefore, those advantageous traits become more common in the overall population of organisms of that type over time in that environment. It is important to note that the advantages provided by traits are dependent on the organism's environment.

Timeline

The time needed to complete this investigation is 270 minutes (4 hours and 30 minutes). The amount of instructional time needed for each stage of the investigation is as follows:

- *Stage 1.* Introduce the task and the guiding question: 35 minutes
- *Stage 2.* Design a method and collect data: 60 minutes
- *Stage 3.* Create a draft argument: 35 minutes
- *Stage 4.* Argumentation session: 30 minutes
- *Stage 5.* Reflective discussion: 15 minutes
- *Stage 6.* Write a draft report: 35 minutes
- *Stage 7.* Peer review: 30 minutes
- *Stage 8.* Revise the report: 30 minutes

This investigation can be completed in one day or over eight days (one day for each stage) during your designated science time in the daily schedule.

Materials and Preparation

The materials needed for this investigation are listed in Table 11.1. The *Natural Selection* simulation, which was developed by PhET Interactive Simulations, University of Colorado (*http://phet.colorado.edu*), is available at *https://phet.colorado.edu/en/simulation/legacy/natural-selection*. It is free to use and can be run online using an internet browser. Students will need a computer or tablet with the capability to run the latest version of Java to use the simulator. At the time of this book's writing, the simulator was not yet available in HTML5.

You should access the website and learn how the simulation works before beginning the investigation. In addition, it is important to check if students can access and use the simulation from a school computer or tablet, because some schools have set up firewalls and other restrictions on web browsing.

The students can use this simulation to investigate the relationship between a rabbit's fur color and its ability to survive and reproduce. The simulation provides students with the choice between two environments: an equatorial environment with a brown background and an arctic environment with a white background. Students are also able to choose a selection factor: food or wolves. A rabbit's fur color has no effect on its ability to eat food in this simulation (though teeth length does). Rabbits in the arctic environment are more likely to survive predation by wolves if they have white fur, whereas rabbits in the equatorial environment are more likely to survive if they have brown fur. Therefore, white fur is an advantageous trait in the arctic environment and a disadvantageous trait in the equatorial environment.

TABLE 11.1
Materials for Investigation 11

Item	Quantity
Computer or tablet with Java capability	1 per group
Whiteboard, 2' × 3'*	1 per group
Investigation Handout	1 per student
Peer-review guide and teacher scoring rubric	1 per student
Checkout Questions (optional)	1 per student

* As an alternative, students can use computer and presentation software such as Microsoft PowerPoint or Apple Keynote to create their arguments.

Safety Precautions

Remind students to follow all normal safety rules.

Lesson Plan by Stage

Stage 1: Introduce the Task and the Guiding Question (35 minutes)

1. Ask the students to sit in six groups, with three or four students in each group.
2. Ask students to clear off their desks except for a pencil (and their *Student Workbook for Argument-Driven Inquiry in Third-Grade Science* if they have one).
3. Pass out an Investigation Handout to each student (or ask students to turn to Investigation Log 11 in their workbook).
4. Read the first paragraph of the "Introduction" aloud to the class. Ask the students to follow along as you read.

5. Remind students of the safety rules for this investigation.
6. Tell students to record their observations and questions about what they see in the simulation in the "OBSERVED/WONDER" chart in the "Introduction" section of their Investigation Handout (or the investigation log in their workbook).
7. Pass out a computer or tablet to each student group. Loading the simulation ahead of time or creating a desktop shortcut on each device will help.
8. Ask students to share *what they observed* when using the simulation.
9. Ask students to share *what questions they have* after using the simulation.
10. Tell the students, "Some of your questions might be answered by reading the rest of the 'Introduction.'"
11. Ask the students to read the rest of the "Introduction" on their own *or* ask them to follow along as you read aloud.
12. Once the students have read the rest of the "Introduction," ask them to fill out the "KNOW/NEED" chart on their Investigation Handout (or in their investigation log) as a group.
13. Ask students to share what they learned from the reading. Add these ideas to a class "know / need to figure out" chart.
14. Ask students to share what they think they will need to figure out based on what they read. Add these ideas to the class "know / need to figure out" chart.
15. Tell the students, "It looks like we have something to figure out. Let's see what we will need to do during our investigation."
16. Read the task and the guiding question aloud.
17. Tell the students, "Take a look at all the options the simulation has. What does it let you do?"
18. Introduce the students to how to set up different environments and change different traits of the rabbits (see Figure 11.1 for a screen shot of the simulation). Be careful not to tell students what to do with the simulation. You are simply helping them see the options available to them.

FIGURE 11.1

PhET simulation screen shot showing equatorial environment and white rabbit

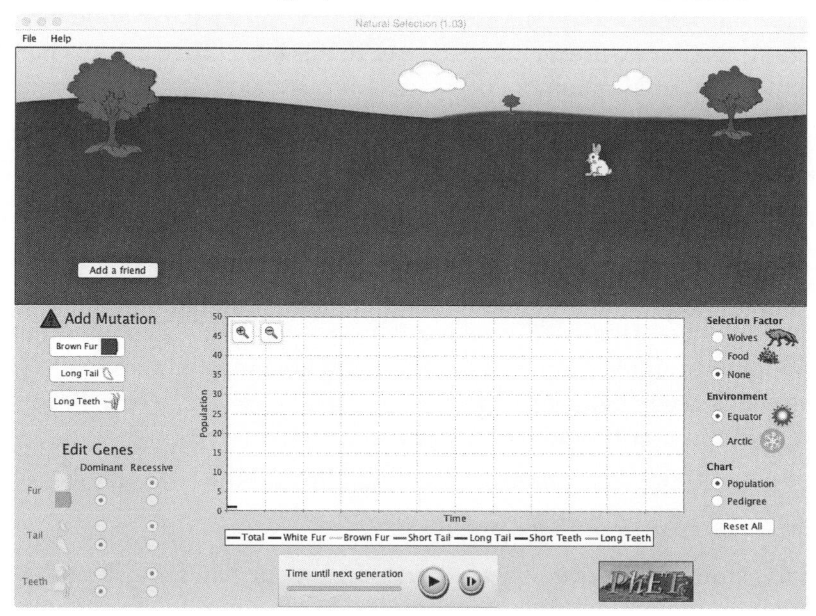

Stage 2: Design a Method and Collect Data (60 minutes: 30 minutes to plan and 30 minutes to collect data)

1. Tell the students, "I am now going to give you and the other members of your group about 30 minutes to plan your investigation. Before you begin, I want you all to take a couple of minutes to discuss the following questions with the rest of your group."
2. Show the following questions on the screen or board:
   - What types of *patterns* might we look for to help answer the guiding question?
   - What information do we need to find a *cause-and-effect relationship?*
3. Tell the students, "Please take a few minutes to come up with an answer to these questions." Give the students two or three minutes to discuss these two questions.
4. Ask two or three different groups to share their answers. Be sure to highlight or write down any important ideas on the board so students can refer to them later.
5. If possible, use a document camera to project an image of the graphic organizer for this investigation on a screen or board (or take a picture of it and project the picture on a screen or board). Tell the students, "I now want you all to plan out

your investigation. To do that, you will need to create an investigation proposal by filling out this graphic organizer."

6. Point to the box labeled "Our guiding question:" and tell the students, "You can put the question we are trying to answer in this box." Then ask, "Where can we find the guiding question?"
7. Wait for a student to answer where to find the guiding question (the answer is "in the handout").
8. Point to the box labeled "We will collect the following data:" and tell the students, "You can list the observations that you will need to collect during the investigation in this box."
9. Point to the box labeled "These are the steps we will follow to collect data:" and tell the students, "You can list what you are going to do to collect the data you need and what you will do with your data once you have it. Be sure to give enough detail that I could do your investigation for you."
10. Ask the students, "Do you have any questions about what you need to do?"
11. Answer any questions that come up.
12. Tell the students, "Once you are done, raise your hand and let me know. I'll then come by and look over your proposal and give you some feedback. You may not begin collecting data until I have approved your proposal by signing it. You need to have your proposal done in the next 20 minutes."
13. For this investigation, you may choose to let students have access to laptops or tablets as they plan their investigations so they can become familiar with the simulation's capabilities.
14. Give the students 20 minutes to work in their groups on their investigation proposal. As they work, move from group to group to check in, ask probing questions, and offer a suggestion if a group gets stuck.

What should a student-designed investigation look like?

The students' investigation proposal should include the following information:

- The guiding question is "How does fur color affect the likelihood that a rabbit will survive?"
- There are a lot of different types of data that students can collect during this investigation. Examples include (1) number of brown rabbits, (2) number of white rabbits, and (3) difference in number between total number of rabbits and white or brown rabbits. Be sure to encourage students to collect quantitative (numerical) data when possible.

- The steps that the students will follow to collect the data should reflect the traits that they decide to examine. However, a procedure might include the following steps:
  1. Set the environment to arctic.
  2. Set the rabbit fur color to white.
  3. Run the simulation.
  4. Record the number of surviving rabbits.
  5. Set the environment to arctic.
  6. Set the rabbit fur color to brown.
  7. Run the simulation.
  8. Record the number of surviving rabbits.
  9. Set the environment to equator.
  10. Set the rabbit fur color to white.
  11. Run the simulation.
  12. Record the number of surviving rabbits.
  13. Set the environment to equator.
  14. Set the rabbit fur color to brown.
  15. Run the simulation.
  16. Record the number of surviving rabbits.

This is just an example of a procedure, and there should be a lot of variation in the student-designed investigations.

15. As each group finishes its investigation proposal, be sure to read it over and determine if it will be productive or not. If you feel the investigation will be productive (not necessarily what you would do or what the other groups are doing), sign your name on the proposal and let the group start collecting data. If the plan needs to be changed, offer some suggestions or ask some probing questions, and have the group make the changes before you approve it.
16. Pass out the tablets or computers again if you haven't done so already, or have one student from each group collect them from a central supply table or cart for the groups that have an approved proposal.
17. Give the students 30 minutes to collect data. Remind the students to collect their data and record their observations or measurements in the "Collect Your Data" box in their Investigation Handout (or the investigation log in their workbook).

Stage 3: Create a Draft Argument (35 minutes)

1. Tell the students, "Now that we have all this data, we need to analyze the data so we can figure out an answer to the guiding question."
2. If possible, project an image of the "Analyze Your Data" section for this investigation on a screen or board using a document camera (or take a picture of it and project the picture on a screen or board). Point to the section and tell the groups of students, "You can create two graphs as a way to analyze your data. You can make your graphs in this section."
3. Ask the students, "What information do we need to include in the graphs?"
4. Tell the students, "Please take a few minutes to discuss this question with your group, and be ready to share."
5. Give the students five minutes to discuss.
6. Ask two or three different groups to share their answers. Be sure to highlight or write down any important ideas on the board so students can refer to them later.
7. Tell the students, "I am now going to give you and the other members of your group about 10 minutes to create your graphs." If the students are having trouble making one or both of the graphs, you can take a few minutes to provide a mini-lesson about how to create a graph from a bunch of observations or measurements (this strategy is called just-in-time instruction because it is offered only when students get stuck).

What should a table or graph look like for this investigation?

There are a number of different ways that students can analyze the observations or measurements they collect during this investigation. One of the most straightforward ways is to create two scaled bar graphs to represent two data sets with several categories. The bar graphs should have the names of conditions on the horizontal or *x*-axis (white fur in the Arctic, white fur on the equator, brown fur in the Arctic, brown fur on the equator). The dependent variable (number of rabbits) should be on the *y*-axis. An example of this type of graph can be seen in Figure 11.2 (p. 408).

8. Give the students 10 minutes to analyze their data. As they work, move from group to group to check in, ask probing questions, and offer suggestions.
9. Tell the students, "I am now going to give you and the other members of your group 15 minutes to create an argument to share what you have learned and convince others that they should believe you. Before you do that, we need to take a few minutes to discuss what you need to include in your argument."
10. If possible, use a document camera to project the "Argument Presentation on a Whiteboard" image from the Investigation Handout (or the investigation log in their workbook) on a screen or board (or take a picture of it and project the picture on a screen or board).
11. Point to the box labeled "The Guiding Question:" and tell the students, "You can put the question we are trying to answer here on your whiteboard."
12. Point to the box labeled "Our Claim:" and tell the students, "You can put your claim here on your whiteboard. The claim is your answer to the guiding question."
13. Point to the box labeled "Our Evidence:" and tell the students, "You can put the evidence that you are using to support your claim here on your whiteboard. Your evidence will need to include the analysis you just did and an explanation of what your analysis means or shows. Scientists always need to support their claims with evidence."
14. Point to the box labeled "Our Justification of the Evidence:" and tell the students, "You can put your justification of your evidence here on your whiteboard. Your justification needs to explain why your evidence is important. Scientists often use core ideas to explain why the evidence they are using matters. Core ideas are important concepts that scientists use to help them make sense of what happens during an investigation."
15. Ask the students, "What are some core ideas that we read about earlier that might help us explain why the evidence we are using is important?"
16. Ask students to share some of the core ideas from the "Introduction" section of the Investigation Handout (or the investigation log in the workbook). List these core ideas on the board.
17. Tell the students, "That is great. I would like to see everyone try to include these core ideas in your justification of the evidence. Your goal is to use these core ideas to help explain why your evidence matters and why the rest of us should pay attention to it."
18. Ask the students, "Do you have any questions about what you need to do?"
19. Answer any questions that come up.
20. Tell the students, "Okay, go ahead and start working on your arguments. You need to have your argument done in the next 15 minutes. It doesn't need to be

perfect. We just need something down on the whiteboards so we can share our ideas."

21. Give the students 15 minutes to work in their groups on their arguments. As they work, move from group to group to check in, ask probing questions, and offer a suggestion if a group gets stuck. Figure 11.2 shows an example of an argument created by students for this investigation.

FIGURE 11.2

Example of an argument

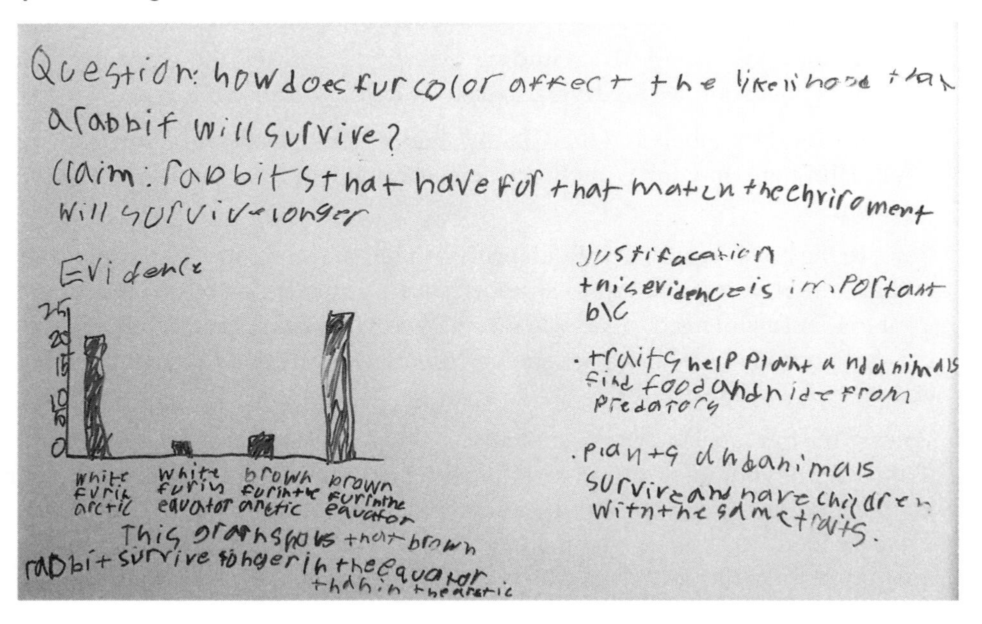

Stage 4: Argumentation Session (30 minutes)

The argumentation session can be conducted in a whole-class presentation format, a gallery walk format, or a modified gallery walk format. We recommend using a whole-class presentation format for the first investigation, but try to transition to either the gallery walk or modified gallery walk format as soon as possible because that will maximize student voice and choice inside the classroom. The following list shows the steps for the three formats; unless otherwise noted, the steps are the same for all three formats.

1. Begin by introducing the use of the whiteboard.
    - *If using the whole-class presentation format,* tell the students, "We are now going to share our arguments. Please set up your whiteboard so everyone can see them."
    - *If using the gallery walk or modified gallery walk format,* tell the students, "We are now going to share our arguments. Please set up your whiteboard so they are facing the walls."

2. Allow the students to set up their whiteboards.
   - *If using the whole-class presentation format,* the whiteboards should be set up on stands or chairs so they are facing toward the center of the room.
   - *If using the gallery walk or modified gallery walk format,* the whiteboards should be set up on stands or chairs so they are facing toward the outside of the room.
3. Give the following instructions to the students:
   - *If using the whole-class presentation format or the modified gallery walk format*, tell the students, "Okay, before we get started I want to explain what we are going to do next. I'm going to ask some of you to present your arguments to your classmates. If you are presenting your argument, your job is to share your group's claim, evidence, and justification of the evidence. The rest of you will be reviewers. If you are a reviewer, your job is to listen to the presenters, ask the presenters questions if you do not understand something, and then offer them some suggestions about ways to make their argument better. After we have a chance to learn from each other, I'm going to give you some time to revise your arguments and make them better."
   - *If using the gallery walk format*, tell the students, "Okay, before we get started I want to explain what we are going to do next. You are going to have an opportunity to read the arguments that were created by other groups. Your group will go to a different group's argument. I'll give you a few minutes to read it and review it. Your job is to offer them some suggestions about ways to make their argument better. You can use sticky notes to give them suggestions. Please be specific about what you want to change and be specific about how you think they should change it. After we have a chance to learn from each other, I'm going to give you some time to revise your arguments and make them better."
4. Use a document camera to project the "Ways to IMPROVE our argument …" box from the Investigation Handout (or the investigation log in their workbook) on a screen or board (or take a picture of it and project the picture on a screen or board).
   - *If using the whole-class presentation format or the modified gallery walk format,* point to the box and tell the students, "If you are a presenter, you can write down the suggestions you get from the reviewers here. If you are a reviewer, and you see a good idea from another group, you can write down that idea here. Once we are done with the presentations, I will give you a chance to use these suggestions or ideas to improve your arguments.
   - *If using the gallery walk format,* point to the box and tell the students, "If you see good ideas from another group, you can write them down here. Once we are done reviewing the different arguments, I will give you a chance to use these ideas to improve your own arguments. It is important to share ideas like this."

Ask the students, "Do you have any questions about what you need to do?"

5. Answer any questions that come up.
6. Give the following instructions:
   - *If using the whole-class presentation format,* tell the students, "Okay. Let's get started."
   - *If using the gallery walk format,* tell the students, "Okay, I'm now going to tell you which argument to go to and review.
   - *If using the modified gallery walk format,* tell the students, "Okay, I'm now going to assign you to be a presenter or a reviewer." Assign one or two students from each group to be presenters and one or two students from each group to be reviewers.
7. Begin the review of the arguments.
   - *If using the whole-class presentation format,* have four or five groups present their argument one at a time. Give each group only two to three minutes to present their argument. Then give the class two to three minutes to ask them questions and offer suggestions. Be sure to encourage as much participation from the students as possible.
   - *If using the gallery walk format,* tell the students, "Okay. Let's get started. Each group, move one argument to the left. Don't move to the next argument until I tell you to move. Once you get there, read the argument and then offer suggestions about how to make it better. I will put some sticky notes next to each argument. You can use the sticky notes to leave your suggestions." Give each group about three to four minutes to read the arguments, talk, and offer suggestions.
     a. Tell the students, "Okay. Let's rotate. Move one group to the left."
     b. Again, give each group three or four minutes to read, talk, and offer suggestions.
     c. Repeat this process for two more rotations.
   - *If using the modified gallery walk format,* tell the students, "Okay. Let's get started. Reviewers, move one group to the left. Don't move to the next group until I tell you to move. Presenters, go ahead and share your argument with the reviewers when they get there." Give each group of presenters and reviewers about three to four minutes to talk.
     a. Tell the students, "Okay. Let's rotate. Reviewers, move one group to the left."
     b. Again, give each group of presenters and reviewers about three or four minutes to talk.
     c. Repeat this process for two more rotations.
8. Tell the students to return to their workstations.

9. Give the following instructions about revising the argument:
   - *If using the whole-class presentation format,* tell the students, "I'm now going to give you about 10 minutes to revise your argument. Take a few minutes to talk in your groups and determine what you want to change to make your argument better. Once you have decided what to change, go ahead and make the changes to your whiteboard."
   - *If using the gallery walk format,* tell the students, "I'm now going to give you about 10 minutes to revise your argument. Take a few minutes to read the suggestions that were left at your argument. Then talk in your groups and determine what you want to change to make your argument better. Once you have decided what to change, go ahead and make the changes to your whiteboard."
   - *If using the modified gallery walk format,* "I'm now going to give you about 10 minutes to revise your argument. Please return to your original groups." Wait for the students to move back into their original groups and then tell the students, "Okay, take a few minutes to talk in your groups and determine what you want to change to make your argument better. Once you have decided what to change, go ahead and make the changes to your whiteboard."

   Ask the students, "Do you have any questions about what you need to do?"
10. Answer any questions that come up.
11. Tell the students, "Okay. Let's get started."
12. Give the students 10 minutes to work in their groups on their arguments. As they work, move from group to group to check in, ask probing questions, and offer a suggestion if a group gets stuck.

Stage 5: Reflective Discussion (15 minutes)

1. Tell the students, "We are now going to take a minute to talk about what we did and what we have learned."
2. Show students several pictures of rabbits, bears, or other animals with fur colors suited to their natural habitats in those habitats, such as those in Figures 11.3–11.5 (p. 412).
3. Ask the students, "What do you all see going on here?"
4. Allow students to share their ideas.
5. Ask the students, "Do you think these animals are well suited to live in their environments?"
6. Allow students to share their ideas.
7. Ask the students, "Do you think animals with different fur colors would be likely to survive in these environments?"

FIGURE 11.3

Arctic fox in its native habitat

FIGURE 11.4

Leopard in South Africa

FIGURE 11.5

Owl in a tree

Full-color versions of these images are available on the book's Extras page at *www.nsta.org/adi-3rd*.

8. Allow students to share their ideas.
9. Ask the students, "Why do many animals' fur, skin, or scales often match the color of their environments?"
10. Allow students to share their ideas. Keep probing until someone mentions that it allows them to hide from predators.
11. Tell the students, "Okay, let's make sure we are on the same page. Animal species have traits that help them survive in the environments in which they live. For some animals, this means that they have the right type of teeth, or the right type of beak, or the right length of tongue to eat the food that grows in their environment. For some animals, it means that they have the right amount of fur or fat to keep them warm in winter. For other animals, like the rabbits in your simulation, it means that their body color allows them to blend in with their environment. This makes it less likely that predators will spot them and be able to catch them, or if the animal is a predator, that its prey will see it and get away before the animal can catch it. This means that animals that can blend in with their environments are more likely to survive. The fact that animals and plants have traits that help them survive in particular environment is a really important core idea in science."
12. Ask the students, "Does anyone have any questions about this core idea?"
13. Answer any questions that come up.
14. Tell the students, "We also looked for a cause-and-effect relationship during our investigation." Then ask, "Can anyone tell me why it is useful to look for cause-and-effect relationships?"
15. Allow students to share their ideas.
16. Tell the students, "Cause-and-effect relationships are important because they allow us to make predictions."
17. Ask the students, "What was the cause and what was the effect that we uncovered today?"
18. Allow students to share their ideas. Keep probing until someone mentions that the cause was the different fur or environment colors and the effect was the survival of the rabbits.
19. Tell the students, "That is great, and if we know that we can use that information to predict how likely it is that an animal will survive in a certain environment."
20. Tell the students, "We are now going take a minute to talk about what went well and what didn't go so well during our investigation. We need to talk about this because you are going to be planning and carrying out your own investigations like this a lot this year, and I want to help you all get better at it."
21. Show an image of the question "What made your investigation scientific?" on the screen. Tell the students, "Take a few minutes to talk about how you would

answer this question with the other people in your group. Be ready to share with the rest of the class." Give the students two to three minutes to talk in their group.

22. Ask the students, "What do you all think? Who would like to share an idea?"
23. Allow students to share their ideas. Be sure to expand on their ideas about what makes an investigation scientific.
24. Show an image of the question "What made your investigation not so scientific?" on the screen. Tell the students, "Take a few minutes to talk about how you would answer this question with the other people in your group. Be ready to share with the rest of the class." Give the students two to three minutes to talk in their group.
25. Ask the students, "What do you all think? Who would like to share an idea?"
26. Allow students to share their ideas. Be sure to expand on their ideas about what makes an investigation less scientific.
27. Show an image of the question "What rules can we put into place to help us make sure our next investigation is more scientific?" on the screen. Tell the students, "Take a few minutes to talk about how you would answer this question with the other people in your group. Be ready to share with the rest of the class." Give the students two to three minutes to talk in their group.
28. Ask the students, "What do you all think? Who would like to share an idea?"
29. Allow students to share their ideas. Once they have shared their ideas, offer a suggestion for a possible class rule.
30. Ask the students, "What do you all think? Should we make this a rule?"
31. If the students agree, write the rule on the board or make a class "Rules for Scientific Investigation" chart so you can refer to it during the next investigation.
32. Tell the students, "We are now going take a minute to talk about what scientists do to investigate the natural world."
33. Show an image of the question "Do all scientists follow the same method?" on the screen. Tell the students, "Take a few minutes to talk about how you would answer this question with the other people in your group. Be ready to share with the rest of the class." Give the students two to three minutes to talk in their group.
34. Ask the students, "What do you all think? Who would like to share an idea?"
35. Allow students to share their ideas.
36. Tell the students, "Okay, let's make sure we are all on the same page. Scientists use lots of different methods to answer different types of questions. Sometimes they need to go out into the field and watch what animals do. Some scientists design experiments, and others analyze data collected by other scientists. Some even create or use computer models to study the way animals behave, just as

you did today. There is no one method that all scientists use, and the method used by scientists depends on what they are studying and what type of question they are asking. This is an important thing to understand about science."

37. Ask the students, "Does anyone have any questions about what scientists do to investigate the natural world?"
38. Answer any questions that come up.

Stage 6: Write a Draft Report (35 minutes)

Your students will use either the Investigation Handout or the investigation log in the student workbook when writing the draft report. When you give the directions shown in quotes in the following steps, substitute "investigation log" (as shown in brackets) for "handout" if they are using the workbook.

1. Tell the students, "You are now going to write an investigation report to share what you have learned. Please take out a pencil and turn to the 'Draft Report' section of your handout [investigation log]."
2. If possible, use a document camera to project the "Introduction" section of the draft report from the Investigation Handout (or the investigation log in their workbook) on a screen or board (or take a picture of it and project the picture on a screen or board).
3. Tell the students, "The first part of the report is called the 'Introduction.' In this section of the report you want to explain to the reader what you were investigating, why you were investigating it, and what question you were trying to answer. All of this information can be found in the text at the beginning of your handout [investigation log]." Point to the image and say, "There are some sentence starters here to help you begin writing the report." Ask the students, "Do you have any questions about what you need to do?"
4. Answer any questions that come up.
5. Tell the students, "Okay. Let's write."
6. Give the students 10 minutes to write the "Introduction" section of the report. As they work, move from student to student to check in, ask probing questions, and offer a suggestion if a student gets stuck.
7. If possible, use a document camera to project the "Method" section of the draft report from the Investigation Handout (or the investigation log in their workbook) on a screen or board (or take a picture of it and project the picture on a screen or board).
8. Tell the students, "The second part of the report is called the 'Method.' In this section of the report you want to explain to the reader what you did during the investigation, what data you collected and why, and how you went about analyzing your data. All of this information can be found in the 'Plan Your

Investigation' section of your handout [investigation log]. Remember that you all planned and carried out different investigations, so do not assume that the reader will know what you did." Point to the image and say, "There are some sentence starters here to help you begin writing this part of the report." Ask the students, "Do you have any questions about what you need to do?"

9. Answer any questions that come up.
10. Tell the students, "Okay. Let's write."
11. Give the students 10 minutes to write the "Method" section of the report. As they work, move from student to student to check in, ask probing questions, and offer a suggestion if a student gets stuck.
12. If possible, use a document camera to project the "Argument" section of the draft report from the Investigation Handout (or the investigation log in their workbook) on a screen or board (or take a picture of it and project the picture on a screen or board).
13. Tell the students, "The last part of the report is called the 'Argument.' In this section of the report you want to share your claim, evidence, and justification of the evidence with the reader. All of this information can be found on your whiteboard." Point to the image and say, "There are some sentence starters here to help you begin writing this part of the report." Ask the students, "Do you have any questions about what you need to do?"
14. Answer any questions that come up.
15. Tell the students, "Okay. Let's write."
16. Give the students 10 minutes to write the "Argument" section of the report. As they work, move from student to student to check in, ask probing questions, and offer a suggestion if a student gets stuck.

Stage 7: Peer Review (30 minutes)

Your students will use either the Investigation Handout or the investigation log in the student workbook when doing the peer review. When you give the directions shown in quotes in the following steps, substitute "workbook" (as shown in brackets) for "Investigation Handout" if they are using the workbook.

1. Tell the students, "We are now going to review our reports to find ways to make them better. I'm going to come around and collect your Investigation Handout [workbook]. While I do that, please take out a pencil."
2. Collect the Investigation Handouts or workbooks from the students.
3. If possible, use a document camera to project the peer-review guide (PRG; see Appendix 4) on a screen or board (or take a picture of it and project the picture on a screen or board).

4. Tell the students, "We are going to use this peer-review guide to give each other feedback." Point to the image.
5. Give the following instructions:
   - *If using the Investigation Handout,* tell the students, "I'm going to ask you to work with a partner to do this. I'm going to give you and your partner a draft report to read and a peer-review guide to fill out. You two will then read the report together. Once you are done reading the report, I want you to answer each of the questions on the peer-review guide." Point to the review questions on the image of the PRG.
   - *If using the workbook,* tell the students, "I'm going to ask you to work with a partner to do this. I'm going to give you and your partner a draft report to read. You two will then read the report together. Once you are done reading the report, I want you to answer each of the questions on the peer-review guide that is right after the report in the investigation log." Point to the review questions on the image of the PRG.
6. Tell the students, "You can check 'yes,' 'almost,' or 'no' after each question." Point to the checkboxes on the image of the PRG.
7. Tell the students, "This will be your rating for this part of the report. Make sure you agree on the rating you give the author. If you mark 'almost' or 'no,' then you need to tell the author what he or she needs to do to get a 'yes.'" Point to the space for the reviewer feedback on the image of the PRG.
8. Tell the students, "It is really important for you to give the authors feedback that is helpful. That means you need to tell them exactly what they need to do to make their reports better." Ask the students, "Do you have any questions about what you need to do?"
9. Answer any questions that come up.
10. Tell the students, "Please sit with a partner who is not in your current group." Allow the students time to sit with a partner.
11. Give the following instructions:
    - *If using the Investigation Handout,* tell the students, "Okay, I am now going to give you one report to read and one peer-review guide to fill out." Pass out one report to each pair. Make sure that the report you give a pair was not written by one of the students in that pair. Give each pair one PRG to fill out as a team.
    - *If using the workbook,* tell the students, "Okay, I am now going to give you one report to read." Pass out a workbook to each pair. Make sure that the workbook you give a pair is not from one of the students in that pair.
12. Tell the students, "Okay, I'm going to give you 15 minutes to read the report I gave you and to fill out the peer-review guide. Go ahead and get started."

13. Give the students 15 minutes to work. As they work, move around from pair to pair to check in and see how things are going, answer questions, and offer advice.
14. After 15 minutes pass, tell the students, "Okay, time is up." *If using the Investigation Handout,* say, "Please give me the report and the peer-review guide that you filled out." *If using the workbook,* say, "Please give me the workbook that you have."
15. Collect the Investigation Handouts and the PRGs, or collect the workbooks if they are being used. Be sure you keep the handout and the PRG together.
16. Give the following instructions:
    - *If using the Investigation Handout,* tell the students, "Okay, I am now going to give you a different report to read and a new peer-review guide to fill out." Pass out another report to each pair. Make sure that this report was not written by one of the students in that pair. Give each pair a new PRG to fill out as a team.
    - *If using the workbook,* tell the students, "Okay, I am now going to give you a different report to read." Pass out a different workbook to each pair. Make sure that the workbook you give a pair is not from one of the students in that pair.
17. Tell the students, "Okay, I'm going to give you 15 minutes to read this new report and to fill out the peer-review guide. Go ahead and get started."
18. Give the students 15 minutes to work. As they work, move around from pair to pair to check in and see how things are going, answer questions, and offer advice.
19. After 15 minutes pass, tell the students, "Okay, time is up." *If using the Investigation Handout,* say, "Please give me the report and the peer-review guide that you filled out." *If using the workbook,* say, "Please give me the workbook that you have."
20. Collect the Investigation Handouts and the PRGs, or collect the workbooks if they are being used. Be sure you keep the handout and the PRG together.

Stage 8: Revise the Report (30 minutes)

Your students will use either the Investigation Handout or the investigation log in the student workbook when revising the report. Except where noted below, the directions are the same whether using the handout or the log.

1. Give the following instructions:
    - *If using the Investigation Handout,* tell the students, "You are now going to revise your investigation report based on the feedback you get from your classmates.

Please take out a pencil while I hand back your draft report and the peer-review guide."

- *If using the investigation log in the student workbook,* tell the students, "You are now going to revise your investigation report based on the feedback you get from your classmates. Please take out a pencil while I hand back your investigation logs."

2. *If using the Investigation Handout,* pass back the handout and the PRG to each student. *If using the investigation log,* pass back the log to each student.
3. Tell the students, "Please take a few minutes to read over the peer-review guide. You should use it to figure out what you need to change in your report and how you will change the report."
4. Allow the students time to read the PRG.
5. *If using the investigation log,* if possible use a document camera to project the "Write Your Final Report" section from the investigation log on a screen or board (or take a picture of it and project the picture on a screen or board).
6. Give the following instructions:
   - *If using the Investigation Handout,* tell the students, "Okay. Let's revise our reports. Please take out a piece of paper. I would like you to rewrite your report. You can use your draft report as a starting point, but use the feedback on the peer-review guide to help make it better."
   - *If using the investigation log,* tell the students, "Okay. Let's revise our reports. I would like you to rewrite your report in the section of the investigation log that says 'Write Your Final Report.'" Point to the image on the screen and tell the students, "You can use your draft report as a starting point, but use the feedback on the peer-review guide to help make your report better."

   Ask the students, "Do you have any questions about what you need to do?"
7. Answer any questions that come up.
8. Tell the students, "Okay. Let's write."
9. Give the students 30 minutes to rewrite their report. As they work, move from student to student to check in, ask probing questions, and offer a suggestion if a student gets stuck.
10. Give the following instructions:
    - *If using the Investigation Handout,* tell the students, "Okay. Time's up. I will now come around and collect your Investigation Handout, the peer-review guide, and your final report."
    - *If using the investigation log,* tell the students, "Okay. Time's up. I will now come around and collect your workbooks."

11. *If using the Investigation Handout,* collect all the Investigation Handouts, PRGs, and final reports. *If using the investigation log,* collect all the workbooks.
12. *If using the Investigation Handout,* use the "Teacher Score" columns in the PRG to grade the final report. *If using the investigation log,* use the "ADI Investigation Report Grading Rubric" in the investigation log to grade the final report. Whether you are using the handout or the log, you can give the students feedback about their writing in the "Teacher Comments" section.

How to Use the Checkout Questions

The Checkout Questions are an optional assessment. We recommend giving them to students one day after they finish stage 8 of the ADI investigation. The Checkout Questions can be used as a formative or summative assessment of student thinking. If you plan to use them as a formative assessment, we recommend that you look over the student answers to determine if you need to reteach the core idea and/or crosscutting concept from the investigation, but do not grade them. If you plan to use them as a summative assessment, we have included a "Teacher Scoring Rubric" at the end of the Checkout Questions that you can use to score a student's ability to apply the core idea in a new scenario and explain their use of a crosscutting concept. The rubric includes a 4-point scale that ranges from 0 (the student cannot apply the core idea correctly in all cases and cannot explain the [crosscutting concept]) to 3 (the student can apply the core idea correctly in all cases and can fully explain the [crosscutting concept]). The Checkout Questions, regardless of how you decide to use them, are a great way to make student thinking visible so you can determine if the students have learned the core idea and the crosscutting concept.

A student who can apply the core idea correctly in all cases and can explain the cause-and-effect relationship would select A for question 1 and B for question 2. He or she should then be able to explain that animals that blend into the environment are more likely to avoid predators when compared with animals that do not blend into the environment.

Connections to Standards

Table 11.2 highlights how the investigation can be used to address specific performance expectations from the *NGSS, Common Core State Standards (CCSS)* in English language arts (ELA) and in mathematics, and *English Language Proficiency (ELP) Standards.*

TABLE 11.2

Investigation 11 alignment with standards

***NGSS* performance expectations**	Strong alignment • 3-LS4-2: Use evidence to construct an explanation for how the variations in characteristics among individuals of the same species may provide advantages in surviving, finding mates, and reproducing. Moderate alignment • 3-LS4-3: Construct an argument with evidence that in a particular habitat some organisms can survive well, some survive less well, and some cannot survive at all.
***CCSS ELA*—Reading: Informational Text**	Key ideas and details • CCSS.ELA-LITERACY.RI.3.1: Ask and answer questions to demonstrate understanding of a text, referring explicitly to the text as the basis for the answers. • CCSS.ELA-LITERACY.RI.3.2: Determine the main idea of a text; recount the key details and explain how they support the main idea. • CCSS.ELA-LITERACY.RI.3.3: Describe the relationship between a series of historical events, scientific ideas or concepts, or steps in technical procedures in a text, using language that pertains to time, sequence, and cause/effect. Craft and structure • CCSS.ELA-LITERACY.RI.3.4: Determine the meaning of general academic and domain-specific words and phrases in a text relevant to a *grade 3 topic or subject area*. • CCSS.ELA-LITERACY.RI.3.5: Use text features and search tools (e.g., key words, sidebars, hyperlinks) to locate information relevant to a given topic efficiently. • CCSS.ELA-LITERACY.RI.3.6: Distinguish their own point of view from that of the author of a text. Integration of knowledge and ideas • CCSS.ELA-LITERACY.RI.3.7: Use information gained from illustrations (e.g., maps, photographs) and the words in a text to demonstrate understanding of the text (e.g., where, when, why, and how key events occur). • CCSS.ELA-LITERACY.RI.3.8: Describe the logical connection between particular sentences and paragraphs in a text (e.g., comparison, cause/effect, first/second/third in a sequence). • CCSS.ELA-LITERACY.RI.3.9: Compare and contrast the most important points and key details presented in two texts on the same topic. Range of reading and level of text complexity • CCSS.ELA-LITERACY.RI.3.10: By the end of the year, read and comprehend informational texts, including history/social studies, science, and technical texts, at the high end of the grades 2–3 text complexity band independently and proficiently.

Continued

Table 11.2 *(continued)*

CCSS ELA—**Writing**	Text types and purposes
	• CCSS.ELA-LITERACY.W.3.1: Write opinion pieces on topics or texts, supporting a point of view with reasons.
	  o CCSS.ELA-LITERACY.W.3.1.A: Introduce the topic or text they are writing about, state an opinion, and create an organizational structure that lists reasons.
	  o CCSS.ELA-LITERACY.W.3.1.B: Provide reasons that support the opinion.
	  o CCSS.ELA-LITERACY.W.3.1.C: Use linking words and phrases (e.g., *because*, *therefore*, *since, for example*) to connect opinion and reasons.
	  o CCSS.ELA-LITERACY.W.3.1.D: Provide a concluding statement or section.
	• CCSS.ELA-LITERACY.W.3.2: Write informative or explanatory texts to examine a topic and convey ideas and information clearly.
	  o CCSS.ELA-LITERACY.W.3.2.A: Introduce a topic and group related information together; include illustrations when useful to aiding comprehension.
	  o CCSS.ELA-LITERACY.W.3.2.B: Develop the topic with facts, definitions, and details.
	  o CCSS.ELA-LITERACY.W.3.2.C: Use linking words and phrases (e.g., *also*, *another*, *and*, *more*, *but*) to connect ideas within categories of information.
	  o CCSS.ELA-LITERACY.W.3.2.D: Provide a concluding statement or section.
	Production and distribution of writing
	• CCSS.ELA-LITERACY.W.3.4: With guidance and support from adults, produce writing in which the development and organization are appropriate to task and purpose.
	• CCSS.ELA-LITERACY.W.3.5: With guidance and support from peers and adults, develop and strengthen writing as needed by planning, revising, and editing.
	• CCSS.ELA-LITERACY.W.3.6: With guidance and support from adults, use technology to produce and publish writing (using keyboarding skills) as well as to interact and collaborate with others.
	Research to build and present knowledge
	• CCSS.ELA-LITERACY.W.3.8: Recall information from experiences or gather information from print and digital sources; take brief notes on sources and sort evidence into provided categories.
	Range of writing
	• CCSS.ELA-LITERACY.W.3.10: Write routinely over extended time frames (time for research, reflection, and revision) and shorter time frames (a single sitting or a day or two) for a range of discipline-specific tasks, purposes, and audiences.

Continued

Table 11.2 (*continued*)

***CCSS ELA—* Speaking and Listening**	Comprehension and collaboration • CCSS.ELA-LITERACY.SL.3.1: Engage effectively in a range of collaborative discussions (one-on-one, in groups, and teacher-led) with diverse partners on *grade 3 topics and texts*, building on others' ideas and expressing their own clearly. o CCSS.ELA-LITERACY.SL.3.1.A: Come to discussions prepared, having read or studied required material; explicitly draw on that preparation and other information known about the topic to explore ideas under discussion. o CCSS.ELA-LITERACY.SL.3.1.B: Follow agreed-upon rules for discussions (e.g., gaining the floor in respectful ways, listening to others with care, speaking one at a time about the topics and texts under discussion). o CCSS.ELA-LITERACY.SL.3.1.C: Ask questions to check understanding of information presented, stay on topic, and link their comments to the remarks of others. o CCSS.ELA-LITERACY.SL.3.1.D: Explain their own ideas and understanding in light of the discussion. • CCSS.ELA-LITERACY.SL.3.2: Determine the main ideas and supporting details of a text read aloud or information presented in diverse media and formats, including visually, quantitatively, and orally. • CCSS.ELA-LITERACY.SL.3.3: Ask and answer questions about information from a speaker, offering appropriate elaboration and detail. Presentation of knowledge and ideas • CCSS.ELA-LITERACY.SL.3.4: Report on a topic or text, tell a story, or recount an experience with appropriate facts and relevant, descriptive details, speaking clearly at an understandable pace. • CCSS.ELA-LITERACY.SL.3.6: Speak in complete sentences when appropriate to task and situation in order to provide requested detail or clarification.

Continued

Table 11.2 (*continued*)

***CCSS Mathematics*—Measurement and Data**	Solve problems involving measurement and estimation. • CCSS.MATH.CONTENT.3.MD.A.2: Measure and estimate liquid volumes and masses of objects using standard units of grams (g), kilograms (kg), and liters (l). Add, subtract, multiply, or divide to solve one-step word problems involving masses or volumes that are given in the same units, e.g., by using drawings (such as a beaker with a measurement scale) to represent the problem. Represent and interpret data • CCSS.MATH.CONTENT.3.MD.B.3: Draw a scaled picture graph and a scaled bar graph to represent a data set with several categories. Solve one- and two-step "how many more" and "how many less" problems using information presented in scaled bar graphs. • CCSS.MATH.CONTENT.3.MD.B.4: Generate measurement data by measuring lengths using rulers marked with halves and fourths of an inch. Show the data by making a line plot, where the horizontal scale is marked off in appropriate units—whole numbers, halves, or quarters.
ELP Standards	Receptive modalities • ELP 1: Construct meaning from oral presentations and literary and informational text through grade-appropriate listening, reading, and viewing. • ELP 8: Determine the meaning of words and phrases in oral presentations and literary and informational text. Productive modalities • ELP 3: Speak and write about grade-appropriate complex literary and informational texts and topics. • ELP 4: Construct grade-appropriate oral and written claims and support them with reasoning and evidence. • ELP 7: Adapt language choices to purpose, task, and audience when speaking and writing. Interactive modalities • ELP 2: Participate in grade-appropriate oral and written exchanges of information, ideas, and analyses, responding to peer, audience, or reader comments and questions. • ELP 5: Conduct research and evaluate and communicate findings to answer questions or solve problems. • ELP 6: Analyze and critique the arguments of others orally and in writing. Linguistic structures of English • ELP 9: Create clear and coherent grade-appropriate speech and text. • ELP 10: Make accurate use of standard English to communicate in grade-appropriate speech and writing.

Investigation 11

Differences in Traits: How Does Fur Color Affect the Likelihood That a Rabbit Will Survive?

Introduction

Animals and plants are adapted to the ecosystems where they live. They eat specific food and are often hunted by a specific predator that lives in the same ecosystem. They also have traits to help them survive. For example, animals who need to hunt at night will often have large ears and use sound to hunt their prey since it is hard to see in the dark. Other traits of animals may also make them more likely to survive.

Take some time to play with a simulation called *Natural Selection*. Keep track of what you observe and what you are wondering about in the boxes below.

Things I OBSERVED …	Things I WONDER about …

Plants and animals are adapted to survive in their environments. Certain traits may help animals find and eat food, hide from predators, or survive harsh weather. For many animals, predators are a major threat to survival. Because of this, prey animals often have *advantageous* (helpful) traits that make them more difficult to see or to catch. These traits can keep prey animals from being discovered and eaten by predators. Plants can also protect themselves against predators. For example, many plants have sharp thorns that can poke animals that try to eat them. When a plant or animal is able to survive, it can reproduce to make new animals or plants. These new plants or animals will often have the same advantageous traits that their parents had. Over time, a whole species can develop these advantageous traits because the animals with those traits pass them on to their offspring, while animals with harmful or less advantageous traits die and are not able to reproduce.

In this investigation you need to figure out how a rabbit's fur color affects how likely it is to survive. To accomplish this task, you will need to use what you know about advantageous (helpful) and less advantageous (harmful) traits and an online simulation to collect data to determine how well rabbits with different colors of fur (a cause) are able to survive (the effect) in different types of environments. Scientists often look for cause-and-effect relationships like this to better understand how animals are able to survive in a specific ecosystem.

Things we KNOW from what we read …	What we will NEED to figure out …

Your Task

Use what you know about traits, ecosystems, and cause-and-effect relationships to design and carry out an investigation to determine how a rabbit's fur color affects the likelihood that the rabbit will survive.

The *guiding question* of this investigation is, ***How does fur color affect the likelihood that a rabbit will survive?***

Materials

You will use a computer or tablet and an online simulation called *Natural Selection* to conduct your investigation; the simulation is available at *https://phet.colorado.edu/en/simulation/legacy/natural-selection.*

Safety Rules

Follow all normal safety rules.

Plan Your Investigation

Prepare a plan for your investigation by filling out the chart that follows; this plan is called an *investigation proposal.* Before you start developing your plan, be sure to discuss the following questions with the other members of your group:

- What types of **patterns** might we look for to help answer the guiding question?
- What information do we need to find a **cause-and-effect relationship?**

Our guiding question:

We will collect the following data:

These are the steps we will follow to collect data:

I approve of this investigation proposal.

Teacher's signature

Date

Collect Your Data

Keep a record of what you measure or observe during your investigation in the space below.

Analyze Your Data

You will need to analyze the data you collected before you can develop an answer to the guiding question. In the space below, create two graphs. One graph should show how many of each type of rabbit survived in the equatorial environment. The second graph should show how many of each type of rabbit survived in the arctic environment.

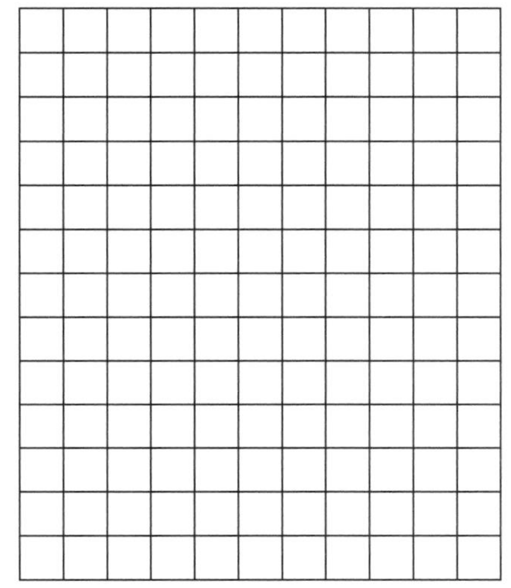

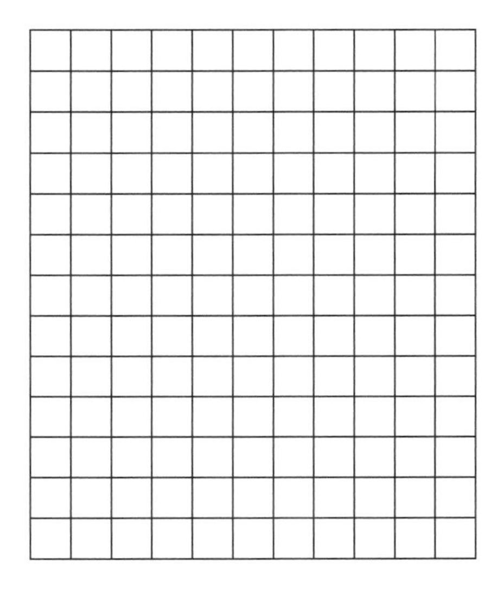

Draft Argument

Develop an argument on a whiteboard. It should include the following parts:

1. A *claim:* Your answer to the guiding question.
2. *Evidence:* An analysis of the data and an explanation of what the analysis means.
3. A *justification of the evidence:* Why your group thinks the evidence is important.

The Guiding Question:

Our Claim:

Our Evidence:

Our Justification of the Evidence:

Argumentation Session

Share your argument with your classmates. Be sure to ask them how to make your draft argument better. Keep track of their suggestions in the space below.

Ways to IMPROVE our argument …

Draft Report

Prepare an *investigation report* to share what you have learned. Use the information in this handout and your group's final argument to write a *draft* of your investigation report.

Introduction

We have been studying ______________________________ in class. Before we started this investigation, we explored ______________________________

____

____

We noticed ______________________________

____

____

My goal for this investigation was to figure out ______________________________

____

____

The guiding question was ______________________________

____

____

Method

To gather the data I needed to answer this question, I ______________________________

____

____

____

____

____

____

I then analyzed the data I collected by ____

____

____

____

Argument

My claim is ____

____

____

____

____

The graphs below show ____

____

____

____

____

____

____

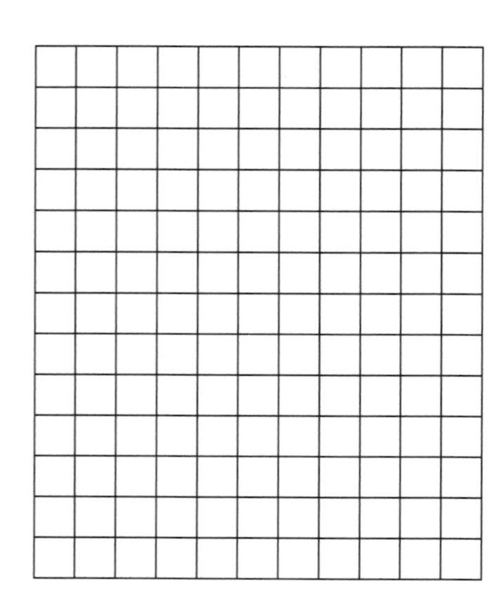

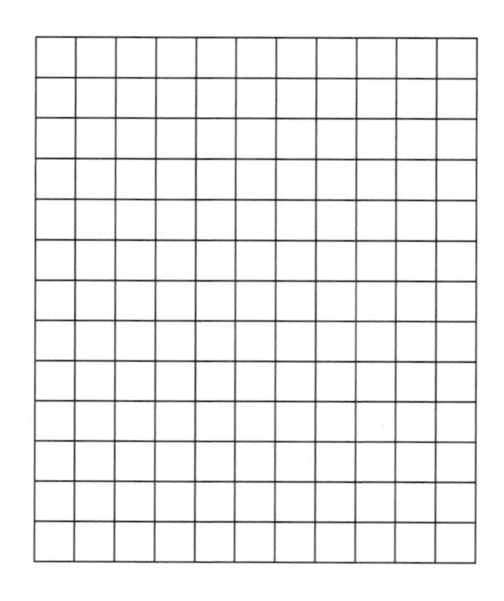

This evidence is important because ____

____

____

____

____

____

____

____

____

____

____

____

____

Review

Your friends need your help! Review the draft of their investigation reports and give them ideas about how to improve. Use the *peer-review guide* when doing your review.

Submit Your Final Report

Once you have received feedback from your friends about your draft report, create your final investigation report and hand it in to your teacher.

Checkout Questions

Investigation 11. Differences in Traits: How Does Fur Color Affect the Likelihood That a Rabbit Will Survive?

The picture below shows a population of mice that live in a sandy environment. Many hawks also live in this environment. Hawks eat mice.

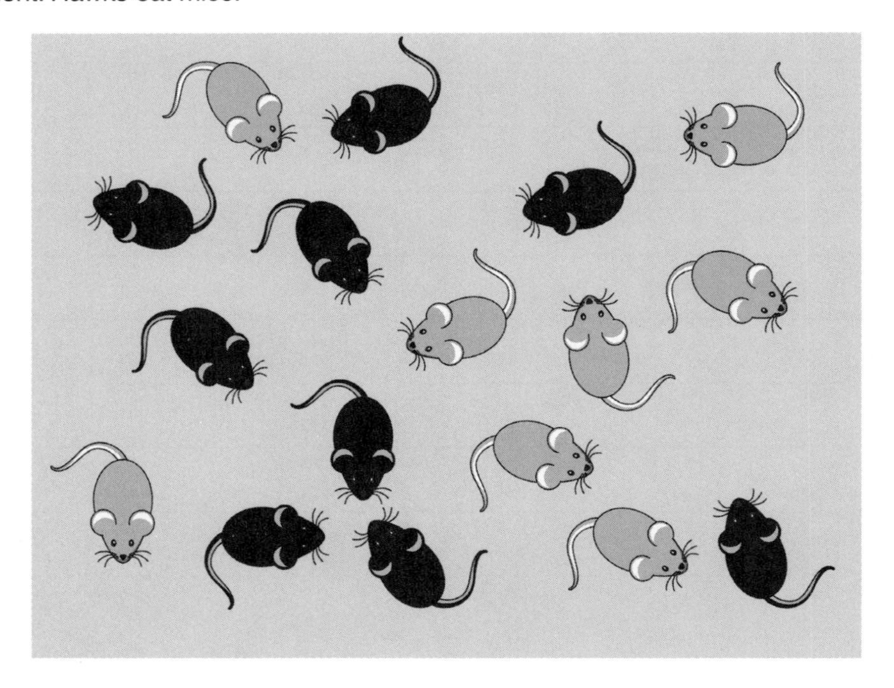

1. Which mice are most likely to become food for the hawks?

   A. Black mice
   B. Brown mice
   C. Both black and brown mice

2. Which fur color is most likely to become more common for mice in this environment over time (after many generations)?

   A. Black mice
   B. Brown mice
   C. Both black and brown mice

Explain your thinking. What *cause-and-effect relationship* did you use to determine what would happen to this population of mice over time?

____

____

____

____

____

____

____

____

Teacher Scoring Rubric for the Checkout Questions

Level	Description
3	The student can apply the core idea correctly in all cases and can fully explain the cause-and-effect relationship.
2	The student can apply the core idea correctly in all cases but cannot fully explain the cause-and-effect relationship.
1	The student cannot apply the core idea correctly in all cases but can fully explain the cause-and-effect relationship.
0	The student cannot apply the core idea correctly in all cases and cannot explain the cause-and-effect relationship.

Investigation 12

Adaptations: Why Do Mammals That Live in the Arctic Ocean Have a Thick Layer of Blubber Under Their Skin?

Purpose

The purpose of this investigation is to give students an opportunity to use the disciplinary core idea (DCI) of LS4.C: Adaptation and the crosscutting concept (CC) of Cause and Effect from *A Framework for K–12 Science Education* (NRC 2012) to figure out why mammals that live in the Arctic Ocean have a thick layer of blubber under their skin. Students will also learn about how scientists use different methods to answer different types of questions during the reflective discussion.

The Disciplinary Core Idea

Students in third grade should understand the following about Adaptation and be able to use this DCI to figure out why mammals that live in the Arctic Ocean have a thick layer of blubber under their skin:

> *Living things can survive only where their needs are met. If some places are too hot or too cold or have too little water or food, plants and animals may not be able to live there. (NRC 2012, p. 165)*

The Crosscutting Concept

Students in third grade should understand the following about the CC Cause and Effect:

> *Repeating patterns in nature, or events that occur together with regularity, are clues that scientists can use to start exploring causal, or cause-and-effect, relationships. … Any application of science, or any engineered solution to a problem, is dependent on understanding the cause-and-effect relationships between events; the quality of the application or solution often can be improved as knowledge of the relevant relationships is improved. (NRC 2012, p. 87)*

Students in third grade should be given opportunities to "begin to look for and analyze patterns—whether in their observations of the world or in the relationships between different quantities in data (e.g., the sizes of plants over time)"; "they can also begin to consider what might be causing these patterns and relationships and design tests that gather more evidence to support or refute their ideas" (NRC 2012, p. 87).

Students should be encouraged to use their developing understanding of cause-and-effect relationships as a tool or a way of thinking about a phenomenon during this investigation to help them figure out why mammals that live in the Arctic Ocean have a thick layer of blubber under their skin.

What Students Figure Out

For any particular environment, some kinds of organisms survive well, some survive less well, and some cannot survive at all. Mammals that live in the Arctic Ocean, such a whales, seals, and walruses, have a thick layer of tissue under their skin called blubber. This tissue is high in fat. Thermal energy does not transfer through blubber easily like it does through tissues such as skin or muscle. This layer of fat therefore acts like an insulator and prevents the mammal from losing too much thermal energy to its surroundings. This layer is an adaptation that allows mammals to live in an environment where other animals cannot survive because it is too cold.

Timeline

The time needed to complete this investigation is 270 minutes (4 hours 30 minutes). The amount of instructional time needed for each stage of the investigation is as follows:

- *Stage 1.* Introduce the task and the guiding question: 35 minutes
- *Stage 2.* Design a method and collect data: 60 minutes
- *Stage 3.* Create a draft argument: 35 minutes
- *Stage 4.* Argumentation session: 30 minutes
- *Stage 5.* Reflective discussion: 15 minutes
- *Stage 6.* Write a draft report: 35 minutes
- *Stage 7.* Peer review: 30 minutes
- *Stage 8.* Revise the report: 30 minutes

This investigation can be completed in one day or over eight days (one day for each stage of ADI) during your designated science time in the daily schedule.

Materials and Preparation

The materials needed for this investigation are listed in Table 12.1 (p. 438). The materials can be purchased from a big-box retail store such as Wal-Mart or Target or through an online retailer such as Amazon. The materials for this investigation can also be purchased as a complete kit (which includes enough materials for 24 students, or six groups) at *www.argumentdriveninquiry.com*.

TABLE 12.1

Materials for Investigation 12

Item	Quantity
Safety goggles	1 per student
Quart-size plastic freezer bags	3 per group
Pint-size plastic freezer bags	3 per group
Vegetable shortening (such as Crisco), 48-ounce can	3 per class
Feathers, 600-count bag	1 per class
Plastic bucket, 2-gallon size	1 per group
Ice	As needed
Thermometer	3 per group
Stopwatch	1 per group
Duct tape	1 roll per class
Whiteboard, 2' × 3'*	1 per group
Investigation Handout	1 per student
Peer-review guide and teacher scoring rubric	1 per student
Checkout Questions (optional)	1 per student

* As an alternative, students can use computer and presentation software such as Microsoft PowerPoint or Apple Keynote to create their arguments.

You will need to make three different plastic-bag gloves for each group before starting this investigation. One plastic-bag glove should be filled with vegetable shortening (to represent blubber), one glove should be filled with feathers (to represent another way to stay warm), and one should be empty (which will serve as the control). Figures 12.1 and 12.2 illustrate how to make the gloves using a 1-quart plastic freezer bag, a 1-pint plastic freezer bag, and some duct tape. To make the glove, add 24 ounces of vegetable shortening (or 100 feathers) to the 1-quart freezer bag. Place the 1-pint freezer bag inside the quart freezer bag so the openings of the two bags are aligned (see Figure 12.2). Use two pieces of duct tape to connect the sides of the 1-pint freezer bag to the sides of the 1-quart freezer bag (see Figure 12.2). Then use two more pieces of duct table to seal the ends of the 1-gallon freezer bag so the vegetable shortening (or feathers) cannot spill out of the glove. Students should be able to put one of their hands inside the 1-pint freezer bag (which is inside the 1-quart freezer bag) without getting vegetable shortening (or feathers) on it. Students can submerge the glove into a bucket of ice water (with a hand inside the glove) to experience the insulating properties of fat, water, and plastic. They can also use a thermometer to monitor how quickly the inside of a glove decreases in temperature when submerged in ice water.

FIGURE 12.1

How to assemble a plastic-bag glove filled with vegetable shortening or feathers (side view)

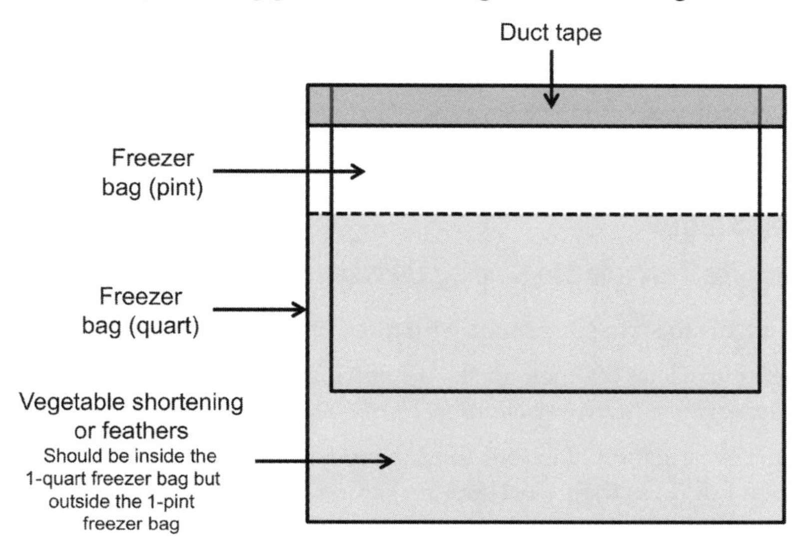

FIGURE 12.2

How to assemble a plastic-bag glove filled with vegetable shortening or feathers (top view)

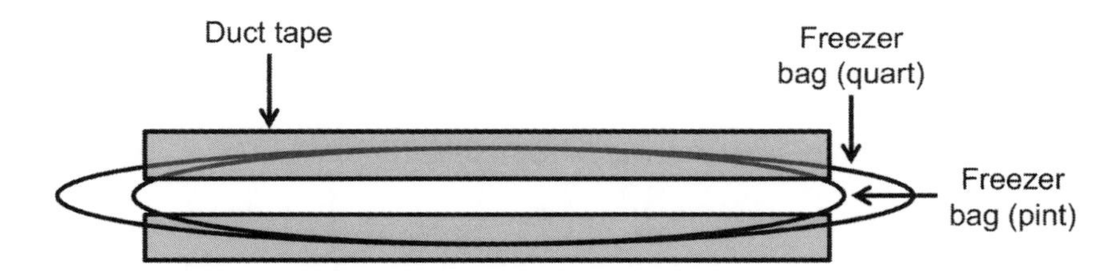

Be sure to use a set routine for distributing and collecting the materials. One option is to set up the materials for each group in a kit that you can deliver to each group. A second option is to have all the materials on a table or cart at a central location. You can then assign a member of each group to be the "materials manager." This individual is responsible for collecting all the materials his or her group needs from the table or cart during class and for returning all the materials at the end of the class.

Safety Precautions

Remind students to follow all normal safety rules. In addition, tell the students to take the following safety precautions:

- Wear sanitized indirectly vented chemical-splash goggles during setup, investigation activity, and cleanup.
- Do not throw objects or put any objects in their mouth.
- Immediately wipe up any slip or fall hazards (such as water).
- Appropriately dispose of materials as directed by the teacher.
- Wash their hands with soap and water when done collecting the data.

Lesson Plan by Stage

Stage 1: Introduce the Task and the Guiding Question (35 minutes)

1. Ask the students to sit in six groups, with three or four students in each group.
2. Ask students to clear off their desks except for a pencil (and their *Student Workbook for Argument-Driven Inquiry in Third-Grade Science* if they have one).
3. Pass out an Investigation Handout to each student (or ask students to turn to Investigation Log 12 in their workbook).
4. Read the first paragraph of the "Introduction" aloud to the class. Ask the students to follow along as you read.
5. Show the video "Arctic Ocean" from National Geographic (available at *https://video.nationalgeographic.com/video/oceans-arctic).*
6. Tell students to watch the video and record their observations and questions in the "OBSERVED/WONDER" chart in the "Introduction" section of their Investigation Handout (or the investigation log in their workbook).
7. Ask students to share *what they observed* in the video.
8. Ask students to share *what questions they have* about the video.
9. Tell the students, "Some of your questions might be answered by reading the rest of the 'Introduction.'"
10. Ask the students to read the rest of the "Introduction" on their own *or* ask them to follow along as you read aloud.
11. Once the students have read the rest of the "Introduction," ask them to fill out the "KNOW/NEED" chart on their Investigation Handout (or in their investigation log) as a group.
12. Ask students to share what they learned from the reading. Add these ideas to a class "know / need to figure out" chart.
13. Ask students to share what they think they will need to figure out based on what they read. Add these ideas to the class "know / need to figure out" chart.
14. Tell the students, "It looks like we have something to figure out. Let's see what we will need to do during our investigation."

15. Read the task and the guiding question aloud.
16. Tell the students, "I have some materials that you can use."
17. Introduce the students to the materials available for them and ask what these materials may be used for. Be sure to explain what is in each plastic-bag glove and what the filling in each glove (shortening, feathers, or air) is intended to represent.
18. Remind students of the safety rules and explain the safety precautions for this investigation.

Stage 2: Design a Method and Collect Data (60 minutes)

1. Tell the students, "I am now going to give you and the other members of your group about 15 minutes to plan your investigation. Before you begin, I want you all to take a couple of minutes to discuss the following questions with the rest of your group."
2. Show the following questions on the screen or board:
   - What information do we need to find a *cause-and-effect relationship?*
   - How might the *structure* of what you are studying relate to its *function?*
3. Tell the students, "Please take a few minutes to come up with an answer to these questions." Give the students two or three minutes to discuss these two questions.
4. Ask two or three different groups to share their answers. Be sure to highlight or write down any important ideas on the board so students can refer to them later.
5. If possible, use a document camera to project an image of the graphic organizer for this investigation on a screen or board (or take a picture of it and project the picture on a screen or board). Tell the students, "I now want you all to plan out your investigation. To do that, you will need to create an investigation proposal by filling out this graphic organizer."
6. Point to the box labeled "Our guiding question:" and tell the students, "You can put the question we are trying to answer in this box." Then ask, "Where can we find the guiding question?"
7. Wait for a student to answer where to find the guiding question (the answer is "in the handout").
8. Point to the box labeled "We will collect the following data:" and tell the students, "You can list the measurements or observations that you will need to collect during the investigation in this box."
9. Point to the box labeled "This is a picture of how we will set up the equipment:" and tell the students, "You can draw a picture of how you plan to set up the equipment during the investigation so you can collect the data you need in this box."

10. Point to the box labeled "These are the steps we will follow to collect data:" and tell the students, "You can list what you are going to do to collect the data you need and what you will do with your data once you have it. Be sure to give enough detail that I could do your investigation for you."
11. Ask the students, "Do you have any questions about what you need to do?"
12. Answer any questions that come up.
13. Tell the students, "Once you are done, raise your hand and let me know. I'll then come by and look over your proposal and give you some feedback. You may not begin collecting data until I have approved your proposal by signing it. You need to have your proposal done in the next 15 minutes."
14. Give the students 15 minutes to work in their groups on their investigation proposal. As they work, move from group to group to check in, ask probing questions, and offer a suggestion if a group gets stuck.

What should a student-designed investigation look like?

The students' investigation proposal should include the following information:

- The guiding question is "Why do mammals that live in the Arctic Ocean have a thick layer of blubber under their skin?"
- The student should collect data on the change in temperature inside the plastic-bag glove.
- To collect the data, the students might use the following procedure:
  1. Place a thermometer inside the plastic-bag glove filled with vegetable shortening.
  2. Record the temperature.
  3. Put the bag (with the thermometer still in it) inside the bucket of ice water.
  4. Wait three minutes.
  5. Measure the temperature inside the bag.
  6. Record the temperature.
  7. Repeat steps 1–6 with other two plastic-bag gloves.
  8. Compare how much the temperature changed in each glove over time.

But there are other ways to design this investigation. Students, for example, can also time how long it takes for the temperature inside the glove to reach a certain temperature. There should be a lot of variation in the student-designed investigations.

15. As each group finishes its investigation proposal, be sure to read it over and determine if it will be productive or not. If you feel the investigation will be productive (not necessarily what you would do or what the other groups are doing), sign your name on the proposal and let the group start collecting data. If the plan needs to be changed, offer some suggestions or ask some probing questions, and have the group make the changes before you approve it.
16. Pass out the materials or have one student from each group collect the materials they need from a central supply table or cart for the groups that have an approved proposal.
17. Remind students of the safety rules and precautions for this investigation.
18. Give the students about 30 minutes to collect data.
19. Tell the students to collect their data and record their observations or measurements in the "Collect Your Data" box in their Investigation Handout (or the investigation log in their workbook).

Stage 3: Create a Draft Argument (35 minutes)

1. Tell the students, "Now that we have all this data, we need to analyze the data so we can figure out an answer to the guiding question."
2. If possible, project an image of the "Analyze Your Data" section for this investigation on a screen or board using a document camera (or take a picture of it and project the picture on a screen or board). Point to the section and tell the students, "You can create a graph as a way to analyze your data. You can make your graph in this section."
3. Ask the students, "What information do we need to include in a graph?"
4. Tell the students, "Please take a few minutes to discuss this question with your group, and be ready to share."
5. Give the students five minutes to discuss.
6. Ask two or three different groups to share their answers. Be sure to highlight or write down any important ideas on the board so students can refer to them later.
7. Tell the students, "I am now going to give you and the other members of your group about 10 minutes to create your graph." If the students are having trouble making a graph, you can take a few minutes to provide a mini-lesson about how to create a graph from a bunch of observations or measurements (this strategy is called just-in-time instruction because it is offered only when students get stuck).

What should a graph look like for this investigation?

There are a number of different ways that students can analyze the observations or measurements they collect during this investigation. One of the most straightforward ways is to create a scaled bar graph. This bar graph should have categories for each type of glove (filled with vegetable shortening, filled with feathers, filled with air) on the horizontal or *x*-axis. The dependent variable (how much the temperature changed or amount of time for the temperature to change by a certain amount) should be on the vertical or *y*-axis. An example of this type of graph can be seen in Figure 12.3 (the students who created this graph used water instead of feathers). There are other options for analyzing the data they collected. Students often come up with some unique ways of analyzing their data, so be sure to give them some voice and choice during this stage.

8. Give the students 10 minutes to analyze their data. As they work, move from group to group to check in, ask probing questions, and offer suggestions.
9. Tell the students, "I am now going to give you and the other members of your group 15 minutes to create an argument to share what you have learned and convince others that they should believe you. Before you do that, we need to take a few minutes to discuss what you need to include in your argument."
10. If possible, use a document camera to project the "Argument Presentation on a Whiteboard" image from the Investigation Handout (or the investigation log in their workbook) on a screen or board (or take a picture of it and project the picture on a screen or board).
11. Point to the box labeled "The Guiding Question:" and tell the students, "You can put the question we are trying to answer here on your whiteboard."
12. Point to the box labeled "Our Claim:" and tell the students, "You can put your claim here on your whiteboard. The claim is your answer to the guiding question."
13. Point to the box labeled "Our Evidence:" and tell the students, "You can put the evidence that you are using to support your claim here on your whiteboard. Your evidence will need to include the analysis you just did and an explanation of what your analysis means or shows. Scientists always need to support their claims with evidence."
14. Point to the box labeled "Our Justification of the Evidence:" and tell the students, "You can put your justification of your evidence here on your whiteboard. Your justification needs to explain why your evidence is important. Scientists often

use core ideas to explain why the evidence they are using matters. Core ideas are important concepts that scientists use to help them make sense of what happens during an investigation."

15. Ask the students, "What are some core ideas that we read about earlier that might help us explain why the evidence we are using is important?"
16. Ask students to share some of the core ideas from the "Introduction" section of the Investigation Handout (or the investigation log in the workbook). List these core ideas on the board.
17. Tell the students, "That is great. I would like to see everyone try to include these core ideas in your justification of the evidence. Your goal is to use these core ideas to help explain why your evidence matters and why the rest of us should pay attention to it."
18. Ask the students, "Do you have any questions about what you need to do?"
19. Answer any questions that come up.
20. Tell the students, "Okay, go ahead and start working on your arguments. You need to have your argument done in the next 15 minutes. It doesn't need to be perfect. We just need something down on the whiteboards so we can share our ideas."
21. Give the students 15 minutes to work in their groups on their arguments. As they work, move from group to group to check in, ask probing questions, and offer a suggestion if a group gets stuck. Figure 12.3 shows an example of an argument created by students for this investigation.

FIGURE 12.3

Example of an argument

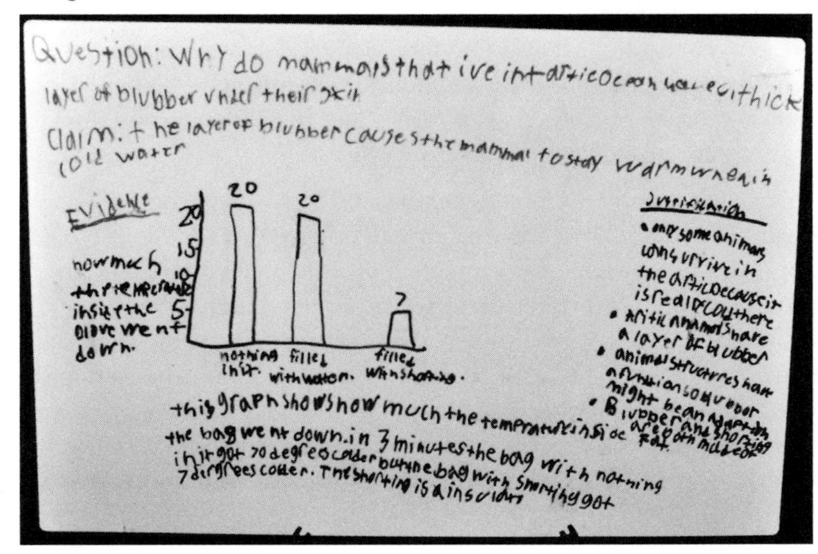

Stage 4: Argumentation Session (30 minutes)

The argumentation session can be conducted in a whole-class presentation format, a gallery walk format, or a modified gallery walk format. We recommend using a whole-class presentation format for the first investigation, but try to transition to either the gallery walk or modified gallery walk format as soon as possible because that will maximize student voice and choice inside the classroom. The following list shows the steps for the three formats; unless otherwise noted, the steps are the same for all three formats.

1. Begin by introducing the use of the whiteboard.
   - *If using the whole-class presentation format,* tell the students, "We are now going to share our arguments. Please set up your whiteboard so everyone can see them."
   - *If using the gallery walk or modified gallery walk format,* tell the students, "We are now going to share our arguments. Please set up your whiteboard so they are facing the walls."
2. Allow the students to set up their whiteboards.
   - *If using the whole-class presentation format,* the whiteboards should be set up on stands or chairs so they are facing toward the center of the room.
   - *If using the gallery walk or modified gallery walk format,* the whiteboards should be set up on stands or chairs so they are facing toward the outside of the room.
3. Give the following instructions to the students:
   - *If using the whole-class presentation format or the modified gallery walk format,* tell the students, "Okay, before we get started I want to explain what we are going to do next. I'm going to ask some of you to present your arguments to your classmates. If you are presenting your argument, your job is to share your group's claim, evidence, and justification of the evidence. The rest of you will be reviewers. If you are a reviewer, your job is to listen to the presenters, ask the presenters questions if you do not understand something, and then offer them some suggestions about ways to make their argument better. After we have a chance to learn from each other, I'm going to give you some time to revise your arguments and make them better."
   - *If using the gallery walk format,* tell the students, "Okay, before we get started I want to explain what we are going to do next. You are going to have an opportunity to read the arguments that were created by other groups. Your group will go to a different group's argument. I'll give you a few minutes to read it and review it. Your job is to offer them some suggestions about ways to make their argument better. You can use sticky notes to give them suggestions. Please be specific about what you want to change and be specific about how you think they should change it. After we have a chance to learn from each other, I'm going to give you some time to revise your arguments and make them better."

4. Use a document camera to project the "Ways to IMPROVE our argument …" box from the Investigation Handout (or the investigation log in their workbook) on a screen or board (or take a picture of it and project the picture on a screen or board).
   - *If using the whole-class presentation format or the modified gallery walk format,* point to the box and tell the students, "If you are a presenter, you can write down the suggestions you get from the reviewers here. If you are a reviewer, and you see a good idea from another group, you can write down that idea here. Once we are done with the presentations, I will give you a chance to use these suggestions or ideas to improve your arguments.
   - *If using the gallery walk format,* point to the box and tell the students, "If you see good ideas from another group, you can write them down here. Once we are done reviewing the different arguments, I will give you a chance to use these ideas to improve your own arguments. It is important to share ideas like this."

   Ask the students, "Do you have any questions about what you need to do?"
5. Answer any questions that come up.
6. Give the following instructions:
   - *If using the whole-class presentation format,* tell the students, "Okay. Let's get started."
   - *If using the gallery walk format,* tell the students, "Okay, I'm now going to tell you which argument to go to and review.
   - *If using the modified gallery walk format,* tell the students, "Okay, I'm now going to assign you to be a presenter or a reviewer." Assign one or two students from each group to be presenters and one or two students from each group to be reviewers.
7. Begin the review of the arguments.
   - *If using the whole-class presentation format,* have four or five groups present their argument one at a time. Give each group only two to three minutes to present their argument. Then give the class two to three minutes to ask them questions and offer suggestions. Be sure to encourage as much participation from the students as possible.
   - *If using the gallery walk format,* tell the students, "Okay. Let's get started. Each group, move one argument to the left. Don't move to the next argument until I tell you to move. Once you get there, read the argument and then offer suggestions about how to make it better. I will put some sticky notes next to each argument. You can use the sticky notes to leave your suggestions." Give each group about three to four minutes to read the arguments, talk, and offer suggestions.

     a. Tell the students, "Okay. Let's rotate. Move one group to the left."

b. Again, give each group three or four minutes to read, talk, and offer suggestions.

c. Repeat this process for two more rotations.

- *If using the modified gallery walk format,* tell the students, "Okay. Let's get started. Reviewers, move one group to the left. Don't move to the next group until I tell you to move. Presenters, go ahead and share your argument with the reviewers when they get there." Give each group of presenters and reviewers about three to four minutes to talk.

  a. Tell the students, "Okay. Let's rotate. Reviewers, move one group to the left."

  b. Again, give each group of presenters and reviewers about three or four minutes to talk.

  c. Repeat this process for two more rotations.

8. Tell the students to return to their workstations.
9. Give the following instructions about revising the argument:
   - *If using the whole-class presentation format,* tell the students, "I'm now going to give you about 10 minutes to revise your argument. Take a few minutes to talk in your groups and determine what you want to change to make your argument better. Once you have decided what to change, go ahead and make the changes to your whiteboard."
   - *If using the gallery walk format,* tell the students, "I'm now going to give you about 10 minutes to revise your argument. Take a few minutes to read the suggestions that were left at your argument. Then talk in your groups and determine what you want to change to make your argument better. Once you have decided what to change, go ahead and make the changes to your whiteboard."
   - *If using the modified gallery walk format,* "I'm now going to give you about 10 minutes to revise your argument. Please return to your original groups." Wait for the students to move back into their original groups and then tell the students, "Okay, take a few minutes to talk in your groups and determine what you want to change to make your argument better. Once you have decided what to change, go ahead and make the changes to your whiteboard."

   Ask the students, "Do you have any questions about what you need to do?"
10. Answer any questions that come up.
11. Tell the students, "Okay. Let's get started."
12. Give the students 10 minutes to work in their groups on their arguments. As they work, move from group to group to check in, ask probing questions, and offer a suggestion if a group gets stuck.

Stage 5: Reflective Discussion (15 minutes)

1. Tell the students, "We are now going to take a minute to talk about what we did and what we have learned."
2. Show Figure 12.4 on a screen. This is an image of two polar bears in the Arctic. Ask the students, "What do you all see going on here?"

FIGURE 12.4

Polar bears in the Arctic

A full-color version of this figure is available on the book's Extras page at *www.nsta.org/adi-3rd*.

3. Allow students to share their ideas.
4. Ask the students, "What types of adaptations allow these animals to survive in this type of environment?"
5. Allow students to share their ideas.
6. Tell the students, "Okay, let's make sure we are on the same page. For any particular environment, some kinds of animals and plants survive well, some survive less well, and some cannot survive at all. Mammals that live in the Arctic, such as whales or polar bears, have a thick layer of tissue under their skin called blubber. This tissue is high in fat. Thermal energy does not transfer through blubber easily like it does through tissues such as skin or muscle. This layer of fat therefore acts like an insulator and prevents the mammal from losing too much thermal energy to its surroundings. This layer is an adaptation that allows mammals to live in an environment where other animals cannot

survive because it is too cold. The fact that all animals and plants have traits that help them survive in a particular environment is a really important core idea in science."

7. Ask the students, "Does anyone have any questions about this core idea?"
8. Answer any questions that come up.
9. Tell the students, "We also looked for a cause-and-effect relationship during our investigation." Then ask, "Can anyone tell me why it is useful to look for a cause-and-effect relationship?"
10. Allow students to share their ideas.
11. Tell the students, "Cause-and-effect relationships are important because they allow us to predict what will happen in the future."
12. Ask the students, "What was the cause and what was the effect that we uncovered today?"
13. Allow students to share their ideas. Keep probing until someone mentions that the cause was what was inside the glove and the effect was the change in temperature.
14. Tell the students, "That is great, and if we know that we can predict what will happen the next time we see an animal that has a layer of blubber under its skin."
15. Tell the students, "We also needed to think about how structure and function are related to each other during our investigation." Then ask, "Can anyone tell me why it is useful to think about the relationship between structure and function?"
16. Allow students to share their ideas.
17. Tell the students, "Plants and animals have structures that have a specific function that allow them to live in a specific type of environment. For example, fish have fins that allow them to swim. Horses have legs that allow them to walk."
18. Ask the students, "What were some other structures of animals that we looked at today, and how did learning about these structures allow us to answer our guiding question?"
19. Allow students to share their ideas. Keep probing until someone mentions that the animals have structures that allow them to live in a specific type of environment.
20. Tell the students, "That is great, and if we know that the structures of a living thing have specific functions, we can try to determine the function of a structure we see when we don't know much about it. For example, if we know that a lot of mammals that live in cold environments have a thick layer of fat under their skin, we can assume that the layer of fat has a function. We can then develop

a possible explanation for the function of that structure, such as that it helps keeps the animal warm. We can then test the possible explanation to see if is supported by our test or not. This is what we did during this investigation."

21. Ask the students, "Can anyone give me another example of the relationship between structure and function?"
22. Allow students to share their ideas.
23. Tell the students, "We are now going take a minute to talk about what went well and what didn't go so well during our investigation. We need to talk about this because you are going to be planning and carrying out your own investigations like this a lot this year, and I want to help you all get better at it."
24. Show an image of the question "What made your investigation scientific?" on the screen. Tell the students, "Take a few minutes to talk about how you would answer this question with the other people in your group. Be ready to share with the rest of the class." Give the students two to three minutes to talk in their group.
25. Ask the students, "What do you all think? Who would like to share an idea?"
26. Allow students to share their ideas. Be sure to expand on their ideas about what makes an investigation scientific.
27. Show an image of the question "What made your investigation not so scientific?" on the screen. Tell the students, "Take a few minutes to talk about how you would answer this question with the other people in your group. Be ready to share with the rest of the class." Give the students two to three minutes to talk in their group.
28. Ask the students, "What do you all think? Who would like to share an idea?"
29. Allow students to share their ideas. Be sure to expand on their ideas about what makes an investigation less scientific.
30. Show an image of the question "What rules can we put into place to help us make sure our next investigation is more scientific?" on the screen. Tell the students, "Take a few minutes to talk about how you would answer this question with the other people in your group. Be ready to share with the rest of the class." Give the students two to three minutes to talk in their group.
31. Ask the students, "What do you all think? Who would like to share an idea?"
32. Allow students to share their ideas. Once they have shared their ideas, offer a suggestion for a possible class rule.
33. Ask the students, "What do you all think? Should we make this a rule?"
34. If the students agree, write the rule on the board or make a class "Rules for Scientific Investigation" chart so you can refer to it during the next investigation.

35. Tell the students, "We are now going take a minute to talk about what scientists do to investigate the natural world."
36. Show an image of the question "Do all scientists follow the same method?" on the screen. Tell the students, "Take a few minutes to talk about how you would answer this question with the other people in your group. Be ready to share with the rest of the class." Give the students two to three minutes to talk in their group.
37. Ask the students, "What do you all think? Who would like to share an idea?"
38. Allow students to share their ideas.
39. Tell the students, "Okay, let's make sure we are all on the same page. Scientists use lots of different methods to answer different types of questions. Sometimes they need to go out into the field and watch what animals do. The video you watched came from scientists working out in the field. Some scientists design experiments, and others analyze data collected by other scientists. There is no one method that all scientists use, and the method used by scientists depends on what they are studying and what type of question they are asking. This is an important thing to understand about science."
40. Ask the students, "Does anyone have any questions about what scientists do to investigate the natural world?"
41. Answer any questions that come up.

Stage 6: Write a Draft Report (35 minutes)

Your students will use either the Investigation Handout or the investigation log in the student workbook when writing the draft report. When you give the directions shown in quotes in the following steps, substitute "investigation log" (as shown in brackets) for "handout" if they are using the workbook.

1. Tell the students, "You are now going to write an investigation report to share what you have learned. Please take out a pencil and turn to the 'Draft Report' section of your handout [investigation log]."
2. If possible, use a document camera to project the "Introduction" section of the draft report from the Investigation Handout (or the investigation log in their workbook) on a screen or board (or take a picture of it and project the picture on a screen or board).
3. Tell the students, "The first part of the report is called the 'Introduction.' In this section of the report you want to explain to the reader what you were investigating, why you were investigating it, and what question you were trying to answer. All of this information can be found in the text at the beginning of your handout [investigation log]." Point to the image and say, "There are some sentence starters here to help you begin writing the report." Ask the students, "Do you have any questions about what you need to do?"

4. Answer any questions that come up.
5. Tell the students, "Okay. Let's write."
6. Give the students 10 minutes to write the "Introduction" section of the report. As they work, move from student to student to check in, ask probing questions, and offer a suggestion if a student gets stuck.
7. If possible, use a document camera to project the "Method" section of the draft report from the Investigation Handout (or the investigation log in their workbook) on a screen or board (or take a picture of it and project the picture on a screen or board).
8. Tell the students, "The second part of the report is called the 'Method.' In this section of the report you want to explain to the reader what you did during the investigation, what data you collected and why, and how you went about analyzing your data. All of this information can be found in the 'Plan Your Investigation' section of your handout [investigation log]. Remember that you all planned and carried out different investigations, so do not assume that the reader will know what you did." Point to the image and say, "There are some sentence starters here to help you begin writing this part of the report." Ask the students, "Do you have any questions about what you need to do?"
9. Answer any questions that come up.
10. Tell the students, "Okay. Let's write."
11. Give the students 10 minutes to write the "Method" section of the report. As they work, move from student to student to check in, ask probing questions, and offer a suggestion if a student gets stuck.
12. If possible, use a document camera to project the "Argument" section of the draft report from the Investigation Handout (or the investigation log in their workbook) on a screen or board (or take a picture of it and project the picture on a screen or board).
13. Tell the students, "The last part of the report is called the 'Argument.' In this section of the report you want to share your claim, evidence, and justification of the evidence with the reader. All of this information can be found on your whiteboard." Point to the image and say, "There are some sentence starters here to help you begin writing this part of the report." Ask the students, "Do you have any questions about what you need to do?"
14. Answer any questions that come up.
15. Tell the students, "Okay. Let's write."
16. Give the students 10 minutes to write the "Argument" section of the report. As they work, move from student to student to check in, ask probing questions, and offer a suggestion if a student gets stuck.

Stage 7: Peer Review (30 minutes)

Your students will use either the Investigation Handout or the investigation log in the student workbook when doing the peer review. When you give the directions shown in quotes in the following steps, substitute "workbook" (as shown in brackets) for "Investigation Handout" if they are using the workbook.

1. Tell the students, "We are now going to review our reports to find ways to make them better. I'm going to come around and collect your Investigation Handout [workbook]. While I do that, please take out a pencil."
2. Collect the Investigation Handouts or workbooks from the students.
3. If possible, use a document camera to project the peer-review guide (PRG; see Appendix 4) on a screen or board (or take a picture of it and project the picture on a screen or board).
4. Tell the students, "We are going to use this peer-review guide to give each other feedback." Point to the image.
5. Give the following instructions:
   - *If using the Investigation Handout,* tell the students, "I'm going to ask you to work with a partner to do this. I'm going to give you and your partner a draft report to read and a peer-review guide to fill out. You two will then read the report together. Once you are done reading the report, I want you to answer each of the questions on the peer-review guide." Point to the review questions on the image of the PRG.
   - *If using the workbook,* tell the students, "I'm going to ask you to work with a partner to do this. I'm going to give you and your partner a draft report to read. You two will then read the report together. Once you are done reading the report, I want you to answer each of the questions on the peer-review guide that is right after the report in the investigation log." Point to the review questions on the image of the PRG.
6. Tell the students, "You can check 'yes,' 'almost,' or 'no' after each question." Point to the checkboxes on the image of the PRG.
7. Tell the students, "This will be your rating for this part of the report. Make sure you agree on the rating you give the author. If you mark 'almost' or 'no,' then you need to tell the author what he or she needs to do to get a 'yes.'" Point to the space for the reviewer feedback on the image of the PRG.
8. Tell the students, "It is really important for you to give the authors feedback that is helpful. That means you need to tell them exactly what they need to do to make their reports better." Ask the students, "Do you have any questions about what you need to do?"
9. Answer any questions that come up.

10. Tell the students, "Please sit with a partner who is not in your current group." Allow the students time to sit with a partner.
11. Give the following instructions:
    - *If using the Investigation Handout,* tell the students, "Okay, I am now going to give you one report to read and one peer-review guide to fill out." Pass out one report to each pair. Make sure that the report you give a pair was not written by one of the students in that pair. Give each pair one PRG to fill out as a team.
    - *If using the workbook,* tell the students, "Okay, I am now going to give you one report to read." Pass out a workbook to each pair. Make sure that the workbook you give a pair is not from one of the students in that pair.
12. Tell the students, "Okay, I'm going to give you 15 minutes to read the report I gave you and to fill out the peer-review guide. Go ahead and get started."
13. Give the students 15 minutes to work. As they work, move around from pair to pair to check in and see how things are going, answer questions, and offer advice.
14. After 15 minutes pass, tell the students, "Okay, time is up." *If using the Investigation Handout,* say, "Please give me the report and the peer-review guide that you filled out." *If using the workbook,* say, "Please give me the workbook that you have."
15. Collect the Investigation Handouts and the PRGs, or collect the workbooks if they are being used. Be sure you keep the handout and the PRG together.
16. Give the following instructions:
    - *If using the Investigation Handout,* tell the students, "Okay, I am now going to give you a different report to read and a new peer-review guide to fill out." Pass out another report to each pair. Make sure that this report was not written by one of the students in that pair. Give each pair a new PRG to fill out as a team.
    - *If using the workbook,* tell the students, "Okay, I am now going to give you a different report to read." Pass out a different workbook to each pair. Make sure that the workbook you give a pair is not from one of the students in that pair.
17. Tell the students, "Okay, I'm going to give you 15 minutes to read this new report and to fill out the peer-review guide. Go ahead and get started."
18. Give the students 15 minutes to work. As they work, move around from pair to pair to check in and see how things are going, answer questions, and offer advice.
19. After 15 minutes pass, tell the students, "Okay, time is up." *If using the Investigation Handout,* say, "Please give me the report and the peer-review guide that you filled out." *If using the workbook,* say, "Please give me the workbook that you have."
20. Collect the Investigation Handouts and the PRGs, or collect the workbooks if they are being used. Be sure you keep the handout and the PRG together.

Stage 8: Revise the Report (30 minutes)

Your students will use either the Investigation Handout or the investigation log in the student workbook when revising the report. Except where noted below, the directions are the same whether using the handout or the log.

1. Give the following instructions:
   - *If using the Investigation Handout,* tell the students, "You are now going to revise your investigation report based on the feedback you get from your classmates. Please take out a pencil while I hand back your draft report and the peer-review guide."
   - *If using the investigation log in the student workbook,* tell the students, "You are now going to revise your investigation report based on the feedback you get from your classmates. Please take out a pencil while I hand back your investigation logs."
2. *If using the Investigation Handout,* pass back the handout and the PRG to each student. *If using the investigation log,* pass back the log to each student.
3. Tell the students, "Please take a few minutes to read over the peer-review guide. You should use it to figure out what you need to change in your report and how you will change the report."
4. Allow the students time to read the PRG.
5. *If using the investigation log,* if possible use a document camera to project the "Write Your Final Report" section from the investigation log on a screen or board (or take a picture of it and project the picture on a screen or board).
6. Give the following instructions:
   - *If using the Investigation Handout,* tell the students, "Okay. Let's revise our reports. Please take out a piece of paper. I would like you to rewrite your report. You can use your draft report as a starting point, but use the feedback on the peer-review guide to help make it better."
   - *If using the investigation log,* tell the students, "Okay. Let's revise our reports. I would like you to rewrite your report in the section of the investigation log that says 'Write Your Final Report.'" Point to the image on the screen and tell the students, "You can use your draft report as a starting point, but use the feedback on the peer-review guide to help make your report better."

   Ask the students, "Do you have any questions about what you need to do?"
7. Answer any questions that come up.
8. Tell the students, "Okay. Let's write."
9. Give the students 30 minutes to rewrite their report. As they work, move from student to student to check in, ask probing questions, and offer a suggestion if a student gets stuck.

10. Give the following instructions:
    - *If using the Investigation Handout,* tell the students, "Okay. Time's up. I will now come around and collect your Investigation Handout, the peer-review guide, and your final report."
    - *If using the investigation log,* tell the students, "Okay. Time's up. I will now come around and collect your workbooks."
11. *If using the Investigation Handout,* collect all the Investigation Handouts, PRGs, and final reports. *If using the investigation log,* collect all the workbooks.
12. *If using the Investigation Handout,* use the "Teacher Score" columns in the PRG to grade the final report. *If using the investigation log,* use the "ADI Investigation Report Grading Rubric" in the investigation log to grade the final report. Whether you are using the handout or the log, you can give the students feedback about their writing in the "Teacher Comments" section.

How to Use the Checkout Questions

The Checkout Questions are an optional assessment. We recommend giving them to students one day after they finish stage 8 of the ADI investigation. The Checkout Questions can be used as a formative or summative assessment of student thinking. If you plan to use them as a formative assessment, we recommend that you look over the student answers to determine if you need to reteach the core idea and/or crosscutting concept from the investigation, but do not grade them. If you plan to use them as a summative assessment, we have included a "Teacher Scoring Rubric" at the end of the Checkout Questions that you can use to score a student's ability to apply the core idea in a new scenario and explain their use of a crosscutting concept. The rubric includes a 4-point scale that ranges from 0 (the student cannot apply the core idea correctly in all cases and cannot explain the [crosscutting concept]) to 3 (the student can apply the core idea correctly in all cases and can fully explain the [crosscutting concept]). The Checkout Questions, regardless of how you decide to use them, are a great way to make student thinking visible so you can determine if the students have learned the core idea and the crosscutting concept.

A student who can apply the core idea correctly in all cases and can explain the relationship between structure and function would select musk ox for question 1. He or she should then be able to explain that animals have specific structures that allows them to live (or function) in specific environments for question 2.

Connections to Standards

Table 12.2 (p. 458) highlights how the investigation can be used to address specific performance expectations from the *NGSS, Common Core State Standards (CCSS)* in English language arts (ELA) and in mathematics, and *English Language Proficiency (ELP) Standards.*

TABLE 12.2

Investigation 12 alignment with standards

***NGSS* performance expectations**	Strong alignment • 3-LS4-3: Construct an argument with evidence that in a particular habitat some organisms can survive well, some survive less well, and some cannot survive at all. Moderate alignment • 3-LS4-4: Make a claim about the merit of a solution to a problem caused when the environment changes and the types of plants and animals that live there may change.
CCSS ELA—Reading: Informational Text	Key ideas and details • CCSS.ELA-LITERACY.RI.3.1: Ask and answer questions to demonstrate understanding of a text, referring explicitly to the text as the basis for the answers. • CCSS.ELA-LITERACY.RI.3.2: Determine the main idea of a text; recount the key details and explain how they support the main idea. • CCSS.ELA-LITERACY.RI.3.3: Describe the relationship between a series of historical events, scientific ideas or concepts, or steps in technical procedures in a text, using language that pertains to time, sequence, and cause/effect. Craft and structure • CCSS.ELA-LITERACY.RI.3.4: Determine the meaning of general academic and domain-specific words and phrases in a text relevant to a *grade 3 topic or subject area*. • CCSS.ELA-LITERACY.RI.3.5: Use text features and search tools (e.g., key words, sidebars, hyperlinks) to locate information relevant to a given topic efficiently. • CCSS.ELA-LITERACY.RI.3.6: Distinguish their own point of view from that of the author of a text. Integration of knowledge and ideas • CCSS.ELA-LITERACY.RI.3.7: Use information gained from illustrations (e.g., maps, photographs) and the words in a text to demonstrate understanding of the text (e.g., where, when, why, and how key events occur). • CCSS.ELA-LITERACY.RI.3.8: Describe the logical connection between particular sentences and paragraphs in a text (e.g., comparison, cause/effect, first/second/third in a sequence). • CCSS.ELA-LITERACY.RI.3.9: Compare and contrast the most important points and key details presented in two texts on the same topic. Range of reading and level of text complexity • CCSS.ELA-LITERACY.RI.3.10: By the end of the year, read and comprehend informational texts, including history/social studies, science, and technical texts, at the high end of the grades 2–3 text complexity band independently and proficiently.

Continued

Table 12.2 (*continued*)

***CCSS ELA*—Writing**	Text types and purposes
	• CCSS.ELA-LITERACY.W.3.1: Write opinion pieces on topics or texts, supporting a point of view with reasons.
	  o CCSS.ELA-LITERACY.W.3.1.A: Introduce the topic or text they are writing about, state an opinion, and create an organizational structure that lists reasons.
	  o CCSS.ELA-LITERACY.W.3.1.B: Provide reasons that support the opinion.
	  o CCSS.ELA-LITERACY.W.3.1.C: Use linking words and phrases (e.g., *because*, *therefore*, *since, for example*) to connect opinion and reasons.
	  o CCSS.ELA-LITERACY.W.3.1.D: Provide a concluding statement or section.
	• CCSS.ELA-LITERACY.W.3.2: Write informative or explanatory texts to examine a topic and convey ideas and information clearly.
	  o CCSS.ELA-LITERACY.W.3.2.A: Introduce a topic and group related information together; include illustrations when useful to aiding comprehension.
	  o CCSS.ELA-LITERACY.W.3.2.B: Develop the topic with facts, definitions, and details.
	  o CCSS.ELA-LITERACY.W.3.2.C: Use linking words and phrases (e.g., *also*, *another*, *and*, *more*, *but*) to connect ideas within categories of information.
	  o CCSS.ELA-LITERACY.W.3.2.D: Provide a concluding statement or section.
	Production and distribution of writing
	• CCSS.ELA-LITERACY.W.3.4: With guidance and support from adults, produce writing in which the development and organization are appropriate to task and purpose.
	• CCSS.ELA-LITERACY.W.3.5: With guidance and support from peers and adults, develop and strengthen writing as needed by planning, revising, and editing.
	• CCSS.ELA-LITERACY.W.3.6: With guidance and support from adults, use technology to produce and publish writing (using keyboarding skills) as well as to interact and collaborate with others.
	Research to build and present knowledge
	• CCSS.ELA-LITERACY.W.3.8: Recall information from experiences or gather information from print and digital sources; take brief notes on sources and sort evidence into provided categories.
	Range of writing
	• CCSS.ELA-LITERACY.W.3.10: Write routinely over extended time frames (time for research, reflection, and revision) and shorter time frames (a single sitting or a day or two) for a range of discipline-specific tasks, purposes, and audiences.

Continued

Table 12.2 (*continued*)

***CCSS ELA—* Speaking and Listening**	Comprehension and collaboration • CCSS.ELA-LITERACY.SL.3.1: Engage effectively in a range of collaborative discussions (one-on-one, in groups, and teacher-led) with diverse partners on *grade 3 topics and texts*, building on others' ideas and expressing their own clearly. o CCSS.ELA-LITERACY.SL.3.1.A: Come to discussions prepared, having read or studied required material; explicitly draw on that preparation and other information known about the topic to explore ideas under discussion. o CCSS.ELA-LITERACY.SL.3.1.B: Follow agreed-upon rules for discussions (e.g., gaining the floor in respectful ways, listening to others with care, speaking one at a time about the topics and texts under discussion). o CCSS.ELA-LITERACY.SL.3.1.C: Ask questions to check understanding of information presented, stay on topic, and link their comments to the remarks of others. o CCSS.ELA-LITERACY.SL.3.1.D: Explain their own ideas and understanding in light of the discussion. • CCSS.ELA-LITERACY.SL.3.2: Determine the main ideas and supporting details of a text read aloud or information presented in diverse media and formats, including visually, quantitatively, and orally. • CCSS.ELA-LITERACY.SL.3.3: Ask and answer questions about information from a speaker, offering appropriate elaboration and detail. Presentation of knowledge and ideas • CCSS.ELA-LITERACY.SL.3.4: Report on a topic or text, tell a story, or recount an experience with appropriate facts and relevant, descriptive details, speaking clearly at an understandable pace. • CCSS.ELA-LITERACY.SL.3.6: Speak in complete sentences when appropriate to task and situation in order to provide requested detail or clarification.
***CCSS Mathematics—* Number and Operations in Base Ten**	Use place value understanding and properties of operations to perform multi-digit arithmetic. • CCSS.MATH.CONTENT.3.NBT.A.1: Use place value understanding to round whole numbers to the nearest 10 or 100. • CCSS.MATH.CONTENT.3.NBT.A.2: Fluently add and subtract within 1,000 using strategies and algorithms based on place value, properties of operations, and/or the relationship between addition and subtraction.

Continued

Table 12.2 *(continued)*

***CCSS Mathematics—* Measurement and Data**	Solve problems involving measurement and estimation. • CCSS.MATH.CONTENT.3.MD.A.1: Tell and write time to the nearest minute and measure time intervals in minutes. Solve word problems involving addition and subtraction of time intervals in minutes. Represent and interpret data. • CCSS.MATH.CONTENT.3.MD.B.3: Draw a scaled picture graph and a scaled bar graph to represent a data set with several categories. Solve one- and two-step "how many more" and "how many less" problems using information presented in scaled bar graphs.
ELP Standards	Receptive modalities • ELP 1: Construct meaning from oral presentations and literary and informational text through grade-appropriate listening, reading, and viewing. • ELP 8: Determine the meaning of words and phrases in oral presentations and literary and informational text. Productive modalities • ELP 3: Speak and write about grade-appropriate complex literary and informational texts and topics. • ELP 4: Construct grade-appropriate oral and written claims and support them with reasoning and evidence. • ELP 7: Adapt language choices to purpose, task, and audience when speaking and writing. Interactive modalities • ELP 2: Participate in grade-appropriate oral and written exchanges of information, ideas, and analyses, responding to peer, audience, or reader comments and questions. • ELP 5: Conduct research and evaluate and communicate findings to answer questions or solve problems. • ELP 6: Analyze and critique the arguments of others orally and in writing. Linguistic structures of English • ELP 9: Create clear and coherent grade-appropriate speech and text. • ELP 10: Make accurate use of standard English to communicate in grade-appropriate speech and writing.

Investigation 12

Adaptations: Why Do Mammals That Live in the Arctic Ocean Have a Thick Layer of Blubber Under Their Skin?

Introduction

There are many different types of ecosystems on Earth. An ecosystem includes all of the living things and non-living things in a given area. The Arctic Ocean is an example of an aquatic ecosystem. Many different types of organisms live in the Arctic Ocean. Watch the video about the Arctic Ocean. As you watch, keep a record what you see (observe) and what you are wondering about in the boxes below.

Things I OBSERVED …	Things I WONDER about …

The Arctic Ocean is a cold and harsh environment. It is covered in ice most of the year. Some kinds of animals survive well in the Arctic Ocean, some survive less well, and some cannot survive at all. All the different types of animals that survive well there, such as bears, whales, birds, and fish, have specific traits that enable them to live in this specific environment. For example, some animals have white fur, feathers, or skin. These animals blend in with the ice that covers the Arctic Ocean. These white animals are able to either hide from predators or sneak up on their prey. Other traits help these animals deal with all the ice that covers the ocean. For example, the narwhal in the video you just watched has a long tusk that it can use to break through the ice that covers the ocean. Beluga whales are able to use sound to find their way around in the dark because it is often very dark under the ice that covers the ocean. All of these different traits make it possible for these animals to survive in the cold and harsh Arctic Ocean.

Sometimes, however, it is difficult to figure out why some animals have a specific trait. For example, all the mammals that live in the Arctic Ocean, such as bears, seals, and whales, have a thick layer of fat under their skin. This thick layer of fat is called *blubber.* In nature, the way an animal's body is shaped or structured determines how it functions and places limits on what it can or cannot do. The thick layer of blubber under the skin of mammals should therefore serve a function that helps these mammals survive in the Arctic Ocean in some way. One possible function is that a thick layer of blubber helps mammals stay warm when they are in very cold water. It is important for mammals that live in the Arctic Ocean to stay warm at all times because mammals will die if their body temperature drops too low.

In this investigation you will need to plan and carry out an experiment to test the hypothesis that a layer of blubber causes mammals to stay warm when they are in cold water. Scientists often investigate possible cause-and-effect relationships like this to explain how animals are able to survive in a specific environment.

Things we KNOW from what we read …	What we will NEED to figure out …

Your Task

Use what you know about the traits of animals, the characteristics of specific environments, cause-and-effect relationships, and how structure determines function in nature to test the idea that a thick layer of fat can help a mammal stay warm in a cold environment.

The *guiding question* of this investigation is, ***Why do mammals that live in the Arctic Ocean have a thick layer of blubber under their skin?***

Materials

You may use any of the following materials during your investigation:

- Safety goggles (required)
- Plastic-bag glove filled with shortening (which is like blubber)
- Plastic-bag glove filled with feathers
- Plastic-bag glove (empty)
- Bucket of ice water
- 3 Thermometers
- Stopwatch

Safety Rules

Follow all normal safety rules. In addition, be sure to follow these rules:

- Wear sanitized indirectly vented chemical-splash goggles during setup, investigation activity, and cleanup.
- Do not throw objects or put any objects in your mouth.
- Immediately wipe up any slip or fall hazards (such as water).
- Appropriately dispose of materials as directed by your teacher
- Wash your hands with soap and water when done collecting the data.

Plan Your Investigation

Prepare a plan for your investigation by filling out the chart that follows; this plan is called an *investigation proposal*. Before you start developing your plan, be sure to discuss the following questions with the other members of your group:

- What information do we need to find a **cause-and-effect relationship?**
- How might the **structure** of what you are studying relate to its **function?**

Our guiding question:

This is a picture of how we will set up the equipment:

We will collect the following data:

These are the steps we will follow to collect data:

I approve of this investigation proposal.

Teacher's signature

Date

Collect Your Data

Keep a record of what you measure or observe during your investigation in the space below.

Analyze Your Data

You will need to analyze the data you collected before you can develop an answer to the guiding question. To do this, create a graph that shows the relationship between the cause (what is inside the glove) and the effect (temperature inside the glove).

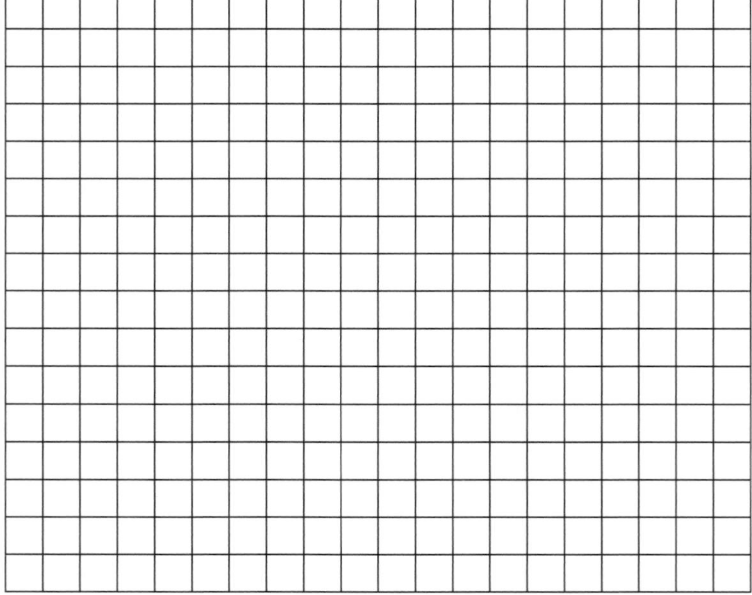

Draft Argument

Develop an argument on a whiteboard. It should include the following parts:

1. A *claim:* Your answer to the guiding question.
2. *Evidence:* An analysis of the data and an explanation of what the analysis means.
3. A *justification of the evidence:* Why your group thinks the evidence is important.

The Guiding Question:	
Our Claim:	
Our Evidence:	Our Justification of the Evidence:

Argumentation Session

Share your argument with your classmates. Be sure to ask them how to make your draft argument better. Keep track of their suggestions in the space below.

Ways to IMPROVE our argument …

Draft Report

Prepare an *investigation report* to share what you have learned. Use the information in this handout and your group's final argument to write a *draft* of your investigation report.

Introduction

We have been studying ______________________________ in class. Before we started this investigation, we explored ______________________________

____

____

We noticed ______________________________

____

____

My goal for this investigation was to figure out ______________________________

____

____

The guiding question was ______________________________

____

____

Method

To gather the data I needed to answer this question, I ______________________________

____

____

____

____

____

____

Argument

My claim is ____

____

____

____

____

The graph below shows ____

____

____

____

____

____

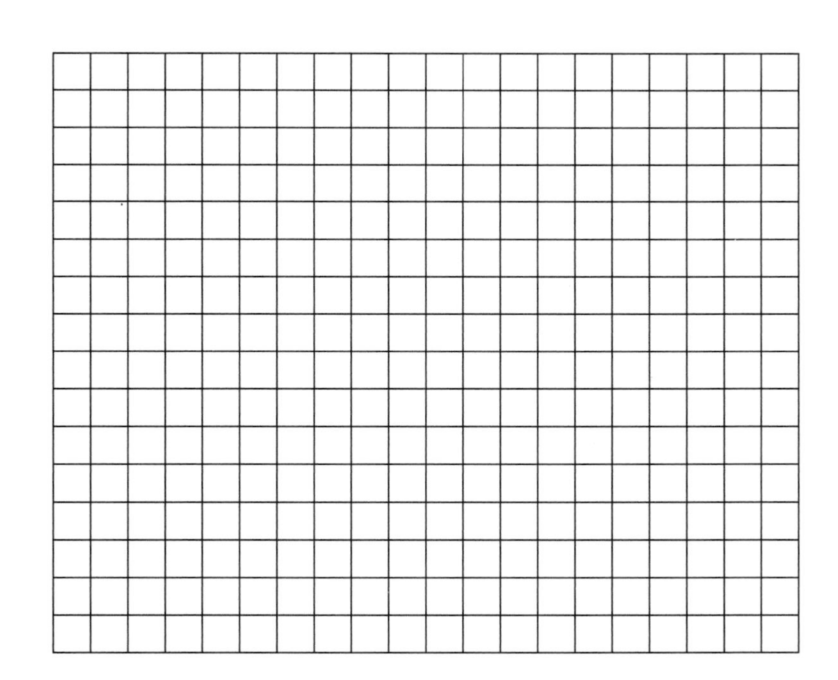

This evidence is important because ______________________________

____

____

____

____

____

____

____

____

____

____

____

Review

Your friends need your help! Review the draft of their investigation reports and give them ideas about how to improve. Use the *peer-review guide* when doing your review.

Submit Your Final Report

Once you have received feedback from your friends about your draft report, create your final investigation report and hand it in to your teacher.

Checkout Questions

Investigation 12. Adaptations: Why Do Mammals That Live in the Arctic Ocean Have a Thick Layer of Blubber Under Their Skin?

Pictured below are two different types of animals. One is called a musk ox and the other is called a gazelle. The animals live in different parts of the world.

Musk ox

Gazelle

1. Which animal do you think is more likely to live in a cold environment?

☐ Musk ox ☐ Gazelle

2. Explain your thinking. How did your understanding of the relationship between *cause and effect* or *structure and function* allow you to determine where the animal is most likely to live?

____

____

____

____

____

____

Teacher Scoring Rubric for the Checkout Questions

Level	Description
3	The student can apply the core idea correctly in all cases and can fully explain the relationship between cause and effect or structure and function.
2	The student can apply the core idea correctly in all cases but cannot fully explain the relationship between cause and effect or structure and function.
1	The student cannot apply the core idea correctly in all cases but can fully explain the relationship between cause and effect or structure and function.
0	The student cannot apply the core idea correctly in all cases and cannot explain the relationship between cause and effect or structure and function.

Section 6
Earth's Systems

Investigation 13

Weather Patterns: What Weather Conditions Can We Expect Here During Each Season?

Purpose

The purpose of this investigation is to give students an opportunity to use the disciplinary core idea (DCI) of ESS2.D: Weather and Climate and the crosscutting concept (CC) of Patterns from *A Framework for K–12 Science Education* (NRC 2012) to figure out how the weather where they live differs by season. Students will also learn about how scientists use different methods to answer different types of questions during the reflective discussion.

The Disciplinary Core Idea

Students in third grade should understand the following about Weather and Climate and be able to use this DCI to figure out how the weather where they live differs by season:

> *Weather is the combination of sunlight, wind, snow or rain, and temperature in a particular region at a particular time. People measure these conditions to describe and record the weather and to notice patterns over time. (NRC 2012, p. 188)*

The Crosscutting Concept

Students in third grade should understand the following about the CC Patterns:

> *Noticing patterns is often a first step to organizing and asking scientific questions about why and how the patterns occur. One major use of pattern recognition is in classification, which depends on careful observation of similarities and differences; objects can be classified into groups on the basis of similarities of visible or microscopic features or on the basis of similarities of function. Such classification is useful in codifying relationships and organizing a multitude of objects or processes into a limited number of groups. (NRC 2012, p. 85)*

Students in third grade should be given opportunities to "describe and predict the patterns in the seasons of the year" (NRC 2012, p. 86).

Students should be encouraged to use their developing understanding of patterns as a tool or a way of thinking about a phenomenon during this investigation to help them figure out how the weather where they live differs by season.

What Students Figure Out

Weather is the minute-by-minute to day-by-day variation of the atmosphere's condition at a specific location. Scientists record the patterns of the weather across different times and areas so that they can make predictions about what kind of weather might happen next. A change in season often brings a marked change in the typical weather at a specific location.

Timeline

The time needed to complete this investigation is 310 minutes (5 hours and 10 minutes). The amount of instructional time needed for each stage of the investigation is as follows:

- *Stage 1.* Introduce the task: 5 minutes; collect data (5 days): 50 minutes (10 minutes per day); introduce the guiding question: 20 minutes
- *Stage 2.* Design a method and collect data: 60 minutes
- *Stage 3.* Create a draft argument: 35 minutes
- *Stage 4.* Argumentation session: 30 minutes
- *Stage 5.* Reflective discussion: 15 minutes
- *Stage 6.* Write a draft report: 35 minutes
- *Stage 7.* Peer review: 30 minutes
- *Stage 8.* Revise the report: 30 minutes

This investigation can be completed over 6 days (5 days for stage 1 and 1 day for the remaining seven stages) or over 12 days (5 days for stage 1 and 7 days for the remaining seven stages) during your designated science time in the daily schedule.

Materials and Preparation

The materials needed for this investigation are listed in Table 13.1 (p. 476). Students will need to collect weather data for five days as part of stage 1. This can be accomplished by hand using weather measurement instruments (such as a thermometer, a barometer, a hygrometer, and an anemometer) if available at your school or by simply using a weather app on a smartphone or tablet.

Students will need access to a computer or tablet with an internet connection to collect data about the typical weather for each season from the World Weather and Climate Information website at *https://weather-and-climate.com*. We recommend at least one computer or tablet per group, but each student can use a computer or tablet on his or her own if there are enough available. Be sure to visit the website and learn how to find the information that the students will need before starting the investigation so you can show students how to use it and help them when they get stuck. In addition, it is important to check if students can access and use the website from a school computer or tablet, because some schools have set up firewalls and other restrictions on web browsing.

TABLE 13.1

Materials for Investigation 13

Item	Quantity
Computer or tablet with internet access	1 per group
Thermometer (optional)	1 per group
Barometer (optional)	1 per group
Hygrometer (optional)	1 per group
Anemometer (optional)	1 per group
Weather app for smartphone or tablet (optional)	1 per group
Whiteboard, 2' × 3'*	1 per group
Investigation Handout	1 per student
Peer-review guide and teacher scoring rubric	1 per student
Checkout Questions (optional)	1 per student

* As an alternative, students can use computer and presentation software such as Microsoft PowerPoint or Apple Keynote to create their arguments.

Safety Precautions

Remind students to follow all normal safety rules.

Lesson Plan by Stage

Stage 1: Introduce the Task and the Guiding Question (75 minutes)

1. Ask the students to sit in six groups, with three or four students in each group.
2. Ask students to clear off their desks except for a pencil (and their *Student Workbook for Argument-Driven Inquiry in Third-Grade Science* if they have one).
3. Pass out an Investigation Handout to each student (or ask students to turn to Investigation Log 13 in their workbook).
4. Read the first paragraph of the "Introduction" aloud to the class. Ask the students to follow along as you read.
5. Show students how to track the weather over the next five school days (see "Materials and Preparation" section).
6. Tell students to record their observations and questions about weather in the "OBSERVED/WONDER" chart in the "Introduction" section of their Investigation Handout (or the investigation log in their workbook).
7. Give the students 10 minutes to record their observations and questions over the next four days.

8. After the students have recorded their measurements on the fifth day, ask students to share *what they observed* about the weather.
9. Ask students to share *what questions they have* about the weather.
10. Tell the students, "Some of your questions might be answered by reading the rest of the 'Introduction.'"
11. Ask the students to read the rest of the "Introduction" on their own *or* ask them to follow along as you read aloud.
12. Once the students have read the rest of the "Introduction," ask them to fill out the "KNOW/NEED" chart on their Investigation Handout (or in their investigation log) as a group.
13. Ask students to share what they learned from the reading. Add these ideas to a class "know / need to figure out" chart.
14. Ask one or two different students to share what they think they will need to figure out based on what they read. Add these ideas to the class "know / need to figure out" chart.
15. Tell the students, "It looks like we have something to figure out. Let's see what we will need to do during our investigation."
16. Read the task and the guiding question aloud.
17. Tell the students, "You will be able to use a website called World Weather and Climate Information during your investigation."
18. Show the students how to use the website by projecting it on the board or on a screen and demonstrating how to select a country and a city in that country and then find the data they need.

Stage 2: Design a Method and Collect Data (60 minutes)

1. Tell the students, "I am now going to give you and the other members of your group about 15 minutes to plan your investigation. Before you begin, I want you all to take a couple of minutes to discuss the following questions with the rest of your group."
2. Show the following questions on the screen or board:
   - What information should we collect so we can *describe* the typical weather here?
   - What types of *patterns* might we look for to help answer the guiding question?
3. Tell the students, "Please take a few minutes to come up with an answer to these questions." Give the students two or three minutes to discuss these two questions.

4. Ask two or three different groups to share their answers. Be sure to highlight or write down any important ideas on the board so students can refer to them later.
5. If possible, use a document camera to project an image of the graphic organizer for this investigation on a screen or board (or take a picture of it and project the picture on a screen or board). Tell the students, "I now want you all to plan out your investigation. To do that, you will need to create an investigation proposal by filling out this graphic organizer."
6. Point to the box labeled "Our guiding question:" and tell the students, "You can put the question we are trying to answer in this box." Then ask, "Where can we find the guiding question?"
7. Wait for a student to answer where to find the guiding question (the answer is "in the handout").
8. Point to the box labeled "We will collect the following data:" and tell the students, "You can list the measurements or observations that you will need to collect during the investigation in this box."
9. Point to the box labeled "These are the steps we will follow to collect data:" and tell the students, "You can list what you are going to do to collect the data you need and what you will do with your data once you have it. Be sure to give enough detail that I could do your investigation for you."
10. Ask the students, "Do you have any questions about what you need to do?"
11. Answer any questions that come up.
12. Tell the students, "Once you are done, raise your hand and let me know. I'll then come by and look over your proposal and give you some feedback. You may not begin collecting data until I have approved your proposal by signing it. You need to have your proposal done in the next 15 minutes."
13. Give the students 15 minutes to work in their groups on their investigation proposal. As they work, move from group to group to check in, ask probing questions, and offer a suggestion if a group gets stuck.

What should a student-designed investigation look like?

The students' investigation proposal should include the following information:

- The guiding question is "What weather conditions can we expect here during each season?"
- There are a lot of different types of data that students can collect during this investigation. Examples include (1) average monthly maximum temperature, (2) average monthly minimum temperature, (3) average monthly precipitation, (4) average monthly relative humidity, and (5) average mean wind speed. At a minimum, each group should collect data about two different weather conditions. This investigation works best if each group selects different conditions.
- The steps that the students will follow to collect the data should reflect the conditions that they decide to examine. However, a procedure might include the following steps:
    1. Identify the city.
    2. Record typical [weather condition 1] for each month.
    3. Record typical [weather condition 2] for each month.
    4. Calculate the seasonal average for each weather condition by adding the value for each month together and then dividing by 3 (the number of months in a season).

There should be a lot of variation in the student-designed investigations.

14. As each group finishes its investigation proposal, be sure to read it over and determine if it will be productive or not. If you feel the investigation will be productive (not necessarily what you would do or what the other groups are doing), sign your name on the proposal and let the group start collecting data. If the plan needs to be changed, offer some suggestions or ask some probing questions, and have the group make the changes before you approve it.
15. Tell the students to collect their data and record their observations or measurements in the "Collect Your Data" box in their Investigation Handout (or the investigation log in their workbook).
16. Give the students 20 minutes to collect their data.

Stage 3: Create a Draft Argument (35 minutes)

1. Tell the students, "Now that we have all this data, we need to analyze the data so we can figure out an answer to the guiding question."
2. If possible, project an image of the "Analyze Your Data" section for this investigation on a screen or board using a document camera (or take a picture of it and project the picture on a screen or board). Point to the section and tell the groups of students, "You can create a couple of graphs as a way to analyze your data. You can make your graphs in this section."
3. Ask the students, "What information do we need to include in these graphs?"
4. Tell the students, "Please take a few minutes to discuss this question with your group, and be ready to share."
5. Give the students five minutes to discuss.
6. Ask two or three different groups to share their answers. Be sure to highlight or write down any important ideas on the board so students can refer to them later.
7. Tell the groups of students, "I am now going to give you and the other members of your group about 10 minutes to create your graphs." If the students are having trouble making a graph, you can take a few minutes to provide a mini-lesson about how to create a graph from a bunch of observations or measurements (this strategy is called just-in-time instruction because it is offered only when students get stuck).

What should a graph look like for this investigation?

There are a number of different ways that students can analyze the observations or measurements they collect during this investigation. One of the most straightforward ways is to create two scaled bar graphs, one for each weather condition (temperature, humidity, wind speed, precipitation). Each bar graph should have categories for each season (winter, spring, summer, fall) on the horizontal or *x*-axis. The average value for a weather condition (average temperature, average humidity, average wind speed, average precipitation) should be on the vertical or *y*-axis. Examples of this type of graph can be seen in Figure 13.1 (p. 482).

8. Give the students 10 minutes to analyze their data. As they work, move from group to group to check in, ask probing questions, and offer suggestions.
9. Tell the students, "I am now going to give you and the other members of your group 15 minutes to create an argument to share what you have learned and

convince others that they should believe you. Before you do that, we need to take a few minutes to discuss what you need to include in your argument."

10. If possible, use a document camera to project the "Argument Presentation on a Whiteboard" image from the Investigation Handout (or the investigation log in their workbook) on a screen or board (or take a picture of it and project the picture on a screen or board).
11. Point to the box labeled "The Guiding Question:" and tell the students, "You can put the question we are trying to answer here on your whiteboard."
12. Point to the box labeled "Our Claim:" and tell the students, "You can put your claim here on your whiteboard. The claim is your answer to the guiding question."
13. Point to the box labeled "Our Evidence:" and tell the students, "You can put the evidence that you are using to support your claim here on your whiteboard. Your evidence will need to include the analysis you just did and an explanation of what your analysis means or shows. Scientists always need to support their claims with evidence."
14. Point to the box labeled "Our Justification of the Evidence:" and tell the students, "You can put your justification of your evidence here on your whiteboard. Your justification needs to explain why your evidence is important. Scientists often use core ideas to explain why the evidence they are using matters. Core ideas are important concepts that scientists use to help them make sense of what happens during an investigation."
15. Ask the students, "What are some core ideas that we read about earlier that might help us explain why the evidence we are using is important?"
16. Ask students to share some of the core ideas from the "Introduction" section of the Investigation Handout (or the investigation log in the workbook). List these core ideas on the board.
17. Tell the students, "That is great. I would like to see everyone try to include these core ideas in your justification of the evidence. Your goal is to use these core ideas to help explain why your evidence matters and why the rest of us should pay attention to it."
18. Ask the students, "Do you have any questions about what you need to do?"
19. Answer any questions that come up.
20. Tell the students, "Okay, go ahead and start working on your arguments. You need to have your argument done in the next 15 minutes. It doesn't need to be perfect. We just need something down on the whiteboards so we can share our ideas."
21. Give the students 15 minutes to work in their groups on their arguments. As they work, move from group to group to check in, ask probing questions, and

offer a suggestion if a group gets stuck. Figure 13.1 shows an example of an argument created by students for this investigation.

FIGURE 13.1

Example of an argument

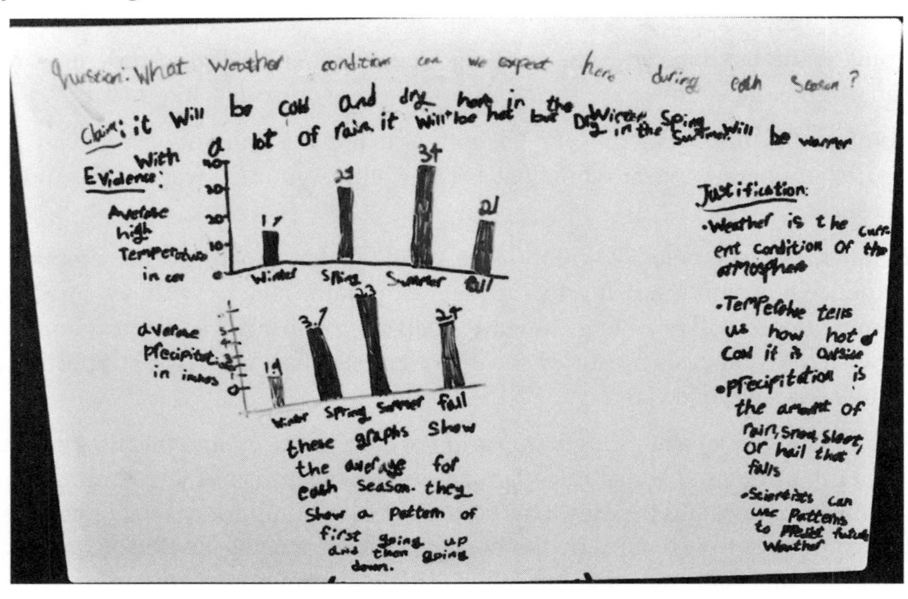

Stage 4: Argumentation Session (30 minutes)

The argumentation session can be conducted in a whole-class presentation format, a gallery walk format, or a modified gallery walk format. We recommend using a whole-class presentation format for the first investigation, but try to transition to either the gallery walk or modified gallery walk format as soon as possible because that will maximize student voice and choice inside the classroom. The following list shows the steps for the three formats; unless otherwise noted, the steps are the same for all three formats.

1. Begin by introducing the use of the whiteboard.
    - *If using the whole-class presentation format,* tell the students, "We are now going to share our arguments. Please set up your whiteboard so everyone can see them."
    - *If using the gallery walk or modified gallery walk format,* tell the students, "We are now going to share our arguments. Please set up your whiteboard so they are facing the walls."
2. Allow the students to set up their whiteboards.
    - *If using the whole-class presentation format,* the whiteboards should be set up on stands or chairs so they are facing toward the center of the room.

- *If using the gallery walk or modified gallery walk format,* the whiteboards should be set up on stands or chairs so they are facing toward the outside of the room.

3. Give the following instructions to the students:
   - *If using the whole-class presentation format or the modified gallery walk format,* tell the students, "Okay, before we get started I want to explain what we are going to do next. I'm going to ask some of you to present your arguments to your classmates. If you are presenting your argument, your job is to share your group's claim, evidence, and justification of the evidence. The rest of you will be reviewers. If you are a reviewer, your job is to listen to the presenters, ask the presenters questions if you do not understand something, and then offer them some suggestions about ways to make their argument better. After we have a chance to learn from each other, I'm going to give you some time to revise your arguments and make them better."
   - *If using the gallery walk format,* tell the students, "Okay, before we get started I want to explain what we are going to do next. You are going to have an opportunity to read the arguments that were created by other groups. Your group will go to a different group's argument. I'll give you a few minutes to read it and review it. Your job is to offer them some suggestions about ways to make their argument better. You can use sticky notes to give them suggestions. Please be specific about what you want to change and be specific about how you think they should change it. After we have a chance to learn from each other, I'm going to give you some time to revise your arguments and make them better."
4. Use a document camera to project the "Ways to IMPROVE our argument …" box from the Investigation Handout (or the investigation log in their workbook) on a screen or board (or take a picture of it and project the picture on a screen or board).
   - *If using the whole-class presentation format or the modified gallery walk format,* point to the box and tell the students, "If you are a presenter, you can write down the suggestions you get from the reviewers here. If you are a reviewer, and you see a good idea from another group, you can write down that idea here. Once we are done with the presentations, I will give you a chance to use these suggestions or ideas to improve your arguments.
   - *If using the gallery walk format,* point to the box and tell the students, "If you see good ideas from another group, you can write them down here. Once we are done reviewing the different arguments, I will give you a chance to use these ideas to improve your own arguments. It is important to share ideas like this."

   Ask the students, "Do you have any questions about what you need to do?"
5. Answer any questions that come up.
6. Give the following instructions:

- *If using the whole-class presentation format,* tell the students, "Okay. Let's get started."
- *If using the gallery walk format,* tell the students, "Okay, I'm now going to tell you which argument to go to and review.
- *If using the modified gallery walk format,* tell the students, "Okay, I'm now going to assign you to be a presenter or a reviewer." Assign one or two students from each group to be presenters and one or two students from each group to be reviewers.

7. Begin the review of the arguments.
   - *If using the whole-class presentation format,* have four or five groups present their argument one at a time. Give each group only two to three minutes to present their argument. Then give the class two to three minutes to ask them questions and offer suggestions. Be sure to encourage as much participation from the students as possible.
   - *If using the gallery walk format,* tell the students, "Okay. Let's get started. Each group, move one argument to the left. Don't move to the next argument until I tell you to move. Once you get there, read the argument and then offer suggestions about how to make it better. I will put some sticky notes next to each argument. You can use the sticky notes to leave your suggestions." Give each group about three to four minutes to read the arguments, talk, and offer suggestions.
     a. Tell the students, "Okay. Let's rotate. Move one group to the left."
     b. Again, give each group three or four minutes to read, talk, and offer suggestions.
     c. Repeat this process for two more rotations.
   - *If using the modified gallery walk format,* tell the students, "Okay. Let's get started. Reviewers, move one group to the left. Don't move to the next group until I tell you to move. Presenters, go ahead and share your argument with the reviewers when they get there." Give each group of presenters and reviewers about three to four minutes to talk.
     a. Tell the students, "Okay. Let's rotate. Reviewers, move one group to the left."
     b. Again, give each group of presenters and reviewers about three or four minutes to talk.
     c. Repeat this process for two more rotations.
8. Tell the students to return to their workstations.
9. Give the following instructions about revising the argument:
   - *If using the whole-class presentation format,* tell the students, "I'm now going to give you about 10 minutes to revise your argument. Take a few minutes to

talk in your groups and determine what you want to change to make your argument better. Once you have decided what to change, go ahead and make the changes to your whiteboard."

- *If using the gallery walk format,* tell the students, "I'm now going to give you about 10 minutes to revise your argument. Take a few minutes to read the suggestions that were left at your argument. Then talk in your groups and determine what you want to change to make your argument better. Once you have decided what to change, go ahead and make the changes to your whiteboard."
- *If using the modified gallery walk format,* "I'm now going to give you about 10 minutes to revise your argument. Please return to your original groups." Wait for the students to move back into their original groups and then tell the students, "Okay, take a few minutes to talk in your groups and determine what you want to change to make your argument better. Once you have decided what to change, go ahead and make the changes to your whiteboard."

Ask the students, "Do you have any questions about what you need to do?"

10. Answer any questions that come up.
11. Tell the students, "Okay. Let's get started."
12. Give the students 10 minutes to work in their groups on their arguments. As they work, move from group to group to check in, ask probing questions, and offer a suggestion if a group gets stuck.

Stage 5: Reflective Discussion (15 minutes)

1. Tell the students, "We are now going to take a minute to talk about what we did and what we have learned."
2. Show Figure 13.2 on a screen. This figure shows images of the same tree in different seasons. Ask the students, "What do you all see going on here?"
3. Allow students to share their ideas.

FIGURE 13.2

A tree during the four seasons

Note: A full-color version of this figure is available on the book's Extras page at *www.nsta.org/adi-3rd.*

4. Ask the students, "How would you describe the weather when each of these pictures was taken?"
5. Allow students to share their ideas.
6. Ask the students, "Which picture do you think was taken during the winter and how do you know?"
7. Allow students to share their ideas.
8. Ask the students, "Which picture do you think was taken during the spring and how do you know?"
9. Allow students to share their ideas.
10. Tell the students, "Okay, let's make sure we are on the same page. Weather is the minute-by-minute to day-by-day changes in the atmosphere's condition at a specific location. Scientists record the patterns of the weather across different times and areas so that they can make predictions about what kind of weather might happen next. A change in season often brings a marked change in the typical weather at a specific location. Our ability to use past weather patterns as a way to predict the weather in the future is a really important core idea in science."
11. Ask the students, "Does anyone have any questions about this core idea?"
12. Answer any questions that come up.
13. Tell the students, "We also looked for patterns during our investigation." Then show Figure 13.3 on the screen. This figure shows the monthly average high temperature for a city in the United States but is missing information for April and October.

FIGURE 13.3

Average high temperature by month for a city in the United States

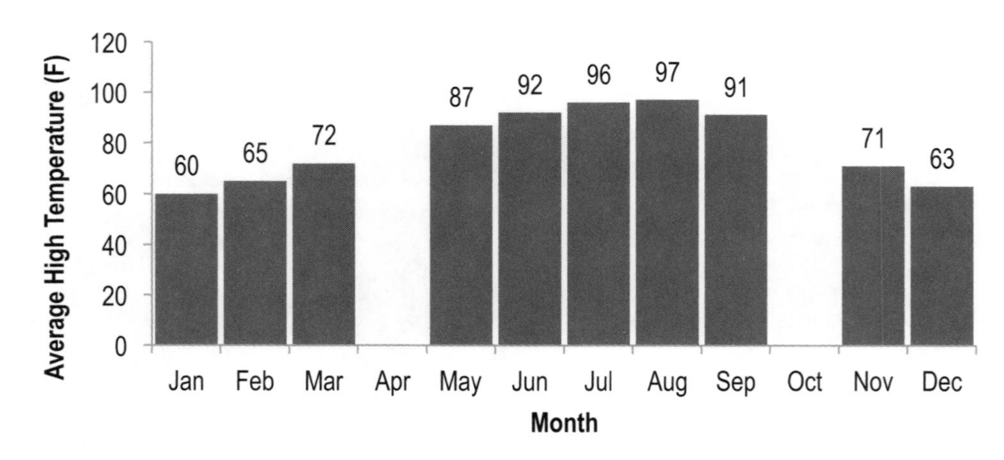

14. Ask the students, "What do you all see going on here?"
15. Allow students to share their ideas.
16. Ask the students, "What pattern do you see here?"
17. Allow students to share their ideas.
18. Ask the students, "What do you think the average temperature will be in April and why?"
19. Allow students to share their ideas.
20. Ask the students, "What do you think the average temperature will be in October and why?"
21. Allow students to share their ideas.
22. Tell the students, "Patterns are really important in science. Scientists look for patterns all the time and use them to make predictions. In fact, they even use patterns to predict what the weather will be like, just like we did."
23. Tell the students, "We are now going take a minute to talk about what went well and what didn't go so well during our investigation. We need to talk about this because you are going to be planning and carrying out your own investigations like this a lot this year, and I want to help you all get better at it."
24. Show an image of the question "What made your investigation scientific?" on the screen. Tell the students, "Take a few minutes to talk about how you would answer this question with the other people in your group. Be ready to share with the rest of the class." Give the students two to three minutes to talk in their group.
25. Ask the students, "What do you all think? Who would like to share an idea?"
26. Allow students to share their ideas. Be sure to expand on their ideas about what makes an investigation scientific.
27. Show an image of the question "What made your investigation not so scientific?" on the screen. Tell the students, "Take a few minutes to talk about how you would answer this question with the other people in your group. Be ready to share with the rest of the class." Give the students two to three minutes to talk in their group.
28. Ask the students, "What do you all think? Who would like to share an idea?"
29. Allow students to share their ideas. Be sure to expand on their ideas about what makes an investigation less scientific.
30. Show an image of the question "What rules can we put into place to help us make sure our next investigation is more scientific?" on the screen. Tell the students, "Take a few minutes to talk about how you would answer this question with the other people in your group. Be ready to share with the rest of the class." Give the students two to three minutes to talk in their group.

31. Ask the students, "What do you all think? Who would like to share an idea?"
32. Allow students to share their ideas. Once they have shared their ideas, offer a suggestion for a possible class rule.
33. Ask the students, "What do you all think? Should we make this a rule?"
34. If the students agree, write the rule on the board or make a class "Rules for Scientific Investigation" chart so you can refer to it during the next investigation.
35. Tell the students, "We are now going take a minute to talk about what makes science different from other subjects."
36. Show an image of the question "What kinds of things do scientists do to answer different types of questions?" on the screen. Tell the students, "Take a few minutes to talk about how you would answer this question with the other people in your group. Be ready to share with the rest of the class." Give the students two to three minutes to talk in their group.
37. Ask the students, "What do you all think? Who would like to share an idea?"
38. Allow students to share their ideas.
39. Tell the students, "Okay, these are all great ideas. Always remember that scientists use different methods to answer different types of questions. Sometimes they conduct experiments, sometimes they use models, and sometimes they make observations over long periods of time and look for patterns. The choice of method depends on what is being studied and the type of question that a scientist wants to answer."
40. Ask the students, "Does anyone have any questions about the different methods that scientists use to answer questions?"
41. Answer any questions that come up.

Stage 6: Write a Draft Report (35 minutes)

Your students will use either the Investigation Handout or the investigation log in the student workbook when writing the draft report. When you give the directions shown in quotes in the following steps, substitute "investigation log" (as shown in brackets) for "handout" if they are using the workbook.

1. Tell the students, "You are now going to write an investigation report to share what you have learned. Please take out a pencil and turn to the 'Draft Report' section of your handout [investigation log]."
2. If possible, use a document camera to project the "Introduction" section of the draft report from the Investigation Handout (or the investigation log in their workbook) on a screen or board (or take a picture of it and project the picture on a screen or board).

3. Tell the students, "The first part of the report is called the 'Introduction.' In this section of the report you want to explain to the reader what you were investigating, why you were investigating it, and what question you were trying to answer. All of this information can be found in the text at the beginning of your handout [investigation log]." Point to the image and say, "There are some sentence starters here to help you begin writing the report." Ask the students, "Do you have any questions about what you need to do?"
4. Answer any questions that come up.
5. Tell the students, "Okay. Let's write."
6. Give the students 10 minutes to write the "Introduction" section of the report. As they work, move from student to student to check in, ask probing questions, and offer a suggestion if a student gets stuck.
7. If possible, use a document camera to project the "Method" section of the draft report from the Investigation Handout (or the investigation log in their workbook) on a screen or board (or take a picture of it and project the picture on a screen or board).
8. Tell the students, "The second part of the report is called the 'Method.' In this section of the report you want to explain to the reader what you did during the investigation, what data you collected and why, and how you went about analyzing your data. All of this information can be found in the 'Plan Your Investigation' section of your handout [investigation log]. Remember that you all planned and carried out different investigations, so do not assume that the reader will know what you did." Point to the image and say, "There are some sentence starters here to help you begin writing this part of the report." Ask the students, "Do you have any questions about what you need to do?"
9. Answer any questions that come up.
10. Tell the students, "Okay. Let's write."
11. Give the students 10 minutes to write the "Method" section of the report. As they work, move from student to student to check in, ask probing questions, and offer a suggestion if a student gets stuck.
12. If possible, use a document camera to project the "Argument" section of the draft report from the Investigation Handout (or the investigation log in their workbook) on a screen or board (or take a picture of it and project the picture on a screen or board).
13. Tell the students, "The last part of the report is called the 'Argument.' In this section of the report you want to share your claim, evidence, and justification of the evidence with the reader. All of this information can be found on your whiteboard." Point to the image and say, "There are some sentence starters here to help you begin writing this part of the report." Ask the students, "Do you have any questions about what you need to do?"

14. Answer any questions that come up.
15. Tell the students, "Okay. Let's write."
16. Give the students 10 minutes to write the "Argument" section of the report. As they work, move from student to student to check in, ask probing questions, and offer a suggestion if a student gets stuck.

Stage 7: Peer Review (30 minutes)

Your students will use either the Investigation Handout or the investigation log in the student workbook when doing the peer review. When you give the directions shown in quotes in the following steps, substitute "workbook" (as shown in brackets) for "Investigation Handout" if they are using the workbook.

1. Tell the students, "We are now going to review our reports to find ways to make them better. I'm going to come around and collect your Investigation Handout [workbook]. While I do that, please take out a pencil."
2. Collect the Investigation Handouts or workbooks from the students.
3. If possible, use a document camera to project the peer-review guide (PRG; see Appendix 4) on a screen or board (or take a picture of it and project the picture on a screen or board).
4. Tell the students, "We are going to use this peer-review guide to give each other feedback." Point to the image.
5. Give the following instructions:
   - *If using the Investigation Handout,* tell the students, "I'm going to ask you to work with a partner to do this. I'm going to give you and your partner a draft report to read and a peer-review guide to fill out. You two will then read the report together. Once you are done reading the report, I want you to answer each of the questions on the peer-review guide." Point to the review questions on the image of the PRG.
   - *If using the workbook,* tell the students, "I'm going to ask you to work with a partner to do this. I'm going to give you and your partner a draft report to read. You two will then read the report together. Once you are done reading the report, I want you to answer each of the questions on the peer-review guide that is right after the report in the investigation log." Point to the review questions on the image of the PRG.
6. Tell the students, "You can check 'yes,' 'almost,' or 'no' after each question." Point to the checkboxes on the image of the PRG.
7. Tell the students, "This will be your rating for this part of the report. Make sure you agree on the rating you give the author. If you mark 'almost' or 'no,' then

you need to tell the author what he or she needs to do to get a 'yes.'" Point to the space for the reviewer feedback on the image of the PRG.

8. Tell the students, "It is really important for you to give the authors feedback that is helpful. That means you need to tell them exactly what they need to do to make their reports better." Ask the students, "Do you have any questions about what you need to do?"
9. Answer any questions that come up.
10. Tell the students, "Please sit with a partner who is not in your current group." Allow the students time to sit with a partner.
11. Give the following instructions:
    - *If using the Investigation Handout,* tell the students, "Okay, I am now going to give you one report to read and one peer-review guide to fill out." Pass out one report to each pair. Make sure that the report you give a pair was not written by one of the students in that pair. Give each pair one PRG to fill out as a team.
    - *If using the workbook,* tell the students, "Okay, I am now going to give you one report to read." Pass out a workbook to each pair. Make sure that the workbook you give a pair is not from one of the students in that pair.
12. Tell the students, "Okay, I'm going to give you 15 minutes to read the report I gave you and to fill out the peer-review guide. Go ahead and get started."
13. Give the students 15 minutes to work. As they work, move around from pair to pair to check in and see how things are going, answer questions, and offer advice.
14. After 15 minutes pass, tell the students, "Okay, time is up." *If using the Investigation Handout,* say, "Please give me the report and the peer-review guide that you filled out." *If using the workbook,* say, "Please give me the workbook that you have."
15. Collect the Investigation Handouts and the PRGs, or collect the workbooks if they are being used. Be sure you keep the handout and the PRG together.
16. Give the following instructions:
    - *If using the Investigation Handout,* tell the students, "Okay, I am now going to give you a different report to read and a new peer-review guide to fill out." Pass out another report to each pair. Make sure that this report was not written by one of the students in that pair. Give each pair a new PRG to fill out as a team.
    - *If using the workbook,* tell the students, "Okay, I am now going to give you a different report to read." Pass out a different workbook to each pair. Make sure that the workbook you give a pair is not from one of the students in that pair.
17. Tell the students, "Okay, I'm going to give you 15 minutes to read this new report and to fill out the peer-review guide. Go ahead and get started."

18. Give the students 15 minutes to work. As they work, move around from pair to pair to check in and see how things are going, answer questions, and offer advice.
19. After 15 minutes pass, tell the students, "Okay, time is up." *If using the Investigation Handout,* say, "Please give me the report and the peer-review guide that you filled out." *If using the workbook,* say, "Please give me the workbook that you have."
20. Collect the Investigation Handouts and the PRGs, or collect the workbooks if they are being used. Be sure you keep the handout and the PRG together.

Stage 8: Revise the Report (30 minutes)

Your students will use either the Investigation Handout or the investigation log in the student workbook when revising the report. Except where noted below, the directions are the same whether using the handout or the log.

1. Give the following instructions:
   - *If using the Investigation Handout,* tell the students, "You are now going to revise your investigation report based on the feedback you get from your classmates. Please take out a pencil while I hand back your draft report and the peer-review guide."
   - *If using the investigation log in the student workbook,* tell the students, "You are now going to revise your investigation report based on the feedback you get from your classmates. Please take out a pencil while I hand back your investigation logs."
2. *If using the Investigation Handout,* pass back the handout and the PRG to each student. *If using the investigation log,* pass back the log to each student.
3. Tell the students, "Please take a few minutes to read over the peer-review guide. You should use it to figure out what you need to change in your report and how you will change the report."
4. Allow the students time to read the PRG.
5. *If using the investigation log,* if possible use a document camera to project the "Write Your Final Report" section from the investigation log on a screen or board (or take a picture of it and project the picture on a screen or board).
6. Give the following instructions:
   - *If using the Investigation Handout,* tell the students, "Okay. Let's revise our reports. Please take out a piece of paper. I would like you to rewrite your report. You can use your draft report as a starting point, but use the feedback on the peer-review guide to help make it better."
   - *If using the investigation log,* tell the students, "Okay. Let's revise our reports. I would like you to rewrite your report in the section of the investigation log that says 'Write Your Final Report.'" Point to the image on the screen and tell

the students, "You can use your draft report as a starting point, but use the feedback on the peer-review guide to help make your report better."

Ask the students, "Do you have any questions about what you need to do?"

7. Answer any questions that come up.
8. Tell the students, "Okay. Let's write."
9. Give the students 30 minutes to rewrite their report. As they work, move from student to student to check in, ask probing questions, and offer a suggestion if a student gets stuck.
10. Give the following instructions:
11. *If using the Investigation Handout,* tell the students, "Okay. Time's up. I will now come around and collect your Investigation Handout, the peer-review guide, and your final report."
    - *If using the investigation log,* tell the students, "Okay. Time's up. I will now come around and collect your workbooks."
    - *If using the Investigation Handout,* collect all the Investigation Handouts, PRGs, and final reports. *If using the investigation log,* collect all the workbooks.
12. *If using the Investigation Handout,* use the "Teacher Score" columns in the PRG to grade the final report. *If using the investigation log,* use the "ADI Investigation Report Grading Rubric" in the investigation log to grade the final report. Whether you are using the handout or the log, you can give the students feedback about their writing in the "Teacher Comments" section.

How to Use the Checkout Questions

The Checkout Questions are an optional assessment. We recommend giving them to students one day after they finish stage 8 of the ADI investigation. The Checkout Questions can be used as a formative or summative assessment of student thinking. If you plan to use them as a formative assessment, we recommend that you look over the student answers to determine if you need to reteach the core idea and/or crosscutting concept from the investigation, but do not grade them. If you plan to use them as a summative assessment, we have included a "Teacher Scoring Rubric" at the end of the Checkout Questions that you can use to score a student's ability to apply the core idea in a new scenario and explain their use of a crosscutting concept. The rubric includes a 4-point scale that ranges from 0 (the student cannot apply the core idea correctly in all cases and cannot explain the [crosscutting concept]) to 3 (the student can apply the core idea correctly in all cases and can fully explain the [crosscutting concept]). The Checkout Questions, regardless of how you decide to use them, are a great way to make student thinking visible so you can determine if the students have learned the core idea and the crosscutting concept.

A student who can apply the core idea correctly in all cases and can explain the pattern would give the following answers to the first two questions:

- Question 1: 2.0, 2.1, or 2.2 (any of these numbers is a correct answer)
- Question 2: a number ranging from 2.4 to 3.9

He or she should then be able to explain that rainfall tends to change with the seasons and that a dry season is often followed by a wetter season.

Connections to Standards

Table 13.2 highlights how the investigation can be used to address specific performance expectations from the *NGSS, Common Core State Standards (CCSS)* in English language arts (ELA) and in mathematics, and *English Language Proficiency (ELP) Standards.*

TABLE 13.2

Investigation 13 alignment with standards

***NGSS* performance expectation**	Strong alignment • 3-ESS2-1: Represent data in tables and graphical displays to describe typical weather conditions expected during a particular season
***CCSS ELA*—Reading: Informational Text**	Key ideas and details • CCSS.ELA-LITERACY.RI.3.1: Ask and answer questions to demonstrate understanding of a text, referring explicitly to the text as the basis for the answers. • CCSS.ELA-LITERACY.RI.3.2: Determine the main idea of a text; recount the key details and explain how they support the main idea. • CCSS.ELA-LITERACY.RI.3.3: Describe the relationship between a series of historical events, scientific ideas or concepts, or steps in technical procedures in a text, using language that pertains to time, sequence, and cause/effect. Craft and structure • CCSS.ELA-LITERACY.RI.3.4: Determine the meaning of general academic and domain-specific words and phrases in a text relevant to a *grade 3 topic or subject area*. • CCSS.ELA-LITERACY.RI.3.5: Use text features and search tools (e.g., key words, sidebars, hyperlinks) to locate information relevant to a given topic efficiently. • CCSS.ELA-LITERACY.RI.3.6: Distinguish their own point of view from that of the author of a text. Integration of knowledge and ideas • CCSS.ELA-LITERACY.RI.3.7: Use information gained from illustrations (e.g., maps, photographs) and the words in a text to demonstrate understanding of the text (e.g., where, when, why, and how key events occur). • CCSS.ELA-LITERACY.RI.3.8: Describe the logical connection between particular sentences and paragraphs in a text (e.g., comparison, cause/effect, first/second/third in a sequence). • CCSS.ELA-LITERACY.RI.3.9: Compare and contrast the most important points and key details presented in two texts on the same topic. Range of reading and level of text complexity • CCSS.ELA-LITERACY.RI.3.10: By the end of the year, read and comprehend informational texts, including history/social studies, science, and technical texts, at the high end of the grades 2–3 text complexity band independently and proficiently.

Continued

Table 13.2 (*continued*)

CCSS ELA—Writing	Text types and purposes • CCSS.ELA-LITERACY.W.3.1: Write opinion pieces on topics or texts, supporting a point of view with reasons. o CCSS.ELA-LITERACY.W.3.1.A: Introduce the topic or text they are writing about, state an opinion, and create an organizational structure that lists reasons. o CCSS.ELA-LITERACY.W.3.1.B: Provide reasons that support the opinion. o CCSS.ELA-LITERACY.W.3.1.C: Use linking words and phrases (e.g., *because*, *therefore*, *since, for example*) to connect opinion and reasons. o CCSS.ELA-LITERACY.W.3.1.D: Provide a concluding statement or section. • CCSS.ELA-LITERACY.W.3.2: Write informative or explanatory texts to examine a topic and convey ideas and information clearly. o CCSS.ELA-LITERACY.W.3.2.A: Introduce a topic and group related information together; include illustrations when useful to aiding comprehension. o CCSS.ELA-LITERACY.W.3.2.B: Develop the topic with facts, definitions, and details. o CCSS.ELA-LITERACY.W.3.2.C: Use linking words and phrases (e.g., *also*, *another*, *and*, *more*, *but*) to connect ideas within categories of information. o CCSS.ELA-LITERACY.W.3.2.D: Provide a concluding statement or section. Production and distribution of writing • CCSS.ELA-LITERACY.W.3.4: With guidance and support from adults, produce writing in which the development and organization are appropriate to task and purpose. • CCSS.ELA-LITERACY.W.3.5: With guidance and support from peers and adults, develop and strengthen writing as needed by planning, revising, and editing. • CCSS.ELA-LITERACY.W.3.6: With guidance and support from adults, use technology to produce and publish writing (using keyboarding skills) as well as to interact and collaborate with others. Research to build and present knowledge • CCSS.ELA-LITERACY.W.3.8: Recall information from experiences or gather information from print and digital sources; take brief notes on sources and sort evidence into provided categories. Range of writing • CCSS.ELA-LITERACY.W.3.10: Write routinely over extended time frames (time for research, reflection, and revision) and shorter time frames (a single sitting or a day or two) for a range of discipline-specific tasks, purposes, and audiences.

Continued

Table 13.2 (*continued*)

<table>
<tr><td>CCSS ELA—Speaking and Listening</td><td>Comprehension and collaboration
• CCSS.ELA-LITERACY.SL.3.1: Engage effectively in a range of collaborative discussions (one-on-one, in groups, and teacher-led) with diverse partners on grade 3 topics and texts, building on others' ideas and expressing their own clearly.
o CCSS.ELA-LITERACY.SL.3.1.A: Come to discussions prepared, having read or studied required material; explicitly draw on that preparation and other information known about the topic to explore ideas under discussion.
o CCSS.ELA-LITERACY.SL.3.1.B: Follow agreed-upon rules for discussions (e.g., gaining the floor in respectful ways, listening to others with care, speaking one at a time about the topics and texts under discussion).
o CCSS.ELA-LITERACY.SL.3.1.C: Ask questions to check understanding of information presented, stay on topic, and link their comments to the remarks of others.
o CCSS.ELA-LITERACY.SL.3.1.D: Explain their own ideas and understanding in light of the discussion.
• CCSS.ELA-LITERACY.SL.3.2: Determine the main ideas and supporting details of a text read aloud or information presented in diverse media and formats, including visually, quantitatively, and orally.
• CCSS.ELA-LITERACY.SL.3.3: Ask and answer questions about information from a speaker, offering appropriate elaboration and detail.
Presentation of knowledge and ideas
• CCSS.ELA-LITERACY.SL.3.4: Report on a topic or text, tell a story, or recount an experience with appropriate facts and relevant, descriptive details, speaking clearly at an understandable pace.
• CCSS.ELA-LITERACY.SL.3.6: Speak in complete sentences when appropriate to task and situation in order to provide requested detail or clarification.</td></tr>
</table>

Continued

Table 13.2 (*continued*)

***CCSS Mathematics*—Operations and Algebraic Thinking**	Represent and solve problems involving multiplication and division. • CCSS.MATH.CONTENT.3.OA.A.2: Interpret whole-number quotients of whole numbers • CCSS.MATH.CONTENT.3.OA.A.3: Use multiplication and division within 100 to solve word problems in situations involving equal groups, arrays, and measurement quantities • CCSS.MATH.CONTENT.3.OA.A.4: Determine the unknown whole number in a multiplication or division equation relating three whole numbers. Understand properties of multiplication and the relationship between multiplication and division. • CCSS.MATH.CONTENT.3.OA.B.5: Apply properties of operations as strategies to multiply and divide. Multiply and divide within 100. • CCSS.MATH.CONTENT.3.OA.C.7: Fluently multiply and divide within 100, using strategies such as the relationship between multiplication and division. Solve problems involving the four operations, and identify and explain patterns in arithmetic. • CCSS.MATH.CONTENT.3.OA.D.8: Solve two-step word problems using the four operations. Represent these problems using equations with a letter standing for the unknown quantity. Assess the reasonableness of answers using mental computation and estimation strategies including rounding.
***CCSS Mathematics*—Number and Operations in Base Ten**	Use place value understanding and properties of operations to perform multi-digit arithmetic. • CCSS.MATH.CONTENT.3.NBT.A.1: Use place value understanding to round whole numbers to the nearest 10 or 100. • CCSS.MATH.CONTENT.3.NBT.A.2: Fluently add and subtract within 1,000 using strategies and algorithms based on place value, properties of operations, and/or the relationship between addition and subtraction. • CCSS.MATH.CONTENT.3.NBT.A.3: Multiply one-digit whole numbers by multiples of 10 in the range 10–90 (e.g., 9 × 80, 5 × 60) using strategies based on place value and properties of operations.
***CCSS Mathematics*—Measurement and Data**	Represent and interpret data. • CCSS.MATH.CONTENT.3.MD.B.3: Draw a scaled picture graph and a scaled bar graph to represent a data set with several categories. Solve one- and two-step "how many more" and "how many less" problems using information presented in scaled bar graphs.

Continued

Table 13.2 (*continued*)

ELP Standards	Receptive modalities • ELP 1: Construct meaning from oral presentations and literary and informational text through grade-appropriate listening, reading, and viewing. • ELP 8: Determine the meaning of words and phrases in oral presentations and literary and informational text. Productive modalities • ELP 3: Speak and write about grade-appropriate complex literary and informational texts and topics. • ELP 4: Construct grade-appropriate oral and written claims and support them with reasoning and evidence. • ELP 7: Adapt language choices to purpose, task, and audience when speaking and writing. Interactive modalities • ELP 2: Participate in grade-appropriate oral and written exchanges of information, ideas, and analyses, responding to peer, audience, or reader comments and questions. • ELP 5: Conduct research and evaluate and communicate findings to answer questions or solve problems. • ELP 6: Analyze and critique the arguments of others orally and in writing. Linguistic structures of English • ELP 9: Create clear and coherent grade-appropriate speech and text. • ELP 10: Make accurate use of standard English to communicate in grade-appropriate speech and writing.

Investigation Handout

Investigation 13

Weather Patterns: What Weather Conditions Can We Expect Here During Each Season?

Introduction

Weather is the current condition of the atmosphere at a specific location. You can also think of weather as what we see outside on a particular day. For example, it may be 75 degrees and sunny outside, or it may be 45 degrees and raining. Weather is important because many of us make plans about where we will go and what we will wear based on current weather conditions. Over the next week, you will have a chance to track how the weather conditions do or do not change. Keep track of what you observe each day and what you are wondering about in the boxes below.

Things I observed on …				
Day 1	Day 2	Day 3	Day 4	Day 5

Things I WONDER about …

Meteorologists are people who study and predict the weather. A meteorologist will often describe the current weather conditions using several different measurements: temperature, humidity, wind speed, precipitation, and cloud cover.

- *Temperature:* Temperature is measured in degrees Fahrenheit or degrees Celsius and tells us how hot or how cold it is outside.
- *Humidity:* Humidity tells us how much water is in the air and is often reported as a percentage (for example, 50%). When it is very humid outside, the air feels wet. People often describe a very humid day as being "muggy" or "sticky."
- *Wind speed:* Wind is the movement of air from one spot to another. Wind is measured in miles per hour (mph). During a major category 5 hurricane, wind speeds can reach over 160 mph!
- *Precipitation:* Precipitation is the amount of rain, snow, sleet, or hail that falls over a specific time period. It is usually reported in millimeters or inches.
- *Cloud cover:* The amount of clouds in the sky can be described as clear (no clouds in the sky), partly cloudy (less than half cloud cover), mostly cloudy (more than half cloud cover but with some breaks), or overcast (complete cloud cover).

Meteorologists and other scientists record patterns of the weather over time and at different locations so that they can make predictions about what kind of weather to expect during each season. This is important to know because people often want to know what the weather tends to be like in different cities during each season so they know what types of plants to grow, when to plant them, and what type of clothes to buy. If someone just moved to your city, they might ask you questions like, "Do you get a lot of rain here?" or "Is it really hot during the summer?" In this investigation you will have an opportunity to figure out what kind of weather conditions can be expected where you live during each season.

Things we KNOW from what we read …	What we will NEED to figure out …

Your Task

Use information about the daily weather conditions over the last year and what you know about the weather, seasons, and patterns to determine the typical weather that can be expected at your school during each season. Spring starts on March 21 and ends on June 20. Summer begins on June 21 and ends on September 20. Fall begins on September 21 and ends on December 20. Winter starts on December 21 and ends on March 20. Be sure to pick at least two measurements to describe the daily weather conditions at your school.

The *guiding question* of this investigation is, ***What weather conditions can we expect here during each season?***

Materials

You may use any of the following materials during your investigation:

Equipment

- Computer or tablet with internet access
- Thermometer (optional)
- Barometer (optional)
- Hygrometer (optional)
- Anemometer (optional)
- Weather app for smartphone or tablet (optional)

You will need to access a website called World Weather and Climate Information. The website is at *https://weather-and-climate.com.*

Safety Rules

Follow all normal safety rules.

Plan Your Investigation

Prepare a plan for your investigation by filling out the chart that follows; this plan is called an *investigation proposal*. Before you start developing your plan, be sure to discuss the following questions with the other members of your group:

- What information should we collect so we can **describe** the typical weather here?
- What types of **patterns** might we look for to help answer the guiding question?

Our guiding question:

We will collect the following data:

These are the steps we will follow to collect data:

I approve of this investigation proposal.

Teacher's signature

Date

Collect Your Data

Keep a record of what you measure or observe during your investigation in the space below.

Analyze Your Data

You will need to analyze the data you decided to use before you can develop an answer to the guiding question. In the space that follows, create two graphs. Each graph should show how a different weather measurement changes by season where you live.

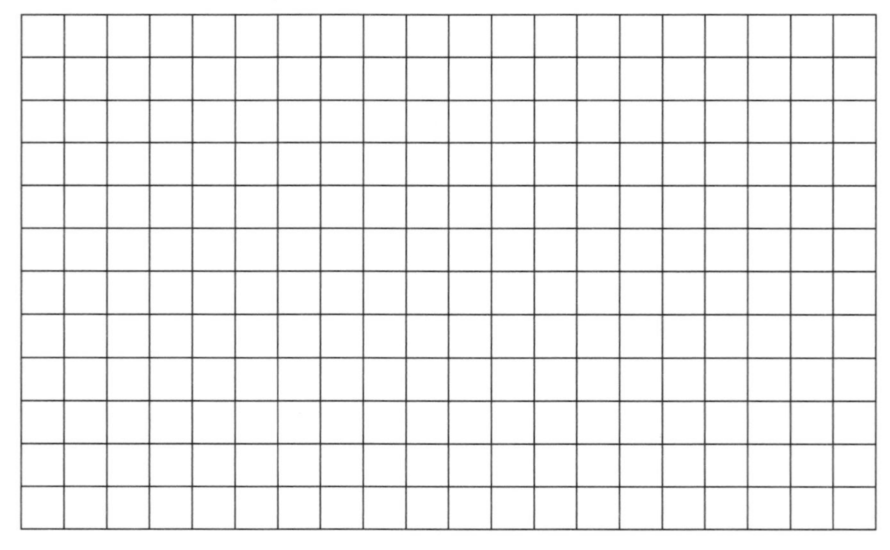

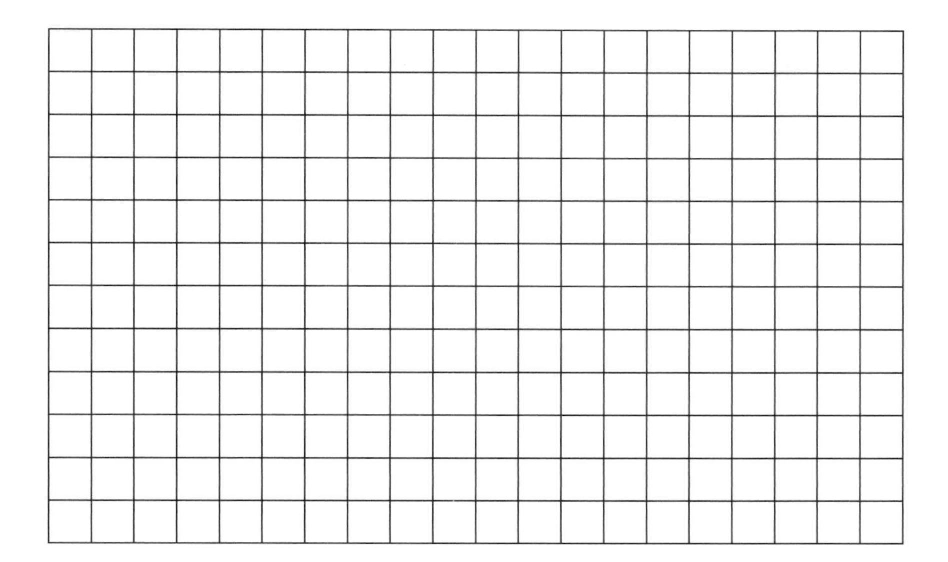

Draft Argument

Develop an argument on a whiteboard. It should include the following parts:

1. A *claim:* Your answer to the guiding question.
2. *Evidence:* An analysis of the data and an explanation of what the analysis means.
3. A *justification of the evidence:* Why your group thinks the evidence is important.

The Guiding Question:	
Our Claim:	
Our Evidence:	Our Justification of the Evidence:

Argumentation Session

Share your argument with your classmates. Be sure to ask them how to make your draft argument better. Keep track of their suggestions in the space below.

Ways to IMPROVE our argument …

Draft Report

Prepare an *investigation report* to share what you have learned. Use the information in this handout and your group's final argument to write a *draft* of your investigation report.

Introduction

We have been studying ______________________________ in class. Before we started this investigation, we explored ______________________________

____

____

We noticed ______________________________

____

____

My goal for this investigation was to figure out ______________________________

____

____

The guiding question was ______________________________

____

____

Method

To gather the data I needed to answer this question, I ______________________________

____

____

____

____

____

____

I then analyzed the data I collected by ____

____

____

____

____

Argument

My claim is ____

____

____

____

____

Figure 1 below shows ____

____

____

____

____

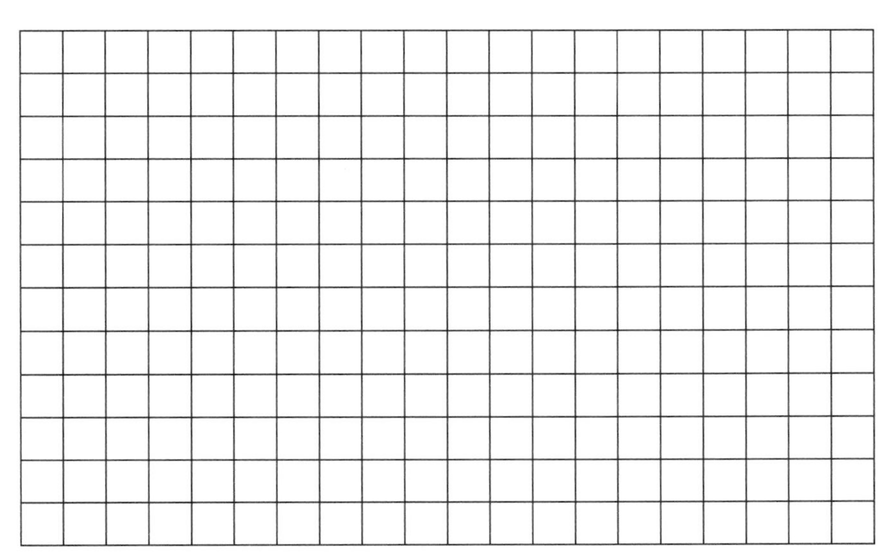

Figure 2 below shows ____

____

____

____

____

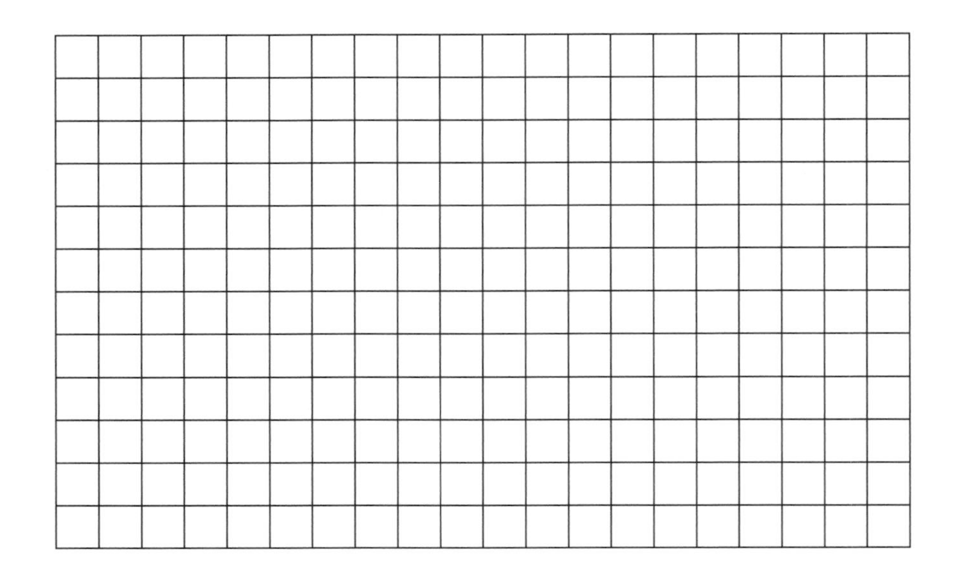

This evidence is important because ____

____

____

____

____

Review

Your friends need your help! Review the draft of their investigation reports and give them ideas about how to improve. Use the *peer-review guide* when doing your review.

Submit Your Final Report

Once you have received feedback from your friends about your draft report, create your final investigation report and hand it in to your teacher.

Checkout Questions

Investigation 13. Weather Patterns: What Weather Conditions Can We Expect Here During Each Season?

The graph below shows the average amount of rainfall in Austin, Texas, by month. This graph is missing the amount of average rainfall for the months of August and October.

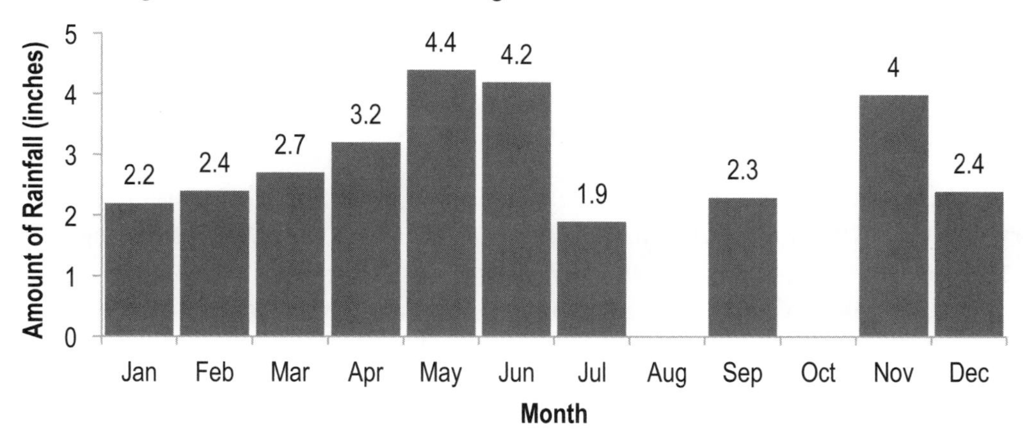

1. How much rainfall would you expect to see in Austin during the month of August?

____

2. How much rainfall would you expect to see in Austin during the month of October?

____

3. Explain your thinking. What *pattern* from your investigation did you use to predict the amount of rainfall in Austin during August and October?

____

____

____

____

____

Teacher Scoring Rubric for the Checkout Questions

Level	Description
3	The student can apply the core idea correctly in all cases and can fully explain the pattern.
2	The student can apply the core idea correctly in all cases but cannot fully explain the pattern.
1	The student cannot apply the core idea correctly in all cases but can fully explain the pattern.
0	The student cannot apply the core idea correctly in all cases and cannot explain the pattern.

Investigation 14

Climate and Location: How Does the Climate Change as One Moves From the Equator Toward the Poles?

Purpose

The purpose of this investigation is to give students an opportunity to use the disciplinary core idea (DCI) of ESS2.D: Weather and Climate along with the crosscutting concept (CC) of Patterns from *A Framework for K–12 Science Education* (NRC 2012) to figure out how latitude affects climate. Students will also learn about how scientific knowledge can change over time during the reflective discussion.

The Disciplinary Core Idea

Students in third grade should understand the following about the Weather and Climate and be able to use this DCI to figure out how latitude affects climate:

> *Weather is the combination of sunlight, wind, snow or rain, and temperature in a particular region at a particular time. People measure these conditions to describe and record the weather and to notice patterns over time. (NRC 2012, p. 188)*

The Crosscutting Concept

Students in third grade should understand the following about the CC Patterns:

> *Noticing patterns is often a first step to organizing and asking scientific questions about why and how the patterns occur. One major use of pattern recognition is in classification, which depends on careful observation of similarities and differences; objects can be classified into groups on the basis of similarities of visible or microscopic features or on the basis of similarities of function. Such classification is useful in codifying relationships and organizing a multitude of objects or processes into a limited number of groups. (NRC 2012, p. 85)*

Students in third grade should be given opportunities to "describe and predict the patterns in the seasons of the year" (NRC 2012, p. 86).

Students should be encouraged to use their developing understanding of patterns as a tool or a way of thinking about a phenomenon during this investigation to help them figure out how latitude affects climate.

What Students Figure Out

Climate describes the range of the typical weather conditions and how much variation there is in the conditions at a specific location over long periods of time. Latitude affects climate. Cities located near the poles have a larger seasonal temperature range than cities located near the equator. Cities near the equator also tend to have higher amounts of annual precipitation.

Timeline

The time needed to complete this investigation is 285 minutes (4 hours and 45 minutes). The amount of instructional time needed for each stage of the investigation is as follows:

- *Stage 1.* Introduce the task and the guiding question: 50 minutes
- *Stage 2.* Design a method and collect data: 60 minutes
- *Stage 3.* Create a draft argument: 35 minutes
- *Stage 4.* Argumentation session: 30 minutes
- *Stage 5.* Reflective discussion: 15 minutes
- *Stage 6.* Write a draft report: 35 minutes
- *Stage 7.* Peer review: 30 minutes
- *Stage 8.* Revise the report: 30 minutes

This investigation can be completed in one day or over eight days (one day for each stage) during your designated science time in the daily schedule.

Materials and Preparation

The materials needed for this investigation are listed in Table 14.1 (p. 512). Students will need access to a computer or tablet with an internet connection to collect data about the typical weather for each city from the World Weather and Climate Information website at *https://weather-and-climate.com*. We recommend at least one computer or tablet per group, but each student can use a computer or tablet on his or her own if there are enough available. Be sure to visit the website and learn how to find the information that the students will need before starting the investigation so you can show students how to use it and help them when they get stuck. In addition, it is important to check if students can access and use the website from a school computer or tablet, because some schools have set up firewalls and other restrictions on web browsing.

TABLE 14.1

Materials for Investigation 14

Item	Quantity
Computer or tablet with internet access	1 per group
Whiteboard, 2' × 3'*	1 per group
Investigation Handout	1 per student
Peer-review guide and teacher scoring rubric	1 per student
Checkout Questions (optional)	1 per student

* As an alternative, students can use computer and presentation software such as Microsoft PowerPoint or Apple Keynote to create their arguments.

Safety Precautions

Remind students to follow all normal safety rules.

Lesson Plan by Stage

Stage 1: Introduce the Task and the Guiding Question (50 minutes)

1. Ask the students to sit in six groups, with three or four students in each group.
2. Ask students to clear off their desks except for a pencil (and their *Student Workbook for Argument-Driven Inquiry in Third-Grade Science* if they have one).
3. Pass out an Investigation Handout to each student (or ask students to turn to Investigation Log 14 in their workbook).
4. Read the first paragraph of the "Introduction" aloud to the class. Ask the students to follow along as you read.
5. Show students how to use the grid lines to identify the *latitude* (the distance north or south from the equator) and *longitude* (the distance east or west of the *prime meridian,* which is an imaginary line running from north to south through Greenwich, England) of one of the cities on the map.
6. Tell students to find the latitude and longitude of the remaining cities on the map and then record their observations and questions about the cities in the "NOTICED/WONDER" chart in the "Introduction" section of their Investigation Handout (or the investigation log in their workbook).
7. Give the students 10 minutes to find the latitude and longitude of the remaining cities and record their observations and questions.
8. After the students have recorded their observations and questions, ask students to share *what they noticed* about the cities.
9. Ask students to share *what questions they have* about the cities.

10. Tell the students, "Some of your questions might be answered by reading the rest of the 'Introduction.'"
11. Ask the students to read the rest of the "Introduction" on their own *or* ask them to follow along as you read aloud.
12. Once the students have read the rest of the "Introduction," ask them to fill out the "KNOW/NEED" chart on their Investigation Handout (or in their investigation log) as a group.
13. Ask students to share what they learned from the reading. Add these ideas to a class "know / need to figure out" chart.
14. Ask students to share what they think they will need to figure out based on what they read. Add these ideas to the class "know / need to figure out" chart.
15. Tell the students, "It looks like we have something to figure out. Let's see what we will need to do during our investigation."
16. Read the task and the guiding question aloud.
17. Tell the students, "You will be able to use a website called World Weather and Climate Information during your investigation."
18. Show the students how to use the website by projecting it on the board or on a screen and demonstrating how to select a country and a city in that country and then find the data they need.
19. Remind students of the safety rules and precautions for this investigation.

Stage 2: Design a Method and Collect Data (60 minutes)

1. Tell the students, "I am now going to give you and the other members of your group about 15 minutes to plan your investigation. Before you begin, I want you all to take a couple of minutes to discuss the following questions with the rest of your group."
2. Show the following questions on the screen or board:
   - What information should we collect so we can *describe* the climate of a city?
   - What types of *patterns* might we look for to help answer the guiding question?
3. Tell the students, "Please take a few minutes to come up with an answer to these questions." Give the students two or three minutes to discuss these two questions.
4. Ask two or three different groups to share their answers. Be sure to highlight or write down any important ideas on the board so students can refer to them later.
5. If possible, use a document camera to project an image of the graphic organizer for this investigation on a screen or board (or take a picture of it and project the picture on a screen or board). Tell the students, "I now want you all to plan out

your investigation. To do that, you will need to create an investigation proposal by filling out this graphic organizer."

6. Point to the box labeled "Our guiding question:" and tell the students, "You can put the question we are trying to answer in this box." Then ask, "Where can we find the guiding question?"
7. Wait for a student to answer where to find the guiding question (the answer is "in the handout").
8. Point to the box labeled "We will collect the following data:" and tell the students, "You can list the measurements or observations that you will need to collect during the investigation in this box."
9. Point to the box labeled "These are the steps we will follow to collect data:" and tell the students, "You can list what you are going to do to collect the data you need and what you will do with your data once you have it. Be sure to give enough detail that I could do your investigation for you."
10. Ask the students, "Do you have any questions about what you need to do?"
11. Answer any questions that come up.
12. Tell the students, "Once you are done, raise your hand and let me know. I'll then come by and look over your proposal and give you some feedback. You may not begin collecting data until I have approved your proposal by signing it. You need to have your proposal done in the next 15 minutes."
13. Give the students 15 minutes to work in their groups on their investigation proposal. As they work, move from group to group to check in, ask probing questions, and offer a suggestion if a group gets stuck.
14. As each group finishes its investigation proposal, be sure to read it over and determine if it will be productive or not. If you feel the investigation will be productive (not necessarily what you would do or what the other groups are doing), sign your name on the proposal and let the group start collecting data. If the plan needs to be changed, offer some suggestions or ask some probing questions, and have the group make the changes before you approve it.
15. Tell the students to collect their data and record their observations or measurements in the "Collect Your Data" box in their Investigation Handout (or the investigation log in their workbook).
16. Give the students 20 minutes to collect their data.

What should a student-designed investigation look like?

The students' investigation proposal should include the following information:

- The guiding question is "How does the climate change as one moves from the equator toward the poles?"
- There are a lot of different types of data that students can collect during this investigation. Students can collect data about (1) average monthly maximum temperature, (2) average monthly minimum temperature, (3) average monthly precipitation, (4) average monthly relative humidity, and (5) average mean wind speed. At a minimum, each group should collect data about two different weather measurements. This investigation works best if each group selects different measurements used to describe the climate of a city.
- The steps that the students will follow to collect the data should reflect the measurements that they decide to examine. However, a procedure might include the following steps: (1) identify two cities, (2) record typical [weather condition 1] for each month in both cities, (3) record typical [weather condition 2] for each month in both cities, and (4) compare how these values change by month in each city. There should be a lot of variation in the student-designed investigations.

Stage 3: Create a Draft Argument (35 minutes)

1. Tell the students, "Now that we have all this data, we need to analyze the data so we can figure out an answer to the guiding question."
2. If possible, project an image of the "Analyze Your Data" section for this investigation on a screen or board using a document camera (or take a picture of it and project the picture on a screen or board). Point to the section and tell the students, "You can create a couple of graphs as a way to analyze your data. You can make your graphs in this section."
3. Ask the students, "What information do we need to include in these graphs?"
4. Tell the students, "Please take a few minutes to discuss this question with your group, and be ready to share."
5. Give the students five minutes to discuss.
6. Ask two or three different groups to share their answers. Be sure to highlight or write down any important ideas on the board so students can refer to them later.
7. Tell the students, "I am now going to give you and the other members of your group about 10 minutes to create your graphs." If the students are having

trouble making a graph, you can take a few minutes to provide a mini-lesson about how to create a graph from a bunch of observations or measurements (this strategy is called just-in-time instruction because it is offered only when students get stuck).

What should a graph look like for this investigation?

There are a number of different ways that students can analyze the observations or measurements they collect during this investigation. One of the most straightforward ways is to create a scaled bar graph, one for each weather measurement (e.g., high temperature, low temperature, humidity, wind speed, precipitation). Each bar graph should have categories for each month on the horizontal or *x*-axis. The average value for a weather condition (e.g., average high temperature, average low temperature, average humidity, average wind speed, average precipitation) should be on the *y*-axis. An example of this type of graph can be seen in Figure 14.1.

8. Give the students 10 minutes to analyze their data. As they work, move from group to group to check in, ask probing questions, and offer suggestions.
9. Tell the students, "I am now going to give you and the other members of your group 15 minutes to create an argument to share what you have learned and convince others that they should believe you. Before you do that, we need to take a few minutes to discuss what you need to include in your argument."
10. If possible, use a document camera to project the "Argument Presentation on a Whiteboard" image from the Investigation Handout (or the investigation log in their workbook) on a screen or board (or take a picture of it and project the picture on a screen or board).
11. Point to the box labeled "The Guiding Question:" and tell the students, "You can put the question we are trying to answer here on your whiteboard."
12. Point to the box labeled "Our Claim:" and tell the students, "You can put your claim here on your whiteboard. The claim is your answer to the guiding question."
13. Point to the box labeled "Our Evidence:" and tell the students, "You can put the evidence that you are using to support your claim here on your whiteboard. Your evidence will need to include the analysis you just did and an explanation of what your analysis means or shows. Scientists always need to support their claims with evidence."
14. Point to the box labeled "Our Justification of the Evidence:" and tell the students, "You can put your justification of your evidence here on your whiteboard. Your

justification needs to explain why your evidence is important. Scientists often use core ideas to explain why the evidence they are using matters. Core ideas are important concepts that scientists use to help them make sense of what happens during an investigation."

15. Ask the students, "What are some core ideas that we read about earlier that might help us explain why the evidence we are using is important?"
16. Ask students to share some of the core ideas from the "Introduction" section of the Investigation Handout (or the investigation log in the workbook). List these core ideas on the board.
17. Tell the students, "That is great. I would like to see everyone try to include these core ideas in your justification of the evidence. Your goal is to use these core ideas to help explain why your evidence matters and why the rest of us should pay attention to it."
18. Ask the students, "Do you have any questions about what you need to do?"
19. Answer any questions that come up.
20. Tell the students, "Okay, go ahead and start working on your arguments. You need to have your argument done in the next 15 minutes. It doesn't need to be perfect. We just need something down on the whiteboards so we can share our ideas."
21. Give the students 15 minutes to work in their groups on their arguments. As they work, move from group to group to check in, ask probing questions, and offer a suggestion if a group gets stuck. Figure 14.1 shows an example of an argument created by students for this investigation.

FIGURE 14.1

Example of an argument

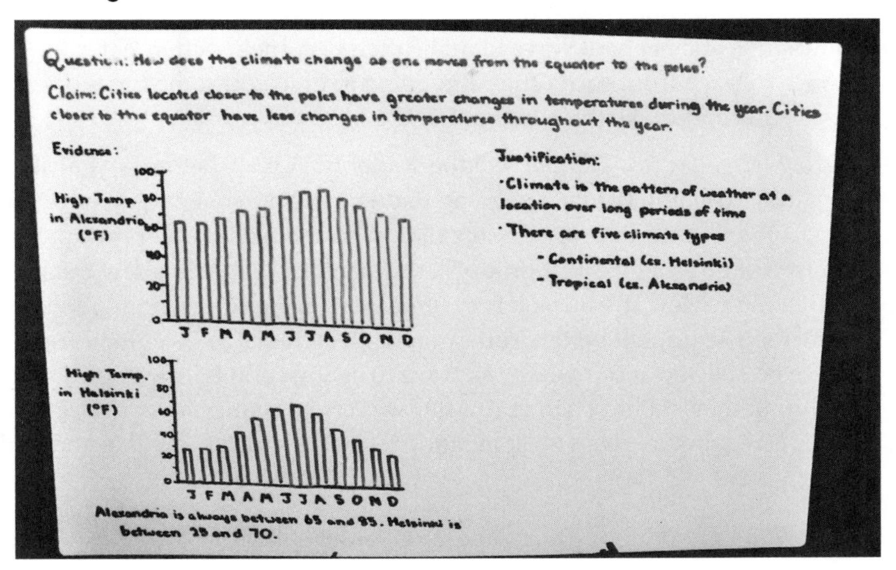

Stage 4: Argumentation Session (30 minutes)

The argumentation session can be conducted in a whole-class presentation format, a gallery walk format, or a modified gallery walk format. We recommend using a whole-class presentation format for the first investigation, but try to transition to either the gallery walk or modified gallery walk format as soon as possible because that will maximize student voice and choice inside the classroom. The following list shows the steps for the three formats; unless otherwise noted, the steps are the same for all three formats.

1. Begin by introducing the use of the whiteboard.
   - *If using the whole-class presentation format,* tell the students, "We are now going to share our arguments. Please set up your whiteboard so everyone can see them."
   - *If using the gallery walk or modified gallery walk format,* tell the students, "We are now going to share our arguments. Please set up your whiteboard so they are facing the walls."
2. Allow the students to set up their whiteboards.
   - *If using the whole-class presentation format,* the whiteboards should be set up on stands or chairs so they are facing toward the center of the room.
   - *If using the gallery walk or modified gallery walk format,* the whiteboards should be set up on stands or chairs so they are facing toward the outside of the room.
3. Give the following instructions to the students:
   - *If using the whole-class presentation format or the modified gallery walk format,* tell the students, "Okay, before we get started I want to explain what we are going to do next. I'm going to ask some of you to present your arguments to your classmates. If you are presenting your argument, your job is to share your group's claim, evidence, and justification of the evidence. The rest of you will be reviewers. If you are a reviewer, your job is to listen to the presenters, ask the presenters questions if you do not understand something, and then offer them some suggestions about ways to make their argument better. After we have a chance to learn from each other, I'm going to give you some time to revise your arguments and make them better."
   - *If using the gallery walk format,* tell the students, "Okay, before we get started I want to explain what we are going to do next. You are going to have an opportunity to read the arguments that were created by other groups. Your group will go to a different group's argument. I'll give you a few minutes to read it and review it. Your job is to offer them some suggestions about ways to make their argument better. You can use sticky notes to give them suggestions. Please be specific about what you want to change and be specific about how you think they should change it. After we have a chance to learn from each other, I'm going to give you some time to revise your arguments and make them better."

4. Use a document camera to project the "Ways to IMPROVE our argument …" box from the Investigation Handout (or the investigation log in their workbook) on a screen or board (or take a picture of it and project the picture on a screen or board).
   - *If using the whole-class presentation format or the modified gallery walk format,* point to the box and tell the students, "If you are a presenter, you can write down the suggestions you get from the reviewers here. If you are a reviewer, and you see a good idea from another group, you can write down that idea here. Once we are done with the presentations, I will give you a chance to use these suggestions or ideas to improve your arguments.
   - *If using the gallery walk format,* point to the box and tell the students, "If you see good ideas from another group, you can write them down here. Once we are done reviewing the different arguments, I will give you a chance to use these ideas to improve your own arguments. It is important to share ideas like this."

   Ask the students, "Do you have any questions about what you need to do?"
5. Answer any questions that come up.
6. Give the following instructions:
   - *If using the whole-class presentation format,* tell the students, "Okay. Let's get started."
   - *If using the gallery walk format,* tell the students, "Okay, I'm now going to tell you which argument to go to and review.
   - *If using the modified gallery walk format,* tell the students, "Okay, I'm now going to assign you to be a presenter or a reviewer." Assign one or two students from each group to be presenters and one or two students from each group to be reviewers.
7. Begin the review of the arguments.
   - *If using the whole-class presentation format,* have four or five groups present their argument one at a time. Give each group only two to three minutes to present their argument. Then give the class two to three minutes to ask them questions and offer suggestions. Be sure to encourage as much participation from the students as possible.
   - *If using the gallery walk format,* tell the students, "Okay. Let's get started. Each group, move one argument to the left. Don't move to the next argument until I tell you to move. Once you get there, read the argument and then offer suggestions about how to make it better. I will put some sticky notes next to each argument. You can use the sticky notes to leave your suggestions." Give each group about three to four minutes to read the arguments, talk, and offer suggestions.

     a. Tell the students, "Okay. Let's rotate. Move one group to the left."

b. Again, give each group three or four minutes to read, talk, and offer suggestions.

c. Repeat this process for two more rotations.

- *If using the modified gallery walk format,* tell the students, "Okay. Let's get started. Reviewers, move one group to the left. Don't move to the next group until I tell you to move. Presenters, go ahead and share your argument with the reviewers when they get there." Give each group of presenters and reviewers about three to four minutes to talk.

  a. Tell the students, "Okay. Let's rotate. Reviewers, move one group to the left."

  b. Again, give each group of presenters and reviewers about three or four minutes to talk.

  c. Repeat this process for two more rotations.

8. Tell the students to return to their workstations.
9. Give the following instructions about revising the argument:
   - *If using the whole-class presentation format,* tell the students, "I'm now going to give you about 10 minutes to revise your argument. Take a few minutes to talk in your groups and determine what you want to change to make your argument better. Once you have decided what to change, go ahead and make the changes to your whiteboard."
   - *If using the gallery walk format,* tell the students, "I'm now going to give you about 10 minutes to revise your argument. Take a few minutes to read the suggestions that were left at your argument. Then talk in your groups and determine what you want to change to make your argument better. Once you have decided what to change, go ahead and make the changes to your whiteboard."
   - *If using the modified gallery walk format,* "I'm now going to give you about 10 minutes to revise your argument. Please return to your original groups." Wait for the students to move back into their original groups and then tell the students, "Okay, take a few minutes to talk in your groups and determine what you want to change to make your argument better. Once you have decided what to change, go ahead and make the changes to your whiteboard."

   Ask the students, "Do you have any questions about what you need to do?"
10. Answer any questions that come up.
11. Tell the students, "Okay. Let's get started."
12. Give the students 10 minutes to work in their groups on their arguments. As they work, move from group to group to check in, ask probing questions, and offer a suggestion if a group gets stuck.

Stage 5: Reflective Discussion (15 minutes)

1. Tell the students, "We are now going to take a minute to talk about what we did and what we have learned."
2. Show Figure 14.2 on a screen. This is a graph showing the average high temperature changes by month in two different cities. Ask the students, "What do you all see going on here?"

FIGURE 14.2

Average high temperature by month in two different cities

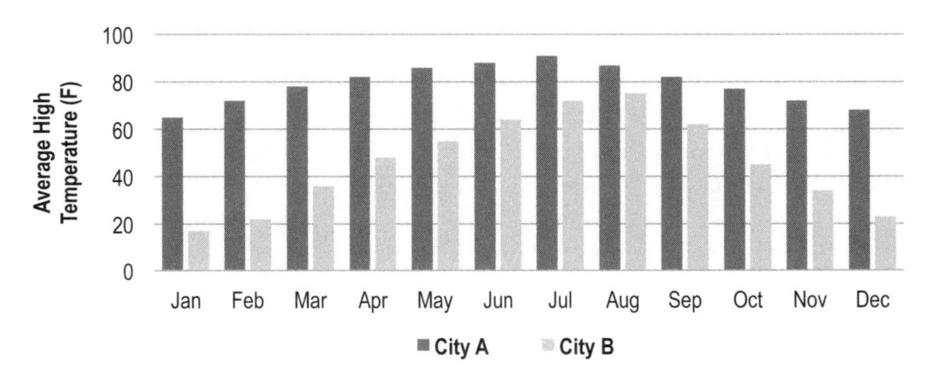

3. Allow students to share their ideas.
4. Ask the students, "Which city do you think is located closer to the equator, and how do you know?"
5. Allow students to share their ideas.
6. Show Figure 14.3 on the screen. This is a graph showing how the average amount of precipitation changes by month in the same two cities.

FIGURE 14.3

Average amount of precipitation by month in two different cities

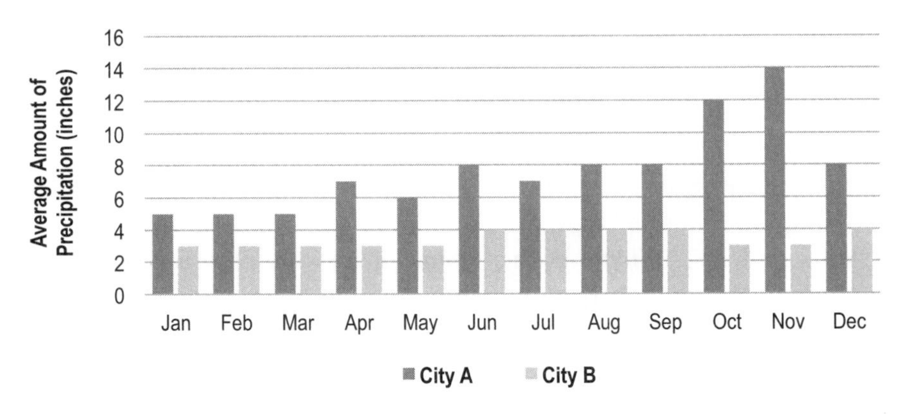

7. Ask the students, "What do you all see going on here?"
8. Allow students to share their ideas.
9. Ask the students, "What type of climate do you think city A has, and how do you know?"
10. Allow students to share their ideas.
11. Ask the students, "What type of climate do you think city B has, and how do you know?"
12. Allow three or four different students to share their ideas.
13. Tell the students, "Okay, let's make sure we are on the same page. Climate describes the range of the typical weather conditions and how much variation there is in the conditions at a specific location over long periods of time. Latitude affects climate. Cities located near the North Pole or the South Pole have a larger seasonal temperature range than cities located near the equator. Cities near the equator also tend to have higher amounts of annual precipitation. This description of climate and how we can predict the climate of different cities based on where they are located is a really important core idea in science."
14. Ask the students, "Does anyone have any questions about this core idea?"
15. Answer any questions that come up.
16. Tell the students, "We also looked for patterns during our investigation." Then show Figures 14.2 and 14.3 (p. 521) on the screen for a second time.
17. Ask the students, "What pattern do you see here?"
18. Allow students to share their ideas.
19. Tell the students, "Patterns are really important in science. Scientists look for patterns all the time and use them to make predictions. In fact, they even use patterns in the weather to describe climates, just like we did."
20. Tell the students, "We are now going take a minute to talk about what went well and what didn't go so well during our investigation. We need to talk about this because you are going to be planning and carrying out your own investigations like this a lot this year, and I want to help you all get better at it."
21. Show an image of the question "What made your investigation scientific?" on the screen. Tell the students, "Take a few minutes to talk about how you would answer this question with the other people in your group. Be ready to share with the rest of the class." Give the students two to three minutes to talk in their group.
22. Ask the students, "What do you all think? Who would like to share an idea?"
23. Allow students to share their ideas. Be sure to expand on their ideas about what makes an investigation scientific.

24. Show an image of the question "What made your investigation not so scientific?" on the screen. Tell the students, "Take a few minutes to talk about how you would answer this question with the other people in your group. Be ready to share with the rest of the class." Give the students two to three minutes to talk in their group.
25. Ask the students, "What do you all think? Who would like to share an idea?"
26. Allow students to share their ideas. Be sure to expand on their ideas about what makes an investigation less scientific.
27. Show an image of the question "What rules can we put into place to help us make sure our next investigation is more scientific?" on the screen. Tell the students, "Take a few minutes to talk about how you would answer this question with the other people in your group. Be ready to share with the rest of the class." Give the students two to three minutes to talk in their group.
28. Ask the students, "What do you all think? Who would like to share an idea?"
29. Allow students to share their ideas. Once they have shared their ideas, offer a suggestion for a possible class rule.
30. Ask the students, "What do you all think? Should we make this a rule?"
31. If the students agree, write the rule on the board or make a class "Rules for Scientific Investigation" chart so you can refer to it during the next investigation.
32. Tell the students, "We are now going take a minute to talk about what makes science different from other subjects."
33. Show an image of the question "Does scientific knowledge ever change?" on the screen. Tell the students, "Take a few minutes to talk about how you would answer this question with the other people in your group. Be ready to share with the rest of the class." Give the students two to three minutes to talk in their group.
34. Ask the students, "What do you all think? Who would like to share an idea?"
35. Allow three or four students to share their ideas.
36. Tell the students, "Okay, these are all great ideas. Always remember that scientific knowledge can change as scientists collect and analyze more data. For example, the way we describe the climate of Helsinki or Izmir may change if the patterns in the weather measurements that they collect change over time. That is another characteristic of scientific knowledge—it is based on evidence. So, if what scientists currently know about something is no longer supported by what they observe or measure, then scientists change what they know."
37. Ask the students, "Does anyone have any questions about how scientific knowledge can change over time?"
38. Answer any questions that come up.

Stage 6: Write a Draft Report (35 minutes)

Your students will use either the Investigation Handout or the investigation log in the student workbook when writing the draft report. When you give the directions shown in quotes in the following steps, substitute "investigation log" (as shown in brackets) for "handout" if they are using the workbook.

1. Tell the students, "You are now going to write an investigation report to share what you have learned. Please take out a pencil and turn to the 'Draft Report' section of your handout [investigation log]."
2. If possible, use a document camera to project the "Introduction" section of the draft report from the Investigation Handout (or the investigation log in their workbook) on a screen or board (or take a picture of it and project the picture on a screen or board).
3. Tell the students, "The first part of the report is called the 'Introduction.' In this section of the report you want to explain to the reader what you were investigating, why you were investigating it, and what question you were trying to answer. All of this information can be found in the text at the beginning of your handout [investigation log]." Point to the image and say, "There are some sentence starters here to help you begin writing the report." Ask the students, "Do you have any questions about what you need to do?"
4. Answer any questions that come up.
5. Tell the students, "Okay. Let's write."
6. Give the students 10 minutes to write the "Introduction" section of the report. As they work, move from student to student to check in, ask probing questions, and offer a suggestion if a student gets stuck.
7. If possible, use a document camera to project the "Method" section of the draft report from the Investigation Handout (or the investigation log in their workbook) on a screen or board (or take a picture of it and project the picture on a screen or board).
8. Tell the students, "The second part of the report is called the 'Method.' In this section of the report you want to explain to the reader what you did during the investigation, what data you collected and why, and how you went about analyzing your data. All of this information can be found in the 'Plan Your Investigation' section of your handout [investigation log]. Remember that you all planned and carried out different investigations, so do not assume that the reader will know what you did." Point to the image and say, "There are some sentence starters here to help you begin writing this part of the report." Ask the students, "Do you have any questions about what you need to do?"
9. Answer any questions that come up.
10. Tell the students, "Okay. Let's write."

11. Give the students 10 minutes to write the "Method" section of the report. As they work, move from student to student to check in, ask probing questions, and offer a suggestion if a student gets stuck.
12. If possible, use a document camera to project the "Argument" section of the draft report from the Investigation Handout (or the investigation log in their workbook) on a screen or board (or take a picture of it and project the picture on a screen or board).
13. Tell the students, "The last part of the report is called the 'Argument.' In this section of the report you want to share your claim, evidence, and justification of the evidence with the reader. All of this information can be found on your whiteboard." Point to the image and say, "There are some sentence starters here to help you begin writing this part of the report." Ask the students, "Do you have any questions about what you need to do?"
14. Answer any questions that come up.
15. Tell the students, "Okay. Let's write."
16. Give the students 10 minutes to write the "Argument" section of the report. As they work, be sure to move from student to student to check in, ask probing questions, and offer a suggestion if a student gets stuck.

Stage 7: Peer Review (30 minutes)

Your students will use either the Investigation Handout or the investigation log in the student workbook when doing the peer review. When you give the directions shown in quotes in the following steps, substitute "workbook" (as shown in brackets) for "Investigation Handout" if they are using the workbook.

1. Tell the students, "We are now going to review our reports to find ways to make them better. I'm going to come around and collect your Investigation Handout [workbook]. While I do that, please take out a pencil."
2. Collect the Investigation Handouts or workbooks from the students.
3. If possible, use a document camera to project the peer-review guide (PRG; see Appendix 4) on a screen or board (or take a picture of it and project the picture on a screen or board).
4. Tell the students, "We are going to use this peer-review guide to give each other feedback." Point to the image.
5. Give the following instructions:
    - *If using the Investigation Handout,* tell the students, "I'm going to ask you to work with a partner to do this. I'm going to give you and your partner a draft report to read and a peer-review guide to fill out. You two will then read the report together. Once you are done reading the report, I want you to answer each of

the questions on the peer-review guide." Point to the review questions on the image of the PRG.

- *If using the workbook,* tell the students, "I'm going to ask you to work with a partner to do this. I'm going to give you and your partner a draft report to read. You two will then read the report together. Once you are done reading the report, I want you to answer each of the questions on the peer-review guide that is right after the report in the investigation log." Point to the review questions on the image of the PRG.

6. Tell the students, "You can check 'yes,' 'almost,' or 'no' after each question." Point to the checkboxes on the image of the PRG.
7. Tell the students, "This will be your rating for this part of the report. Make sure you agree on the rating you give the author. If you mark 'almost' or 'no,' then you need to tell the author what he or she needs to do to get a 'yes.'" Point to the space for the reviewer feedback on the image of the PRG.
8. Tell the students, "It is really important for you to give the authors feedback that is helpful. That means you need to tell them exactly what they need to do to make their reports better." Ask the students, "Do you have any questions about what you need to do?"
9. Answer any questions that come up.
10. Tell the students, "Please sit with a partner who is not in your current group." Allow the students time to sit with a partner.
11. Give the following instructions:
    - *If using the Investigation Handout,* tell the students, "Okay, I am now going to give you one report to read and one peer-review guide to fill out." Pass out one report to each pair. Make sure that the report you give a pair was not written by one of the students in that pair. Give each pair one PRG to fill out as a team.
    - *If using the workbook,* tell the students, "Okay, I am now going to give you one report to read." Pass out a workbook to each pair. Make sure that the workbook you give a pair is not from one of the students in that pair.
12. Tell the students, "Okay, I'm going to give you 15 minutes to read the report I gave you and to fill out the peer-review guide. Go ahead and get started."
13. Give the students 15 minutes to work. As they work, move around from pair to pair to check in and see how things are going, answer questions, and offer advice.
14. After 15 minutes pass, tell the students, "Okay, time is up." *If using the Investigation Handout,* say, "Please give me the report and the peer-review guide that you filled out." *If using the workbook,* say, "Please give me the workbook that you have."

15. Collect the Investigation Handouts and the PRGs, or collect the workbooks if they are being used. Be sure you keep the handout and the PRG together.
16. Give the following instructions:
    - *If using the Investigation Handout,* tell the students, "Okay, I am now going to give you a different report to read and a new peer-review guide to fill out." Pass out another report to each pair. Make sure that this report was not written by one of the students in that pair. Give each pair a new PRG to fill out as a team.
    - *If using the workbook,* tell the students, "Okay, I am now going to give you a different report to read." Pass out a different workbook to each pair. Make sure that the workbook you give a pair is not from one of the students in that pair.
17. Tell the students, "Okay, I'm going to give you 15 minutes to read this new report and to fill out the peer-review guide. Go ahead and get started."
18. Give the students 15 minutes to work. As they work, move around from pair to pair to check in and see how things are going, answer questions, and offer advice.
19. After 15 minutes pass, tell the students, "Okay, time is up." *If using the Investigation Handout,* say, "Please give me the report and the peer-review guide that you filled out." *If using the workbook,* say, "Please give me the workbook that you have."
20. Collect the Investigation Handouts and the PRGs, or collect the workbooks if they are being used. Be sure you keep the handout and the PRG together.

Stage 8: Revise the Report (30 minutes)

Your students will use either the Investigation Handout or the investigation log in the student workbook when revising the report. Except where noted below, the directions are the same whether using the handout or the log.

1. Give the following instructions:
    - *If using the Investigation Handout,* tell the students, "You are now going to revise your investigation report based on the feedback you get from your classmates. Please take out a pencil while I hand back your draft report and the peer-review guide."
    - *If using the investigation log in the student workbook,* tell the students, "You are now going to revise your investigation report based on the feedback you get from your classmates. Please take out a pencil while I hand back your investigation logs."
2. *If using the Investigation Handout,* pass back the handout and the PRG to each student. *If using the investigation log,* pass back the log to each student.

3. Tell the students, "Please take a few minutes to read over the peer-review guide. You should use it to figure out what you need to change in your report and how you will change the report."
4. Allow the students time to read the PRG.
5. *If using the investigation log,* if possible use a document camera to project the "Write Your Final Report" section from the investigation log on a screen or board (or take a picture of it and project the picture on a screen or board).
6. Give the following instructions:
   - *If using the Investigation Handout,* tell the students, "Okay. Let's revise our reports. Please take out a piece of paper. I would like you to rewrite your report. You can use your draft report as a starting point, but use the feedback on the peer-review guide to help make it better."
   - *If using the investigation log,* tell the students, "Okay. Let's revise our reports. I would like you to rewrite your report in the section of the investigation log that says 'Write Your Final Report.'" Point to the image on the screen and tell the students, "You can use your draft report as a starting point, but use the feedback on the peer-review guide to help make your report better."

   Ask the students, "Do you have any questions about what you need to do?"
7. Answer any questions that come up.
8. Tell the students, "Okay. Let's write."
9. Give the students 30 minutes to rewrite their report. As they work, move from student to student to check in, ask probing questions, and offer a suggestion if a student gets stuck.
10. Give the following instructions:
    - *If using the Investigation Handout,* tell the students, "Okay. Time's up. I will now come around and collect your Investigation Handout, the peer-review guide, and your final report."
    - *If using the investigation log,* tell the students, "Okay. Time's up. I will now come around and collect your workbooks."
11. *If using the Investigation Handout,* collect all the Investigation Handouts, PRGs, and final reports. *If using the investigation log,* collect all the workbooks.
12. *If using the Investigation Handout,* use the "Teacher Score" columns in the PRG to grade the final report. *If using the investigation log,* use the "ADI Investigation Report Grading Rubric" in the investigation log to grade the final report. Whether you are using the handout or the log, you can give the students feedback about their writing in the "Teacher Comments" section.

How to Use the Checkout Questions

The Checkout Questions are an optional assessment. We recommend giving them to students one day after they finish stage 8 of the ADI investigation. The Checkout Questions can be used as a formative or summative assessment of student thinking. If you plan to use them as a formative assessment, we recommend that you look over the student answers to determine if you need to reteach the core idea and/or crosscutting concept from the investigation, but do not grade them. If you plan to use them as a summative assessment, we have included a "Teacher Scoring Rubric" at the end of the Checkout Questions that you can use to score a student's ability to apply the core idea in a new scenario and explain their use of a crosscutting concept. The rubric includes a 4-point scale that ranges from 0 (the student cannot apply the core idea correctly in all cases and cannot explain the [crosscutting concept]) to 3 (the student can apply the core idea correctly in all cases and can fully explain the [crosscutting concept]). The Checkout Questions, regardless of how you decide to use them, are a great way to make student thinking visible so you can determine if the students have learned the core idea and the crosscutting concept.

A student who can apply the core idea correctly in all cases and can explain the pattern would give the following answers: question 1, city A; question 2, city A; question 3, continental; and question 4, tropical. He or she should then be able to explain that latitude affects climate, so cities located near the equator tend to be warmer all year round and have more annual precipitation than cities located farther from the equator.

Connections to Standards

Table 14.2 (p. 530) highlights how the investigation can be used to address specific performance expectations from the *NGSS, Common Core State Standards (CCSS)* in English language arts (ELA) and in mathematics, and *English Language Proficiency (ELP) Standards.*

TABLE 14.2

Investigation 14 alignment with standards

***NGSS* performance expectation**	Strong alignment • 3-ESS2-2. Obtain and combine information to describe climates in different regions of the world.
***CCSS ELA*—Reading: Informational Text**	Key ideas and details • CCSS.ELA-LITERACY.RI.3.1: Ask and answer questions to demonstrate understanding of a text, referring explicitly to the text as the basis for the answers. • CCSS.ELA-LITERACY.RI.3.2: Determine the main idea of a text; recount the key details and explain how they support the main idea. • CCSS.ELA-LITERACY.RI.3.3: Describe the relationship between a series of historical events, scientific ideas or concepts, or steps in technical procedures in a text, using language that pertains to time, sequence, and cause/effect. Craft and structure • CCSS.ELA-LITERACY.RI.3.4: Determine the meaning of general academic and domain-specific words and phrases in a text relevant to a *grade 3 topic or subject area*. • CCSS.ELA-LITERACY.RI.3.5: Use text features and search tools (e.g., key words, sidebars, hyperlinks) to locate information relevant to a given topic efficiently. • CCSS.ELA-LITERACY.RI.3.6: Distinguish their own point of view from that of the author of a text. Integration of knowledge and ideas • CCSS.ELA-LITERACY.RI.3.7: Use information gained from illustrations (e.g., maps, photographs) and the words in a text to demonstrate understanding of the text (e.g., where, when, why, and how key events occur). • CCSS.ELA-LITERACY.RI.3.8: Describe the logical connection between particular sentences and paragraphs in a text (e.g., comparison, cause/effect, first/second/third in a sequence). • CCSS.ELA-LITERACY.RI.3.9: Compare and contrast the most important points and key details presented in two texts on the same topic. Range of reading and level of text complexity • CCSS.ELA-LITERACY.RI.3.10: By the end of the year, read and comprehend informational texts, including history/social studies, science, and technical texts, at the high end of the grades 2–3 text complexity band independently and proficiently.

Continued

Table 14.2 *(continued)*

***CCSS ELA*—Writing**	Text types and purposes
	• CCSS.ELA-LITERACY.W.3.1: Write opinion pieces on topics or texts, supporting a point of view with reasons.
	  o CCSS.ELA-LITERACY.W.3.1.A: Introduce the topic or text they are writing about, state an opinion, and create an organizational structure that lists reasons.
	  o CCSS.ELA-LITERACY.W.3.1.B: Provide reasons that support the opinion.
	  o CCSS.ELA-LITERACY.W.3.1.C: Use linking words and phrases (e.g., *because*, *therefore*, *since, for example*) to connect opinion and reasons.
	  o CCSS.ELA-LITERACY.W.3.1.D: Provide a concluding statement or section.
	• CCSS.ELA-LITERACY.W.3.2: Write informative or explanatory texts to examine a topic and convey ideas and information clearly.
	  o CCSS.ELA-LITERACY.W.3.2.A: Introduce a topic and group related information together; include illustrations when useful to aiding comprehension.
	  o CCSS.ELA-LITERACY.W.3.2.B: Develop the topic with facts, definitions, and details.
	  o CCSS.ELA-LITERACY.W.3.2.C: Use linking words and phrases (e.g., *also*, *another*, *and*, *more*, *but*) to connect ideas within categories of information.
	  o CCSS.ELA-LITERACY.W.3.2.D: Provide a concluding statement or section.
	Production and distribution of writing
	• CCSS.ELA-LITERACY.W.3.4: With guidance and support from adults, produce writing in which the development and organization are appropriate to task and purpose.
	• CCSS.ELA-LITERACY.W.3.5: With guidance and support from peers and adults, develop and strengthen writing as needed by planning, revising, and editing.
	• CCSS.ELA-LITERACY.W.3.6: With guidance and support from adults, use technology to produce and publish writing (using keyboarding skills) as well as to interact and collaborate with others.
	Research to build and present knowledge
	• CCSS.ELA-LITERACY.W.3.8: Recall information from experiences or gather information from print and digital sources; take brief notes on sources and sort evidence into provided categories.
	Range of writing
	• CCSS.ELA-LITERACY.W.3.10: Write routinely over extended time frames (time for research, reflection, and revision) and shorter time frames (a single sitting or a day or two) for a range of discipline-specific tasks, purposes, and audiences.

Continued

Table 14.2 (*continued*)

***CCSS ELA—* Speaking and Listening**	Comprehension and collaboration • CCSS.ELA-LITERACY.SL.3.1: Engage effectively in a range of collaborative discussions (one-on-one, in groups, and teacher-led) with diverse partners on *grade 3 topics and texts*, building on others' ideas and expressing their own clearly. o CCSS.ELA-LITERACY.SL.3.1.A: Come to discussions prepared, having read or studied required material; explicitly draw on that preparation and other information known about the topic to explore ideas under discussion. o CCSS.ELA-LITERACY.SL.3.1.B: Follow agreed-upon rules for discussions (e.g., gaining the floor in respectful ways, listening to others with care, speaking one at a time about the topics and texts under discussion). o CCSS.ELA-LITERACY.SL.3.1.C: Ask questions to check understanding of information presented, stay on topic, and link their comments to the remarks of others. o CCSS.ELA-LITERACY.SL.3.1.D: Explain their own ideas and understanding in light of the discussion. • CCSS.ELA-LITERACY.SL.3.2: Determine the main ideas and supporting details of a text read aloud or information presented in diverse media and formats, including visually, quantitatively, and orally. • CCSS.ELA-LITERACY.SL.3.3: Ask and answer questions about information from a speaker, offering appropriate elaboration and detail. Presentation of knowledge and ideas • CCSS.ELA-LITERACY.SL.3.4: Report on a topic or text, tell a story, or recount an experience with appropriate facts and relevant, descriptive details, speaking clearly at an understandable pace. • CCSS.ELA-LITERACY.SL.3.6: Speak in complete sentences when appropriate to task and situation in order to provide requested detail or clarification.
***CCSS Mathematics—* Measurement and Data**	Represent and interpret data • CCSS.MATH.CONTENT.3.MD.B.3: Draw a scaled picture graph and a scaled bar graph to represent a data set with several categories. Solve one- and two-step "how many more" and "how many less" problems using information presented in scaled bar graphs.

Continued

Table 14.2 (*continued*)

ELP Standards	Receptive modalities • ELP 1: Construct meaning from oral presentations and literary and informational text through grade-appropriate listening, reading, and viewing. • ELP 8: Determine the meaning of words and phrases in oral presentations and literary and informational text. Productive modalities • ELP 3: Speak and write about grade-appropriate complex literary and informational texts and topics. • ELP 4: Construct grade-appropriate oral and written claims and support them with reasoning and evidence. • ELP 7: Adapt language choices to purpose, task, and audience when speaking and writing. Interactive modalities • ELP 2: Participate in grade-appropriate oral and written exchanges of information, ideas, and analyses, responding to peer, audience, or reader comments and questions. • ELP 5: Conduct research and evaluate and communicate findings to answer questions or solve problems. • ELP 6: Analyze and critique the arguments of others orally and in writing. Linguistic structures of English • ELP 9: Create clear and coherent grade-appropriate speech and text. • ELP 10: Make accurate use of standard English to communicate in grade-appropriate speech and writing.

Investigation 14

Climate and Location: How Does the Climate Change as One Moves From the Equator Toward the Poles?

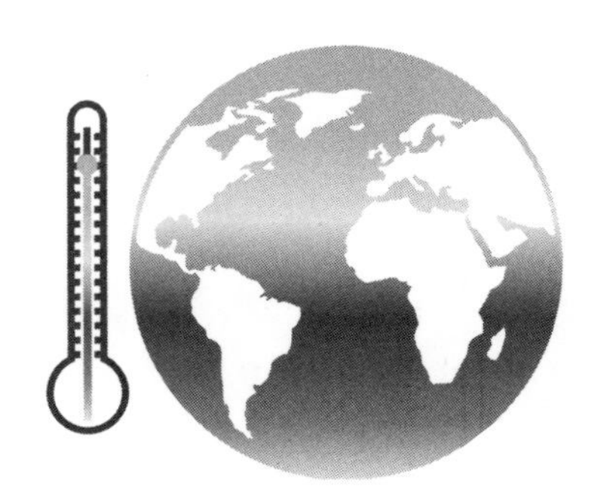

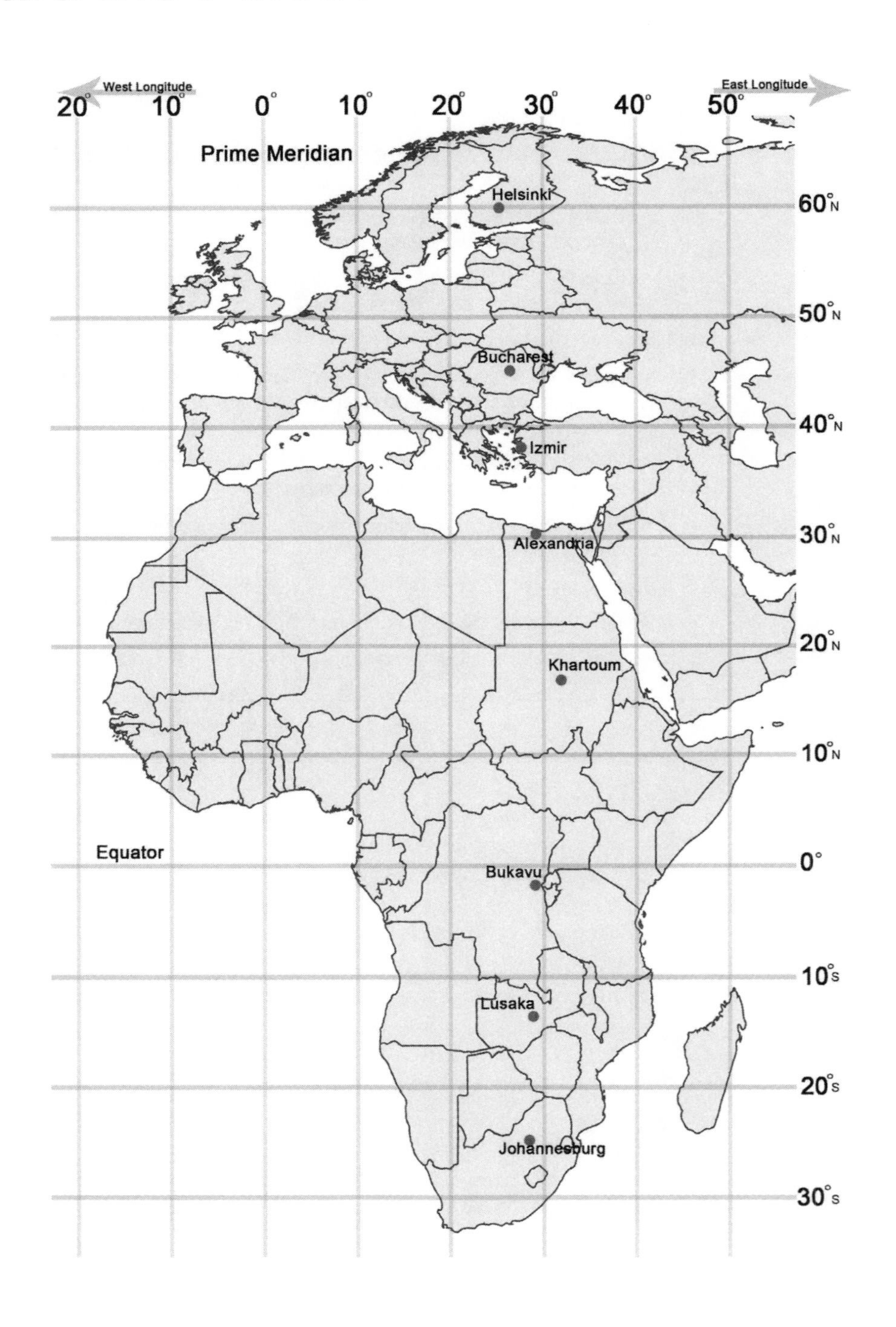

Introduction

There are many cities all over the world. All of these cities have characteristics that make them special. Take a minute to find the latitude and longitude of the eight different cities labeled on the map on the previous page. Keep track of what you notice and what you are wondering about in the boxes below.

City	Longitude	Latitude
Helsinki, Finland		
Bucharest, Romania		
Izmir, Turkey		
Alexandria, Egypt		

City	Longitude	Latitude
Khartoum, Sudan		
Bukavu, Democratic Republic of the Congo		
Lusaka, Zambia		
Johannesburg, South Africa		

Things I NOTICED …	Things I WONDER about …

People often want to know about the climate and the current weather in a city before they travel to that city so they know what type of clothing to bring with them. *Climate* is a pattern of weather in a particular region over a long period of time. *Weather* is the current condition of the atmosphere at a specific place. We describe the weather by measuring the air temperature, humidity, wind speed, precipitation, and cloud cover. Weather can change from hour to hour or day to day.

A region's weather patterns, tracked for more than 30 years, are used to describe the climate of that region. There are five main climate types:

- *Tropical*—a region that is warm all year and gets a lot of rain
- *Dry*—a region that gets very little rain; it can be hot or cool
- *Mild*—a region with warm and dry summers and short, cool, but rainy winters
- *Continental*—a region with short summers and long winters that are cold with a lot of snow
- *Polar*—a region with temperatures that are cold all year

There are many reasons why different regions have different climates. One cause that may or may not affect the climate of a specific region is *latitude,* or how far that specific region is from the equator. Think about the eight cities shown on the map. These cities are all located at between about 25 and 30 East longitude, but each city is at a different latitude. *Longitude* is the distance east or west of the *prime meridian* (an imaginary line running from north to south through Greenwich, England). Some of the cities, in other words, are close to the equator and some are far away from the equator, even though all these cities are found on the same side of the Earth.

Your goal in this investigation is to first determine if these cities have different climates and then use this information to figure out if the climate at a specific location is related to how far it is from the equator. To accomplish this task, you will need to compare how the typical weather in at least two of these cities changes by month over an entire year. You can then use this information to look for a pattern. If you can find a pattern, you will be able to figure out how latitude and climate are related.

Things we KNOW from what we read …	What we will NEED to figure out …

Your Task

Use what you know about weather, climate, and patterns to determine the climate of at least two different cities that are located at a similar longitude but different latitudes. Then determine if there is a relationship between latitude and climate.

The *guiding question* of this investigation is, ***How does the climate change as one moves from the equator toward the poles?***

Materials

You will use a computer or tablet with internet access and a website called World Weather and Climate Information during your investigation. The website is at *https://weather-and-climate.com.*

Safety Rules

Follow all normal safety rules.

Plan Your Investigation

Prepare a plan for your investigation by filling out the chart that follows; this plan is called an *investigation proposal.* Before you start developing your plan, be sure to discuss the following questions with the other members of your group:

- What information should we collect so we can **describe** the climate of a city?
- What types of **patterns** might we look for to help answer the guiding question?

Our guiding question:

We will collect the following data:

These are the steps we will follow to collect data:

I approve of this investigation proposal.

Teacher's signature

Date

Collect Your Data

Keep a record of what you measure or observe during your investigation in the space below.

Analyze Your Data

You will need to analyze the data you collected before you can develop an answer to the guiding question. In the space that follows, create two graphs to illustrate how the climate of each city is different.

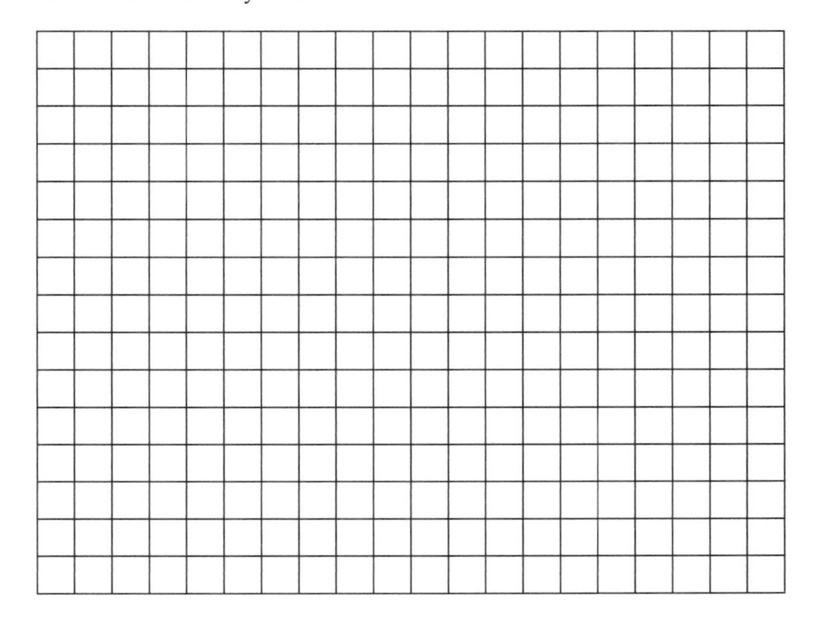

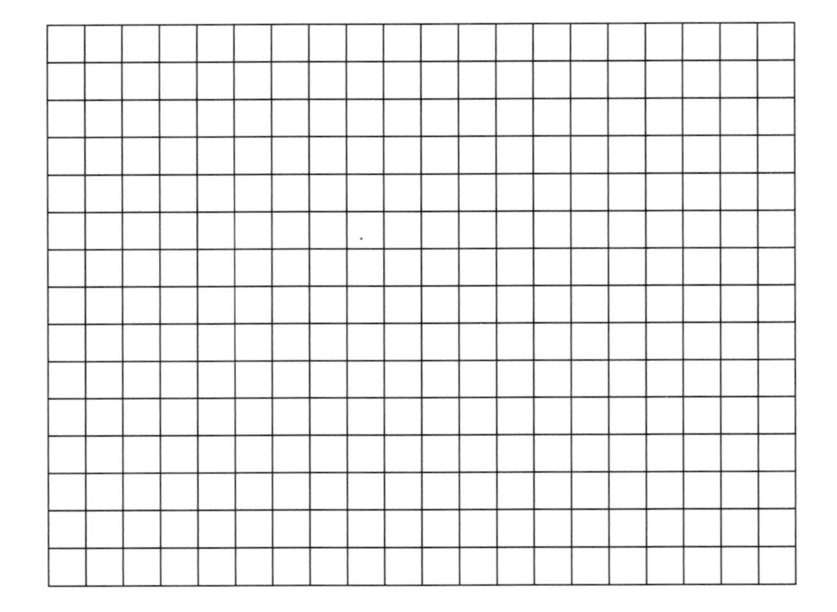

Draft Argument

Develop an argument on a whiteboard. It should include the following parts:

1. A claim: Your answer to the guiding question.
2. Evidence: An analysis of the data and an explanation of what the analysis means.
3. A justification of the evidence: Why your group thinks the evidence is important.

The Guiding Question:	
Our Claim:	
Our Evidence:	Our Justification of the Evidence:

Argumentation Session

Share your argument with your classmates. Be sure to ask them how to make your draft argument better. Keep track of their suggestions in the space below.

Ways to IMPROVE our argument ...

Draft Report

Prepare an investigation report to share what you have learned. Use the information in this handout and your group's final argument to write a draft of your investigation report.

Introduction

We have been studying ______________________________ in class. Before we started this investigation, we explored ______________________________

____

____

We noticed ______________________________

____

____

My goal for this investigation was to figure out ______________________________

____

____

The guiding question was ______________________________

____

____

Method

To gather the data I needed to answer this question, I ______________________________

____

____

____

____

____

____

____

____

____

I then analyzed the data I collected by ______________________________

____

____

____

Argument

My claim is ______________________________

____

____

____

Figure 1 below shows ______________________________

____

____

____

____

____

____

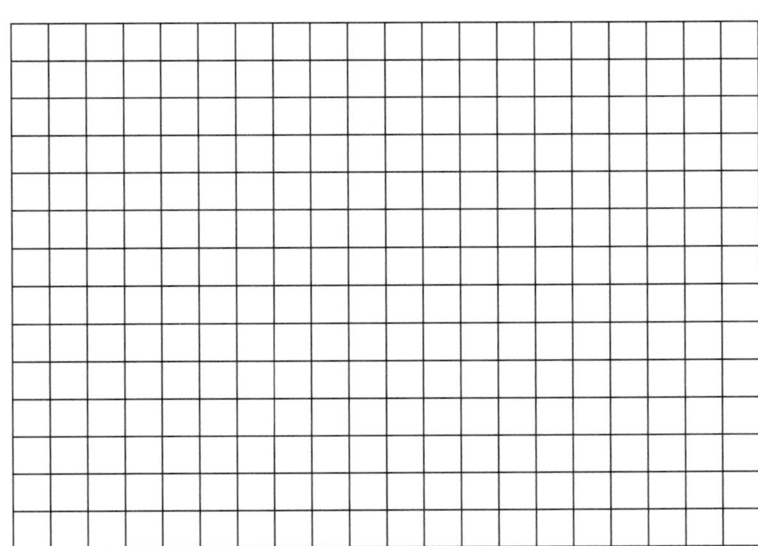

Figure 2 below shows ______________________________

This evidence is important because ______________________________

Review

Your friends need your help! Review the draft of their investigation reports and give them ideas about how to improve. Use the *peer-review guide* when doing your review.

Submit Your Final Report

Once you have received feedback from your friends about your draft report, create your final investigation report and hand it in to your teacher.

Checkout Questions

Investigation 14. Climate and Location: How Does the Climate Change as One Moves From the Equator Toward the Poles?

The graph below shows the average high temperature by month in two different cities. Both cities are located on the same line of longitude in the Western Hemisphere.

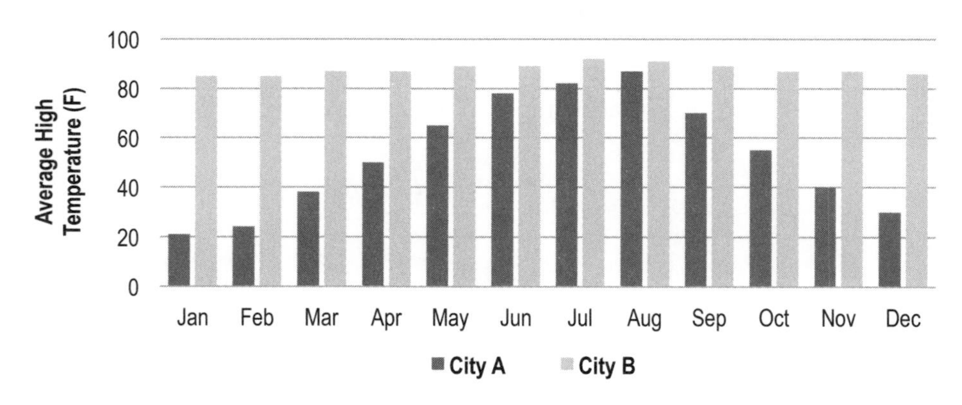

1. Which city has the greatest change in seasonal temperature?

☐ City A ☐ City B

2. Which city is most likely located farthest from the equator?

☐ City A ☐ City B

The graph below shows the average amount of precipitation by month in these two cities.

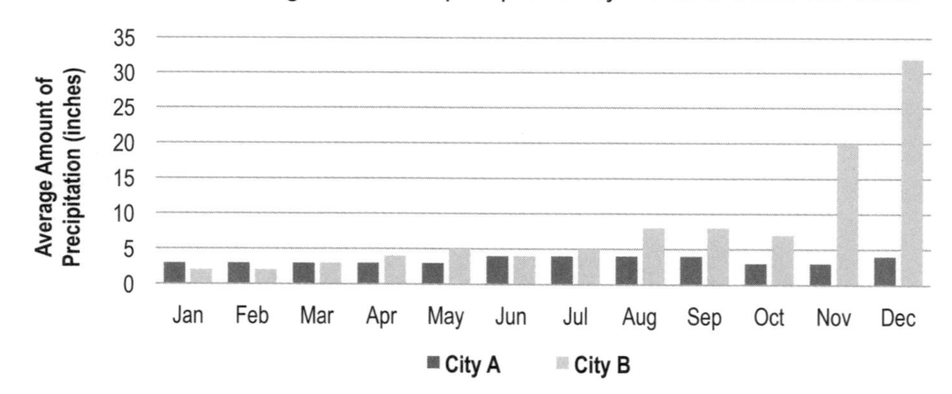

3. How would you describe the climate in city A based on all the information available?

☐ Tropical ☐ Continental

☐ Dry ☐ Polar

☐ Mild

4. How would you describe the climate in city B based on all the information available?

   ☐ Tropical ☐ Continental

   ☐ Dry ☐ Polar

   ☐ Mild

5. Explain your thinking. What *pattern* from your investigation did you use to determine the location and climate of these two cities?

   ___
   ___
   ___
   ___
   ___
   ___
   ___
   ___
   ___
   ___

Teacher Scoring Rubric for the Checkout Questions

Level	Description
3	The student can apply the core idea correctly in all cases and can fully explain the pattern.
2	The student can apply the core idea correctly in all cases but cannot fully explain the pattern.
1	The student cannot apply the core idea correctly in all cases but can fully explain the pattern.
0	The student cannot apply the core idea correctly in all cases and cannot explain the pattern.

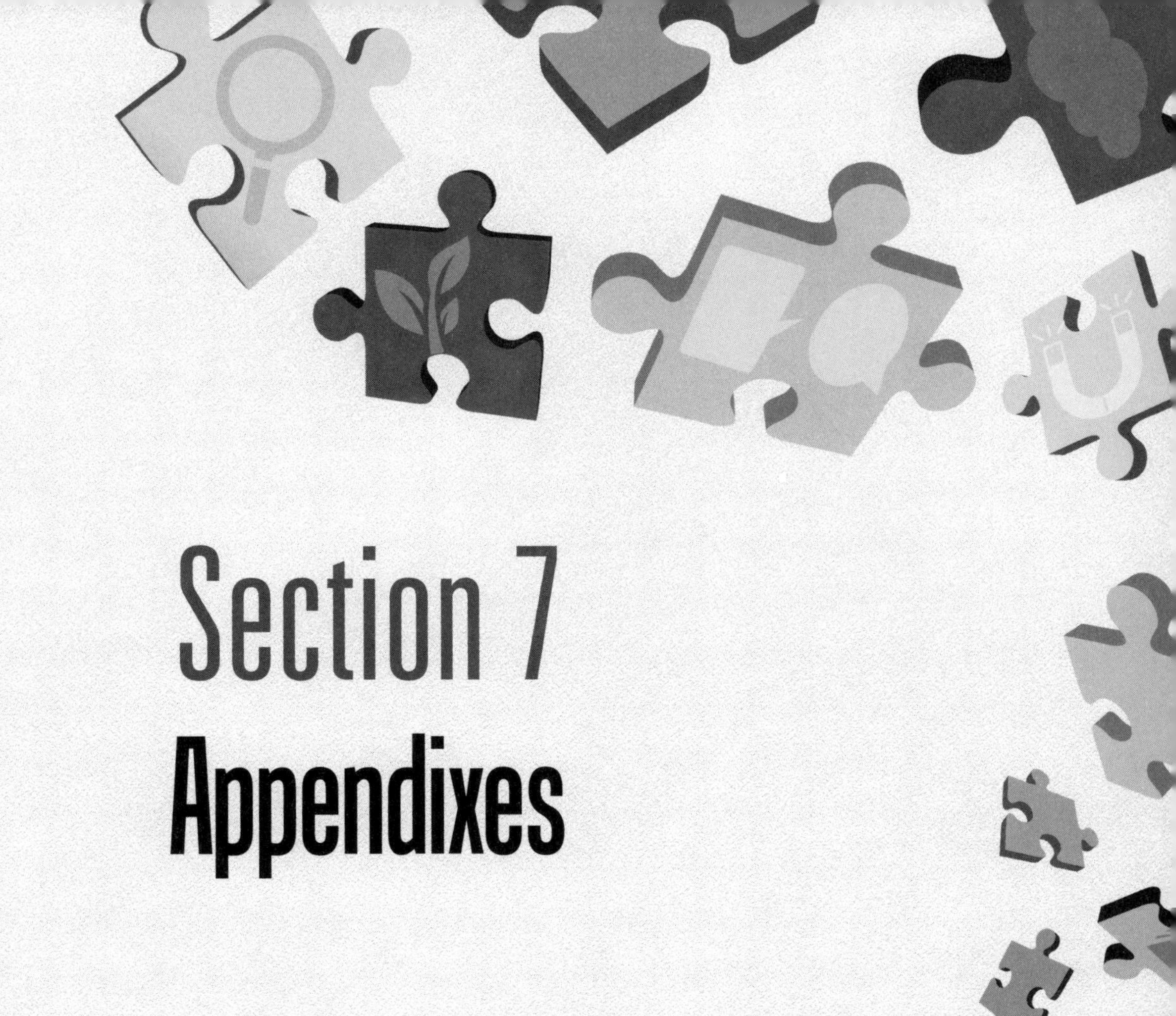

Section 7
Appendixes

APPENDIX 1

Standards Alignment Matrixes

Standards Matrix A: Alignment of the Argument-Driven Inquiry (ADI) Investigations With the Scientific and Engineering Practices (SEPs), Crosscutting Concepts (CCs), and Disciplinary Core Ideas (DCIs) in *A Framework for K–12 Science Education* (NRC 2012)

	ADI investigation													
Aspect of the *Framework*	Investigation 1. Magnetic Attraction	Investigation 2. Magnetic Force	Investigation 3. Changes in Motion	Investigation 4. Balanced and Unbalanced Forces	Investigation 5. Life Cycles	Investigation 6. Life in Groups	Investigation 7. Variation Within a Species	Investigation 8. Inheritance of Traits	Investigation 9. Traits and the Environment	Investigation 10. Fossils	Investigation 11. Differences in Traits	Investigation 12. Adaptations	Investigation 13. Weather Patterns	Investigation 14. Climate and Location
Scientific and engineering practices														
SEP 1. Asking Questions and Defining Problems	■	■	■	■	■	■	■	■	■	■	■	■	■	■
SEP 2. Developing and Using Models				□	■						■	■		
SEP 3. Planning and Carrying Out Investigations	■	■	■	■	■	■	■	■	■	■	■	■	■	■
SEP 4. Analyzing and Interpreting Data	■	■	■	■	■	■	■	■	■	■	■	■	■	■
SEP 5. Using Mathematics and Computational Thinking	■	■	■	■	■	■	■	■	■	■	■	■	■	■
SEP 6. Constructing Explanations and Designing Solutions	■	■	■	■	■	■	■	■	■	■	■	■	■	■
SEP 7. Engaging in Argument From Evidence	■	■	■	■	■	■	■	■	■	■	■	■	■	■
SEP 8. Obtaining, Evaluating, and Communicating Information	■	■	■	■	■	■	■	■	■	■	■	■	■	■
Crosscutting concepts														
CC 1. Patterns	■		■		■	□	■	■		□			■	■
CC 2. Cause and Effect: Mechanism and Explanation		■		■		■			■		■	■		
CC 3. Scale, Proportion, and Quantity	□		□											
CC 4. Systems and System Models														
CC 5. Energy and Matter: Flows, Cycles, and Conservation		□												
CC 6. Structure and Function					□		□	□	□	■	□	□		
CC 7. Stability and Change													□	□

Key: ■ = strong alignment; □ = moderate alignment.

Continued

Standards Matrix A *(continued)*

Aspect of the *Framework*	ADI investigation													
	Investigation 1. Magnetic Attraction	Investigation 2. Magnetic Force	Investigation 3. Changes in Motion	Investigation 4. Balanced and Unbalanced Forces	Investigation 5. Life Cycles	Investigation 6. Life in Groups	Investigation 7. Variation Within a Species	Investigation 8. Inheritance of Traits	Investigation 9. Traits and the Environment	Investigation 10. Fossils	Investigation 11. Differences in Traits	Investigation 12. Adaptations	Investigation 13. Weather Patterns	Investigation 14. Climate and Location
Disciplinary core ideas														
PS1. Matter and Its Interactions	■													
PS2. Motion and Stability: Forces and Interactions	□	■	■	■										
PS3. Energy			□											
LS1. From Molecules to Organisms: Structures and Process					■									
LS2. Ecosystems: Interactions, Energy, and Dynamics						■								
LS3. Heredity: Inheritance and Variation of Traits							■	■	■					
LS4. Biological Evolution: Unity and Diversity										■	■	■		
ESS2. Earth's Systems													■	■

Key: ■ = strong alignment; □ = moderate alignment.

Standards Matrix B: Alignment of the Argument-Driven Inquiry (ADI) Investigations With the *NGSS* Performance Expectations for Third-Grade Science (NGSS Lead States 2013)

NGSS performance expectations	ADI investigation													
	Investigation 1. Magnetic Attraction	Investigation 2. Magnetic Force	Investigation 3. Changes in Motion	Investigation 4. Balanced and Unbalanced Forces	Investigation 5. Life Cycles	Investigation 6. Life in Groups	Investigation 7. Variation Within a Species	Investigation 8. Inheritance of Traits	Investigation 9. Traits and the Environment	Investigation 10. Fossils	Investigation 11. Differences in Traits	Investigation 12. Adaptations	Investigation 13. Weather Patterns	Investigation 14. Climate and Location
Motion and Stability: Forces and Interactions														
3-PS2-1: Plan and conduct an investigation to provide evidence of the effects of balanced and unbalanced forces on the motion of an object.			□	■										
3-PS2-2: Make observations and/or measurements of an object's motion to provide evidence that a pattern can be used to predict future motion.			■											
3-PS2-3: Ask questions to determine cause and effect relationships of electric or magnetic interactions between two objects not in contact with each other.	□	■												
3-PS2-4: Define a simple design problem that can be solved by applying scientific ideas about magnets.	□	□												
From Molecules to Organisms: Structures and Process														
3-LS1-1: Develop models to describe that organisms have unique and diverse life cycles but all have in common birth, growth, reproduction, and death.					■									
Ecosystems: Interactions, Energy, and Dynamics														
3-LS2-1: Construct an argument that some animals form groups that help members survive.						■								
Heredity: Inheritance and Variation of Traits														
3-LS3-1: Analyze and interpret data to provide evidence that plants and animals have traits inherited from parents and that variation of these traits exists in a group of similar organisms							■	■	□					

Key: ■ = strong alignment; □ = moderate alignment.

Continued

Standards Matrix B (*continued*)

NGSS performance expectations	ADI investigation													
	Investigation 1. Magnetic Attraction	Investigation 2. Magnetic Force	Investigation 3. Changes in Motion	Investigation 4. Balanced and Unbalanced Forces	Investigation 5. Life Cycles	Investigation 6. Life in Groups	Investigation 7. Variation Within a Species	Investigation 8. Inheritance of Traits	Investigation 9. Traits and the Environment	Investigation 10. Fossils	Investigation 11. Differences in Traits	Investigation 12. Adaptations	Investigation 13. Weather Patterns	Investigation 14. Climate and Location
3-LS3-2: Use evidence to support the explanation that traits can be influenced by the environment.									■					
Biological Evolution: Unity and Diversity														
3-LS4-1: Analyze and interpret data from fossils to provide evidence of the organisms and the environments in which they lived long ago.										■				
3-LS4-2: Use evidence to construct an explanation for how the variations in characteristics among individuals of the same species may provide advantages in surviving, finding mates, and reproducing.											■			
3-LS4-3: Construct an argument with evidence that in a particular habitat some organisms can survive well, some survive less well, and some cannot survive at all.											□	■		
3-LS4-4: Make a claim about the merit of a solution to a problem caused when the environment changes and the types of plants and animals that live there may change.												□		
Earth's Systems														
3-ESS2-1: Represent data in tables and graphical displays to describe typical weather conditions expected during a particular season.													■	
3-ESS2-2: Obtain and combine information to describe climates in different regions of the world.														■

Key: ■ = strong alignment; □ = moderate alignment.

Standards Matrix C: Alignment of the Argument-Driven Inquiry (ADI) Investigations With the Nature of Scientific Knowledge (NOSK) and Nature of Scientific Inquiry (NOSI) Concepts*

	ADI Investigation													
NOSK and NOSI concepts	Investigation 1. Magnetic Attraction	Investigation 2. Magnetic Force	Investigation 3. Changes in Motion	Investigation 4. Balanced and Unbalanced Forces	Investigation 5. Life Cycles	Investigation 6. Life in Groups	Investigation 7. Variation Within a Species	Investigation 8. Inheritance of Traits	Investigation 9. Traits and the Environment	Investigation 10. Fossils	Investigation 11. Differences in Traits	Investigation 12. Adaptations	Investigation 13. Weather Patterns	Investigation 14. Climate and Location
NOSK														
How scientific knowledge changes over time														■
The difference between laws and theories in science									■					
The difference between data and evidence in science		■					■							
The difference between observations and inferences in science	■							■						
NOSI														
The types of questions that scientists can investigate					■									
How scientists use different methods to answer different types of questions						■					■	■	■	
Science as a way of knowing			■											
The assumptions made by scientists about order and consistency in nature				■						■				

Key: ■ = strong alignment.

*The NOSK/NOSI concepts listed in this matrix are based on the work of Abd-El-Khalick and Lederman 2000; Akerson, Abd-El-Khalick, and Lederman 2000; Lederman et al. 2002, 2014; Schwartz, Lederman, and Crawford 2004; and NGSS Lead States 2013.

Standards Matrix D: Alignment of the Argument-Driven Inquiry (ADI) Investigations With the *Common Core State Standards for English Language Arts* (*CCSS ELA*; NGAC and CCSSO 2010)

	ADI investigation													
***CCSS ELA* for third grade**	Investigation 1. Magnetic Attraction	Investigation 2. Magnetic Force	Investigation 3. Changes in Motion	Investigation 4. Balanced and Unbalanced Forces	Investigation 5. Life Cycles	Investigation 6. Life in Groups	Investigation 7. Variation Within a Species	Investigation 8. Inheritance of Traits	Investigation 9. Traits and the Environment	Investigation 10. Fossils	Investigation 11. Differences in Traits	Investigation 12. Adaptations	Investigation 13. Weather Patterns	Investigation 14. Climate and Location
Reading: Informational Text														
Key ideas and details (CCSS.ELA-LITERACY.RI.3.1–3)	■	■	■	■	■	■	■	■	■	■	■	■	■	■
Craft and structure (CCSS.ELA-LITERACY.RI.3.4–6)	■	■	■	■	■	■	■	■	■	■	■	■	■	■
Integration of knowledge and ideas (CCSS.ELA-LITERACY.RI.3.7–9)	■	■	■	■	■	■	■	■	■	■	■	■	■	■
Range of reading and level of text complexity (CCSS.ELA-LITERACY.RI.3.10)	■	■	■	■	■	■	■	■	■	■	■	■	■	■
Writing														
Text types and purposes (CCSS.ELA-LITERACY.W.3.1–2)	■	■	■	■	■	■	■	■	■	■	■	■	■	■
Production and distribution of writing (CCSS.ELA-LITERACY.W.3.4–6)	■	■	■	■	■	■	■	■	■	■	■	■	■	■
Research to build and present knowledge (CCSS.ELA-LITERACY.W.3.8)	■	■	■	■	■	■	■	■	■	■	■	■	■	■
Range of writing (CCSS.ELA-LITERACY.W.3.10)	■	■	■	■	■	■	■	■	■	■	■	■	■	■
Speaking and Listening														
Comprehension and collaboration (CCSS.ELA-LITERACY.SL.3.1–3)	■	■	■	■	■	■	■	■	■	■	■	■	■	■
Presentation of knowledge and ideas (CCSS.ELA-LITERACY.SL.3.4,6)	■	■	■	■	■	■	■	■	■	■	■	■	■	■

Key: ■ = strong alignment.

Standards Matrix E: Alignment of the Argument-Driven Inquiry (ADI) Investigations With the *Common Core State Standards for Mathematics* (*CCSS Mathematics;* NGAC and CCSSO 2010)

CCSS Mathematics for third grade	ADI Investigation													
	Investigation 1. Magnetic Attraction	Investigation 2. Magnetic Force	Investigation 3. Changes in Motion	Investigation 4. Balanced and Unbalanced Forces	Investigation 5. Life Cycles	Investigation 6. Life in Groups	Investigation 7. Variation Within a Species	Investigation 8. Inheritance of Traits	Investigation 9. Traits and the Environment	Investigation 10. Fossils	Investigation 11. Differences in Traits	Investigation 12. Adaptations	Investigation 13. Weather Patterns	Investigation 14. Climate and Location
Mathematical Practices														
Make sense of problems and persevere in solving them. (CCSS.MATH.PRACTICE.MP1)		■	■	■					■				■	
Reason abstractly and quantitatively. (CCSS.MATH.PRACTICE.MP2)	■	■	■	■	■	■	■	■	■	□	■	□	■	■
Construct viable arguments and critique the reasoning of others. (CCSS.MATH.PRACTICE.MP3)		■	■	■										
Model with mathematics. (CCSS.MATH.PRACTICE.MP4)		□	□	□			□	□		□		□	■	■
Use appropriate tools strategically. (CCSS.MATH.PRACTICE.MP5)		■	■	■	□		■		■				■	
Attend to precision. (CCSS.MATH.PRACTICE.MP6)		■	■	■	□		■	□	■	□	■	□	■	■
Look for and make use of structure. (CCSS.MATH.PRACTICE.MP7)													■	■
Look for and express regularity in repeated reasoning. (CCSS.MATH.PRACTICE.MP8)		□	□	□	□			□	□	□		□	■	■

Key: ■ = strong alignment; □ = moderate alignment.

Continued

Standards Matrix E *(continued)*

CCSS Mathematics for third grade	ADI Investigation													
	Investigation 1. Magnetic Attraction	Investigation 2. Magnetic Force	Investigation 3. Changes in Motion	Investigation 4. Balanced and Unbalanced Forces	Investigation 5. Life Cycles	Investigation 6. Life in Groups	Investigation 7. Variation Within a Species	Investigation 8. Inheritance of Traits	Investigation 9. Traits and the Environment	Investigation 10. Fossils	Investigation 11. Differences in Traits	Investigation 12. Adaptations	Investigation 13. Weather Patterns	Investigation 14. Climate and Location
Operations and Algebraic Thinking														
Represent and solve problems involving multiplication and division. (CCSS.MATH.CONTENT.3.OA.A.1–4)			■										■	
Understand properties of multiplication and the relationship between multiplication and division. (CCSS.MATH.CONTENT.3.OA.B.5–6)			■										■	
Multiple and divide within 100. (CCSS.MATH.CONTENT.3.OA.C.7)			□										■	
Solve problems involving the four operations, and identify and explain patterns in arithmetic. (CCSS.MATH.CONTENT.3.OA.D.8–9)			■	■			■		■				■	
Number and Operations in Base Ten														
Use place value understanding and properties of operations to perform multi-digit arithmetic. (CCSS.MATH.CONTENT.3.NBT.A.1–3)		■	■	■			■		■			■	■	
Measurement and Data														
Solve problems involving measurement and estimation. (CCSS.MATH.CONTENT.3.MD.A.1–2)			■	■			■		■		□	■		
Represent and interpret data. (CCSS.MATH.CONTENT.3.MD.B.3–4)	■	■	■	■	■	□	■	□	■	■	■	■	■	■

Key: ■ = strong alignment; □ = moderate alignment.

Standards Matrix F: Alignment of the Argument-Driven Inquiry (ADI) Investigations With the *English Language Proficiency (ELP) Standards* (CCSSO 2014)

ELP Standards for grades 2–3	ADI Investigation													
	Investigation 1. Magnetic Attraction	Investigation 2. Magnetic Force	Investigation 3. Changes in Motion	Investigation 4. Balanced and Unbalanced Forces	Investigation 5. Life Cycles	Investigation 6. Life in Groups	Investigation 7. Variation Within a Species	Investigation 8. Inheritance of Traits	Investigation 9. Traits and the Environment	Investigation 10. Fossils	Investigation 11. Differences in Traits	Investigation 12. Adaptations	Investigation 13. Weather Patterns	Investigation 14. Climate and Location
Receptive Modalities														
ELP 1: Construct meaning from oral presentations and informational text through grade-appropriate listening, reading, and viewing.	■	■	■	■	■	■	■	■	■	■	■	■	■	■
ELP 8: Determine the meaning of words and phrases in oral presentations and literary and informational text.	■	■	■	■	■	■	■	■	■	■	■	■	■	■
Productive Modalities														
ELP 3: Speak and write about grade-appropriate complex literary and informational texts and topics.	■	■	■	■	■	■	■	■	■	■	■	■	■	■
ELP 4: Construct grade-appropriate oral and written claims and support them with reasoning and evidence.	■	■	■	■	■	■	■	■	■	■	■	■	■	■
ELP 7: Adapt language choices to purpose, task, and audience when speaking and writing.	■	■	■	■	■	■	■	■	■	■	■	■	■	■
Interactive Modalities														
ELP 2: Participate in grade-appropriate oral and written exchanges of information, ideas, and analyses, responding to peer, audience, or reader comments and questions.	■	■	■	■	■	■	■	■	■	■	■	■	■	■
ELP 5: Conduct research and evaluate and communicate findings to answer questions or solve problems.	■	■	■	■	■	■	■	■	■	■	■	■	■	■
ELP 6: Analyze and critique the arguments of others orally and in writing.	■	■	■	■	■	■	■	■	■	■	■	■	■	■
Linguistic Structures of English														
ELP 9: Create clear and coherent grade-appropriate speech and text.	■	■	■	■	■	■	■	■	■	■	■	■	■	■
ELP 10: Make accurate use of standard English to communicate in grade-appropriate speech and writing.	■	■	■	■	■	■	■	■	■	■	■	■	■	■

Key: ■ = strong alignment.

References

Abd-El-Khalick, F., and N. G. Lederman. 2000. Improving science teachers' conceptions of nature of science: A critical review of the literature. *International Journal of Science Education* 22: 665–701.

Akerson, V., F. Abd-El-Khalick, and N. Lederman. 2000. Influence of a reflective explicit activity-based approach on elementary teachers' conception of nature of science. *Journal of Research in Science Teaching* 37 (4): 295–317.

Council of Chief State School Officers (CCSSO). 2014. *English language proficiency (ELP) standards.* Washington, DC: NGAC and CCSSO. *www.ccsso.org/resource-library/english-language-proficiency-elp-standards.*

Lederman, N. G., F. Abd-El-Khalick, R. L. Bell, and R. S. Schwartz. 2002. Views of nature of science questionnaire: Toward a valid and meaningful assessment of learners' conceptions of nature of science. *Journal of Research in Science Teaching* 39 (6): 497–521.

Lederman, J., N. Lederman, S. Bartos, S. Bartels, A. Meyer, and R. Schwartz. 2014. Meaningful assessment of learners' understanding about scientific inquiry: The Views About Scientific Inquiry (VASI) questionnaire. *Journal of Research in Science Teaching* 51 (1): 65–83.

National Governors Association Center for Best Practices and Council of Chief State School Officers (NGAC and CCSSO). 2010. *Common core state standards.* Washington, DC: NGAC and CCSSO.

NGSS Lead States. 2013. *Next Generation Science Standards: For states, by states.* Washington, DC: National Academies Press. *www.nextgenscience.org/next-generation-science-standards.*

National Research Council (NRC). 2012. *A framework for K–12 science education: Practices, crosscutting concepts, and core ideas.* Washington, DC: National Academies Press.

Schwartz, R. S., N. Lederman, and B. Crawford. 2004. Developing views of nature of science in an authentic context: An explicit approach to bridging the gap between nature of science and scientific inquiry. *Science Education* 88: 610–645.

APPENDIX 2

OVERVIEW OF *NGSS* CROSSCUTTING CONCEPTS AND NATURE OF SCIENTIFIC KNOWLEDGE AND SCIENTIFIC INQUIRY CONCEPTS

Overview of *NGSS* Crosscutting Concepts

Patterns

Scientists look for patterns in nature and attempt to understand the underlying cause of these patterns. For example, scientists often collect data and then look for patterns to identify a relationship between two variables, a trend over time, or a difference between groups.

Cause and Effect: Mechanism and Explanation

Natural phenomena have causes, and uncovering causal relationships (e.g., how changes in x affect y) is a major activity of science. Scientists also need to understand that correlation does not imply causation, some effects can have more than one cause, and some cause-and-effect relationships in systems can only be described using probability.

Scale, Proportion, and Quantity

It is critical for scientists to be able to recognize what is relevant at different sizes, times, and scales. An understanding of scale involves not only understanding how systems and processes vary in size, time span, and energy, but also how different mechanisms operate at different scales. Scientists must also be able to recognize proportional relationships between categories, groups, or quantities.

Systems and System Models

Scientists often need to define a system under study, and making a model of the system is a tool for developing a better understanding of natural phenomena in science. Scientists also need to understand that a system may interact with other systems and a system might include several different subsystems. Scientists often describe a system in terms of inputs and outputs or processes and interactions. All models of a system have limitations because they only represent certain aspects of the system under study.

Energy and Matter: Flows, Cycles, and Conservation

It is important to track how energy and matter move into, out of, and within systems during investigations. Scientists understand that the total amount of energy and matter remains the same in a closed system and that energy cannot be created or destroyed; it only moves between objects and/or fields, between one place and another place, or between systems. Energy drives the cycling and transformation of matter within and between systems.

Structure and Function

The way an object or a material is structured or shaped determines how it functions and places limits on what it can and cannot do. Scientists can make inferences about the function of an object or system by making observations about the structure or shape of its component parts and how these components interact with each other.

Stability and Change

It is critical to understand what makes a system stable or unstable and what controls rates of change in a system. Scientists understand that changes in one part of a system might cause large changes in another part. They also understand that systems in dynamic equilibrium are stable due to a balance of feedback mechanisms, but the stability of these systems can be disturbed by a sudden change in the system or a series of gradual changes that accumulate over time.

Overview of Nature of Scientific Knowledge and Scientific Inquiry Concepts

Nature of Scientific Knowledge Concepts

How scientific knowledge changes over time

A person can have confidence in the validity of scientific knowledge but must also accept that scientific knowledge may be abandoned or modified in light of new evidence or because existing evidence has been reconceptualized by scientists. There are many examples in the history of science of both *evolutionary changes* (i.e., the slow or gradual refinement of ideas) and *revolutionary changes* (i.e., the rapid abandonment of a well-established idea) in scientific knowledge.

The difference between laws and theories in science

A *scientific law* describes the behavior of a natural phenomenon or a generalized relationship under certain conditions; a *scientific theory* is a well-substantiated explanation of some aspect of the natural world. Theories do not become laws even with additional evidence; they explain laws. However, not all scientific laws have an accompanying explanatory theory. It is also important for students to understand that scientists do not discover laws or theories; the scientific community develops them over time.

The use of models as tools for reasoning about natural phenomena

Scientists use conceptual models as tools to understand natural phenomena and to make predictions. A *conceptual model* is a representation of a set of ideas about how something works or why something happens. Models can take the form of diagrams, mathematical relationships, analogies, or simulations. Scientists often develop, use, test, and refine models as part of an investigation. All models are based on a set of assumptions and include approximations that limit how a model can be used and its overall predictive power.

The difference between data and evidence in science

Data are measurements, observations, and findings from other studies that are collected as part of an investigation. *Evidence,* in contrast, is analyzed data and an interpretation of the analysis. Scientists do not collect evidence; they collect data and then transform the data they collect into evidence through a process of analysis and interpretation.

The difference between observations and inferences in science

An *observation* is a descriptive statement about a natural phenomenon, whereas an *inference* is an interpretation of an observation. Students should also understand that current scientific knowledge and the perspectives of individual scientists guide both observations and inferences. Thus, different scientists can have different but equally valid interpretations of the same observations due to differences in their perspectives and background knowledge.

Nature of Scientific Inquiry Concepts

The types of questions that scientists can investigate

Scientists answer questions about the natural or material world, but not all questions can be answered by science. Science and technology may raise ethical issues for which science, by itself, does not provide answers and solutions. Scientists attempt to answer questions about what can happen in natural systems, why things happen, or how things happen. Scientists do not attempt to answer questions about what should happen. To answer questions about what should happen requires consideration of issues related to ethics, morals, values, politics, and economics.

How scientists use different methods to answer different types of questions

Examples of methods include experiments, systematic observations of a phenomenon, literature reviews, and analysis of existing data sets; the choice of method depends on the objectives of the research. There is no universal step-by step scientific method that all scientists follow; rather, different scientific disciplines (e.g., geoscience vs. chemistry) and fields within a discipline (e.g., geophysics vs. paleontology) use different types of methods, use different core theories, and rely on different standards to develop scientific knowledge.

Science as a way of knowing

Science can help us figure out how or why things happen in the world around us. Science is both a body of knowledge (which includes core ideas and crosscutting concepts) and a set of practices (such as asking questions, planning and carrying out investigations, constructing explanations, and arguing from evidence) that people use to add, revise, and refine ideas. It is a way of making sense of the world that is used by many people, not just scientists.

The assumptions made by scientists about order and consistency in nature

Scientific investigations are designed based on the assumptions that natural laws operate today as they did in the past and that they will continue to do so in the future. Scientists also assume that the universe is a vast single system in which basic laws are consistent.

APPENDIX 3

SOME FREQUENTLY ASKED QUESTIONS ABOUT ARGUMENT-DRIVEN INQUIRY (ADI)

What are some things I can do to encourage productive talk among my students during ADI?

We suggest that you …

- Choose a **talk format** that will maximize participation (we recommend small-group or partner format during most stages).
- Establish some **class talk norms** (agreed-upon rules for discussions) or remind your students about the existing ones.
- Make the **goal of the discussion** explicit (e.g., identify important or useful ideas in a text, share ideas, improve an argument, help others improve their argument, reach consensus).
- Post different **conversation starters** in the classroom. Conservation starters are general question stems (e.g., "I disagree with that because ...."; "Could you tell me more about ...?"; "Can you explain what you mean by ...?") that encourage students to think, reason, and collaborate in academically productive ways.
- Make and use a set of **back-pocket question cards**. You can carry a set of cards with you as you move from group to group. Each card has a different question written on it. When you see a group of students no longer talking, hand one of the cards to a student. The student then asks the question written on the card. This is a great way to get a conversation started and to encourage students to engage in more productive talk.
- **Model** how to argue from evidence, critique ideas, or offer useful feedback.

What can I do to help my students comprehend more of what they read?

We suggest that you …

- **Read the text aloud** so your students can listen to a fluent adult.
- If you are reading the text aloud, stop when you read a sentence that includes an important or useful idea and **tell students to highlight** or "star" that part in the text (e.g., "I think that will be really useful for you later, so you might want to put a star next to that sentence.").
- Encourage your students to **annotate** the text as they read if you ask them to read the text on their own.
- Remind students to **talk with their peers** about what they read.
- Break the text into more **manageable chunks** by telling the students to read only one paragraph at a time and add to their "KNOW/NEED" chart before moving on to the next paragraph of text.

What are some things I can do to support my emerging bilingual students during ADI?

We suggest that you …

- **Group students strategically** to maximize productive interaction between students of varying language proficiencies. For example, you may want to pair a student who is bilingual with one who is learning English (if they share the same first language) so these two students can work together when you ask them to present or critique arguments.

- Encourage students to **use the language they have** to express their ideas (e.g., everyday language, imperfect English, native language).
- Provide a **visual representation** of the key terms and ideas.
- Provide a **copy of the handout** in the students' native language. Ask your school's ELL (English-language learner) support staff for assistance in translation.
- Encourage students to **draw** what they are thinking.
- **Model** language expectations for a task (e.g., demonstrate what critiquing looks like, demonstrate what giving good feedback looks like, highlight how a section of a report is organized, demonstrate how to write a section of a report).
- Allow the students to **work with a peer** as they write.
- Provide **extended time** to complete tasks that require writing.

What are some things that I can do if my students get stuck when they are developing their draft argument?

We suggest that you …

- Provide students with **sentence starters** for each component of the argument. Examples of sentence starters that will help students provide a good interpretation of the data analysis include the following:
  - "Our analysis suggests …"
  - "Our analysis shows …"
  - "This table shows …"

  Examples of sentence starters that will help students provide a good justification of the evidence include the following:
  - "This evidence matters because …"
  - "We included this evidence in our argument because …"
  - "We think this evidence is important because …"
- Provide them with **strong and weak examples** of each component of an argument (claim, evidence, and justification of evidence) and then discuss what makes the component strong or weak.
- Send one or two students from each group to a different group to **learn something from their classmates.** When students visit another group, encourage those students to ask questions about what the other group is doing and why. They can then return to their own group and share what they have learned.
- Encourage students to **revisit the "Introduction" section of the handout** (or the investigation log) and look for useful ideas to include in the argument.

What are some things that I can do if my students get stuck as they start writing their report?

We suggest that you …

- Provide **strong and weak examples** of each section of the report and then discuss what makes that section strong or weak.
- Allow students to talk to each other as they write so they can **learn something from their classmates**. When the students talk, encourage them to ask questions about what they are writing and why.
- Encourage students to **revisit the "Introduction" section of the handout** (or the investigation log) and look for useful ideas to include in the "Introduction" section of the report.
- Encourage students to **revisit the investigation proposal section of the handout** (or the investigation log) and look for useful ideas to include in the "Method" section of the report.
- Encourage students to **use their group's argument as a foundation** when they write the "Argument" section of their report. All the basic information should be on their whiteboard. All they need to do is write a paragraph or two that includes this information.

What are some things that I can do to help my students give better feedback to each other during the peer-review process?

We suggest that you …

- Provide **strong and weak examples** of each section of the report and review them together as a class. Then **model** how to give good feedback.
- Provide students with **feedback sentence starters** for different types of feedback. Examples of feedback sentence starters include the following:
  - "We suggest adding …"
  - "We suggest making the following changes …"
  - "We think you can make your writing more clear by …"
  - "We think you should change ___ to ___."
- Be sure to establish some **class feedback norms** (agreed-upon rules for critiquing the work of others) or remind them about the existing ones.
- Make the **goal of the activity** explicit (in this case the goal is to help others improve the quality of their report).

APPENDIX 4

ADI INVESTIGATION REPORT PEER-REVIEW GUIDE: ELEMENTARY SCHOOL VERSION

Report By: ______________________ Date: ______________
Author

Reviewed By: ______________________ ______________________
Reviewer 1 Reviewer 2

Section 1: The Investigation	Reviewer Rating			Teacher Score
1. Did the author do a good job of explaining what the investigation was about?	☐ No	☐ Almost	☐ Yes	0 1 2
2. Did the author do a good job of making the **guiding question** clear?	☐ No	☐ Almost	☐ Yes	0 1 2
3. Did the author do a good job of describing what he or she did to **collect data?**	☐ No	☐ Almost	☐ Yes	0 1 2
4. Did the author do a good job describing **how** he or she **analyzed** the data?	☐ No	☐ Almost	☐ Yes	0 1 2

Reviewers: If your group gave the author any "No" or "Almost" ratings, please give the author some advice about what to do to improve this part of his or her investigation report.

Section 2: The Argument	Reviewer Rating			Teacher Score
1. Does the author's claim provide a clear and detailed **answer** to the guiding question?	☐ No	☐ Almost	☐ Yes	0 1 2
2. Did the author support his or her claim with **scientific evidence?** Scientific evidence includes analyzed data and an explanation of the analysis.	☐ No	☐ Almost	☐ Yes	0 1 2
3. Does the **evidence** that the author uses in his or her argument **support the claim?**	☐ No	☐ Almost	☐ Yes	0 1 2
4. Did the author include enough **evidence** in his or her argument?	☐ No	☐ Almost	☐ Yes	0 1 2
5. Did the author do a good job of **explaining why the evidence** is important (why it matters)?	☐ No	☐ Almost	☐ Yes	0 1 2

6. Is the content of the argument **correct** based on the science concepts we talked about in class?	☐ No	☐ Almost	☐ Yes	0 1 2
Reviewers: If your group gave the author any "No" or "Almost" ratings, please give the author some advice about what to do to improve this part of his or her investigation report.				

Section 3: Mechanics	**Reviewer Rating**			**Teacher Score**
1. ***Grammar:*** Are the sentences complete? Is there proper subject-verb agreement in each sentence? Are there no run-on sentences?	☐ No	☐ Almost	☐ Yes	0 1 2
2. ***Conventions:*** Did the author use proper spelling, punctuation, and capitalization?	☐ No	☐ Almost	☐ Yes	0 1 2
3. **Word Choice:** Did the author use the right words in each sentence (for example, *there* vs. *their, to* vs. *too, then* vs. *than*)?	☐ No	☐ Almost	☐ Yes	0 1 2
Reviewers: If your group gave the author any "No" or "Almost" ratings, please give the author some advice about what to do to improve the writing mechanics of his or her investigation report.				

General Reviewer Comments	**Teacher Comments**
We liked … We wonder …	

Total: ____ /26

APPENDIX 5

SAFETY ACKNOWLEDGMENT FORM

I know that it is very important to be as safe as I can during an investigation. My teacher has told me how to be safer in science. I agree to follow these 11 safety rules when I am working with my classmates to figure things out in science:

1. I will act in a responsible manner at all times. I will not run around the classroom, throw things, play jokes on my classmates, or be careless.
2. I will never eat, drink, or chew gum.
3. I will never touch, taste, or smell any materials, tools, or chemicals without permission.
4. I will wear my safety goggles at all times during the activity setup, hands-on work, and cleanup.
5. I will do my best to take care of the materials and tools that my teacher allows me to use.
6. I will always tell my teacher about any accidents as soon as they happen.
7. I will always dress in a way that will help keep me safer. I will wear closed-toed shoes and pants. My clothes will not be loose, baggy, or bulky. I will also use hair ties to keep my hair out of the way while I am working if my hair is long.
8. I will keep my work area clean and neat at all times. I will put my backpack, books, and other personal items where my teacher tells me to put them and I will not get them out unless my teacher tells me that it is okay.
9. I will clean my work area and the materials or tools that I use.
10. I will wash my hands with soap and water at the end of the activity.
11. I will follow my teacher's directions at all times.

____________________	____________________	__________
Print Name	Signature	Date

I have read and reviewed the 11 investigation safety rules with my child. He or she understands how important it is to follow safety rules in science and has agreed to follow these safety rules at all times. I give my permission for my child to participate in the investigations this year.

____________________	____________________	__________
Parent or Guardian Name	Parent or Guardian Signature	Date

IMAGE CREDITS

All images in this book are stock photographs or courtesy of the authors unless otherwise noted below.

Investigation 5

Figure 5.2: Username1927, Wikimedia Commons, CC BY-SA 4.0, *https://commons.wikimedia.org/wiki/File:Holometabolous_vs._Hemimetabolous.svg*

Figure 5.3: Brocken Inaglory, Wikimedia Commons, GFDL 1.2, *https://commons.wikimedia.org/wiki/File:Sunset_at_ocean_in_san_francisco.jpg*

Investigation 6

Figure 6.2: Doug Smith, National Park Service, Public domain. *https://commons.wikimedia.org/wiki/File:Wolves_and_elk.jpg*

Investigation 7

Figure 7.2: SuSanA Secretariat, Wikimedia Commons, CC BY-SA 3.0, *https://commons.wikimedia.org/wiki/File:Compost_with_earthworms_(8261572027).jpg*

Figure 7.3: Rulexip, Wikimedia Commons, CC BY-SA 3.0, *https://commons.wikimedia.org/wiki/File:Ladybugs_on_Jurmala_beach.jpg*

Figure 7.4: Brian Gratwicke, CC BY 2.0, *www.flickr.com/photos/briangratwicke/14324430306*

Earthworm illustration in checkout question 1: Modified from Pearson Scott Foresman, Wikimedia Commons, Public domain. *https://commons.wikimedia.org/wiki/File:Earthworm_1_(PSF).png*

Ladybug illustration in checkout question 2: Modified from Pearson Scott Foresman, Wikimedia Commons, Public domain. *https://commons.wikimedia.org/wiki/File:Ladybug_(PSF).svg*

Investigation 8

Figure 8.2: Cartman0052007, Wikimedia Commons, GFDL 1.2, *https://commons.wikimedia.org/wiki/File:Havanese_Litter.png*

Figure 8.3: IDS.photos, Wikimedia Commons, CC BY-SA 2.0, *https://upload.wikimedia.org/wikipedia/commons/f/f2/Yellow_Labrador_puppies_%284166519466%29.jpg*

Figure 8.4: Moheen Reeyad, Wikimedia Commons, CC BY-SA 4.0, *https://upload.wikimedia.org/wikipedia/commons/8/8a/Bangladeshi_Puppies_behavior_%2802%29.jpg*

Investigation 10

Figure 10.2: National Park Service, Public domain. *www.nps.gov/fobu/learn/nature/fossil-fish.htm*

Figure 10.3: National Park Service, Public domain. *www.nps.gov/fobu/learn/nature/fossil-mammals.htm*

Frog skeleton illustration in checkout question 1: *https://pixabay.com/en/amphibian-animal-bone-dead-frog-2028309*

Investigation 11

Figure 11.3: Algkalv, Wikimedia Commons, CC BY-SA 3.0, *https://upload.wikimedia.org/wikipedia/commons/e/e4/Terianniaq-Qaqortaq-arctic-fox.jpg*

Figure 11.4: Lukas Kaffer, Wikimedia Commons, CC BY-SA 3.0, *https://upload.wikimedia.org/wikipedia/commons/7/7a/Great_male_Leopard_in_South_Afrika-JD.JPG*

Investigation 12

Figure 12.4: AWeith, Wikimedia Commons, CC BY-SA 4.0, *https://commons.wikimedia.org/wiki/File:Arctic_ocean_drift_ice,_the_realm_of_the_polar_bear.jpg*

Musk ox image above checkout question 1: Jeangagnon, Wikimedia Commons, CC BY-SA 2.0, *https://commons.wikimedia.org/wiki/File:Musk_ox.jpg*

Gazelle image above checkout question 1: Charles J. Sharp, Wikimedia Commons, CC BY-SA 3.0, *https://commons.wikimedia.org/wiki/File:Mountain_gazelle_(gazella_gazella).jpg*

Investigation 13

Figure 13.2: Cherubino, Wikimedia Commons, Public domain. *https://commons.wikimedia.org/wiki/File:Four_Poplars_in_four_seasons.JPG*

INDEX

Page numbers printed in **boldface type** refer to figures or tables.

Index

D

M